PRODUCTION AND NEUTRALIZATION OF NEGATIVE IONS AND BEAMS

Related Titles from AIP Conference Proceedings

To learn more about these titles, or the AIP Conference Proceedings Series, please visit the webpage **http://proceedings.aip.org/proceedings**

PRODUCTION AND NEUTRALIZATION OF NEGATIVE IONS AND BEAMS

10th International Symposium on Production and Neutralization of Negative Ions and Beams

Kiev, Ukraine   14 –17 September 2004

EDITORS

Joseph D. Sherman
Los Alamos National Laboratory
Los Alamos, New Mexico

Yuri I. Belchenko
Budker Institute of Nuclear Physics
Novosibirsk, Russia

SPONSORING ORGANIZATIONS
National Academy of Sciences, Ukraine
Los Alamos National Laboratory, USA

Melville, New York, 2005
AIP CONFERENCE PROCEEDINGS ■ VOLUME 763

Editors:

Joseph D. Sherman
Los Alamos National Laboratory
MS H838 LANSCE-2
Los Alamos, NM 87545
USA
E-mail: jsherman@lanl.gov

Yuri I. Belchenko
Budker Institute of Nuclear Physics
11 Lavrentiev Avenue
Novosibirsk 630090
RUSSIA
E-mail: belchenko@inp.nsk.su

L.C. Catalog Card No. 2005923944
ISBN 0-7354-0248-5
ISSN 0094-243X
Printed in the United States of America

CONTENTS

NEUTRAL BEAM SYSTEMS

FUNDAMENTAL PROCESSES

EXTRACTION, TRANSPORT, AND ACCELERATION

H⁻ AND D⁻ SOURCES

NEUTRALIZATION

APPLICATIONS

PREFACE

The 10th International Symposium on Production and Neutralization of Negative Ions and Beams was held in Kiev (Ukraine), on September 13-17, 2004. The sequence of these symposia was initiated by Brookhaven National Laboratory (USA) in 1977. The initial seven meetings were held at Brookhaven with an interval of three years between them. The eighth meeting was arranged in Villagium of Giens (France) in 1997, and the ninth one, in CEA Saclay, close to Paris (France) in 2002.

The decision about arranging the 10th Symposium in Kiev was made at the previous Conference on Ion Sources (ICIS03) in 2003. In the Former Soviet Union, Kiev was a traditional place of holding annual workshops on physics and techniques of ion sources and ion beams. An outstanding number of specialists from Russia and Ukraine were able to attend the 10th Symposium. 34 reports on research in areas of knowledge traditional for this symposium (fundamental processes; hydrogen and deuterium negative ion sources; extraction, acceleration, transportation and neutralization of ions; the applications) have been presented. Research in those areas was stimulated mainly by two problems: 1) plasma heating in thermonuclear setups by means powerful beams of neutral particles, and 2) creation of spallation source based on high-power accelerators. A solution of the first problem requires high-current beams with current value up to 20 A which can be formed only from large plasma surface by means of multi-aperture extracting systems; and for solving of the second problem bright beams with current value up to 60 mA are required. We hope that the 10th Symposium has made an essential contribution to solving of two those problems.

We are grateful to all organizations and persons assisting us in the arrangement of this Symposium; National Academy of Sciences of Ukraine and Los Alamos National Laboratory (USA) that were sponsors of the Symposium; the Local Organizing Committee which arranged appropriate conditions for work of the Symposium; the International Organizing Committee which formed an interesting program for the Symposium; Mr. Joseph Sherman and Mr. Yuri Belchenko who agreed to edit the Proceedings of the Symposium; and, finally, all participants for their active work on the presentation and consideration of the reports.

We hope that the next Symposium will be arranged by Los Alamos National Laboratory in one of the cities of New Mexico (USA).

Igor Soloshenko
Institute of Physics of
National Academy of
Sciences of Ukraine,
Kiev, Ukraine

SYMPOSIUM ORGANIZATION

CHAIR

Igor Soloshenko (Chair)   Institute of Physics, National Academy of Sciences of Ukraine, Kiev, Ukraine
Martin Stockli (Co-chair)   Spallation Neutron Source, Oak Ridge National Lab, Oak Ridge, Tennessee

INTERNATIONAL PROGRAM COMMITTEE

Igor Soloshenko   IP NASU (Ukraine)
Martin Stockli   SNS ORNL (USA)
Marthe Bacal   Ecole Polytechnique (France)
Reinard Becker   University of Frankfurt (Germany)
Yuri Belchenko   BINP, Novosibirsk (Russia)
Raphael Gobin   CEA/Saclay (France)
Ronald Hemsworth   CEA/Cadarache (France)
Jens Peters   DESY (Germany)
Joseph Sherman   LANL (USA)

LOCAL ORGANIZING COMMITTEE
(Institute of Physics of NAS Ukraine, Kiev, Ukraine)

Igor Soloshenko (Chair)
Vladimir Lukashenko
Vladimir Bazhenov
Anna Terentyeva

NEUTRAL BEAM SYSTEMS

Development of the Long Pulse Negative Ion Source for ITER

R. S. Hemsworth, D. Boilson*, U. Fanz**, L. Svensson, H. P. L. de Esch, A. Krylov, P. Massmann, and B. Zaniol***

*Association EURATOM-CEA, CEA/DSM/DRFC, CEA-Cadarache,
13108 St Paul-lez-Durance (France)*
*Association EURATOM-DCU , PRL/NCPST, Glasnevin, Dublin 13, Ireland
** Association EURATOM-IPP, Max-Planck–Institut fuer Plasmaphysik,D-85748 Garching, Germany
*** CONSORZIO RFX Association EURATOM-ENEA, Corso Stati Uniti 4, I-35127 Padova, Italy

Abstract. A model of the ion source designed for the neutral beam injectors of the International Thermonuclear Experimental Reactor (ITER), the KAMABOKO III ion source, is being tested on the MANTIS test stand at the DRFC Cadarache in collaboration with JAERI, Japan, who designed and supplied the ion source. The ion source is attached to a 3 grid 30 keV accelerator (also supplied by JAERI) and the accelerated negative ion current is determined from the energy deposited on a calorimeter located 1.6 m from the source.

During experiments on MANTIS three adverse effects of long pulse operation were found:

- The negative ion current to the calorimeter is $\approx 50\%$ of that obtained from short pulse operation
- Increasing the plasma grid (PG) temperature results in $\leq 40\%$ enhancement in negative ion yield, substantially below that reported for short pulse operation, $\geq 100\%$.
- The caesium "consumption" is up to 1500 times that expected.

Results presented here indicate that each of these is, at least partially, explained by thermal effects. Additionally presented are the results of a detailed characterisation of the source, which enable the most efficient mode of operation to be identified.

BACKGROUND

Each of the neutral beam injectors of ITER is designed to deliver 17 MW of 1 MeV D^0 to the ITER plasma throughout the ITER pulse, i.e. for up to 3600 s [1]. The design of the injectors is based on the acceleration and neutralisation of D^-, assuming an accelerated D^- current density of 200 A/m^2 with <1 electron extracted per accelerated D^- ion. The reference design of the negative ion source is a caesiated, filamented, multi-pole, arc discharge source requiring an arc power of about 1.5 kW per litre of source volume, at a source pressure of 0.3 Pa. The KAMABOKO III source has operated at these parameters for 5 s pulses [2], and the aim of the ongoing experiments is to achieve them during long pulse operation, when the system reaches thermal equilibrium.

CP763, *Production and Neutralization of Negative Ions and Beams*, edited by J. D. Sherman and Y. I. Belchenko
© 2005 American Institute of Physics 0-7354-0248-5/05/$22.50

EXPERIMENTAL APPARATUS

The source has been described in the past [3] and only some features are briefly summarised here. The source is a 30 l quasi-cylindrical chamber of machined oxygen free copper, which forms the anode; with 12 filaments mounted on water cooled co-axial feedthroughs. The primary electron confinement (and some plasma confinement) is achieved by 16 magnetic line cusps generated by SmCo permanent magnets (width x height = 10 x 20 mm^2) arranged in machined vertical channels on the outside of the chamber. The cooling of the source is achieved by water lines brazed into channels on the outside of the source next to the columns of magnets.

The source is directly attached to, but electrically insulated from, the accelerator. The end wall of the ion source is the "plasma grid" (PG), the first grid of the accelerator, which is biased with respect to the source walls (the anode) to +5V. Columns of large (30 x 30 x 20 mm^3 or 50 x 30 x 20 mm^3, length by width by height) permanent magnets mounted inside the source flange produce a horizontal magnetic field across the front of the plasma grid. This field forms the "standard" magnetic filter with a strength of $\approx$900 G cm, defined as $S = \int_0^{\ell_{fil}} B.d\ell$ where B is the local field strength, ℓ is the distance from the PG along a line between the PG and the filaments, and ℓ_{fil} is the position of the filaments. Electrons and negative ions arriving at the apertures in the PG are extracted and accelerated to $\leq$30 keV. As they have the same charge it is not possible to distinguish between electron and negative ion extraction and acceleration electrically. Therefore the accelerated negative ion current is usually determined from the energy deposited by the ions on a calorimeter located 1.6 m downstream of the accelerator which electrons are unable to reach as they are deflected out of the beam path by a small transverse magnetic field.

BEAM TRANSMISSION

Fig. 1 shows the accelerated current and the current to the calorimeter as a function of the arc power: only part of the accelerated current reaches the calorimeter. A possible reasons for the poor transmission is that the "lost" beam is accelerated electrons as the magnetic field between the accelerator and the calorimeter deflects such electrons out of the beam path. Accelerated electrons can arise from either accelerated extracted electrons or electrons created by stripping in the accelerator.

Extracted electrons: Extracted electrons are deflected onto the surface of the extraction grid by the magnetic field from the filter in the ion source and the field from permanent magnets buried in the extraction grid, but some electrons escape to the acceleration region. The fraction of the extracted electrons that are accelerated has been measured by operating the source in pure argon. In this situation no negative ions are produced in the discharge, but a high electron current (assumed equal to the current to the extraction grid) of 2.8 A was extracted. No power was recorded on the calorimeter, and the accelerated current, was 50 mA, i.e.<2% of the extracted current. Furthermore the current to the acceleration grid was equal (within the measurement error) to the current drain from the high voltage power supply, which means that most of the accelerated electrons were collected on that grid. As the extracted electron current during H_2 operation is <20% of the accelerated current, and approximately equal to the accelerated current in D_2 operation, extracted electrons cannot explain an accelerated electron current that is ≈50% of the total accelerated current

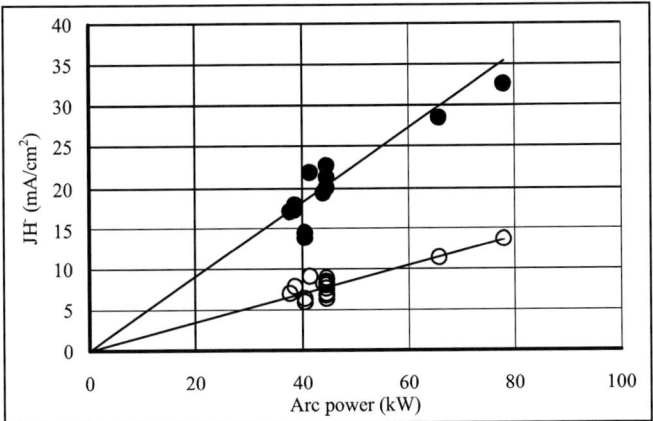

FIGURE 1. Accelerated current and the current to the calorimeter as a function of the arc power. The filled circles are the H⁻ current density calculated by assuming the drain current from the high voltage power supply is entirely accelerated H⁻. The open circles are the H⁻ current density derived from the energy falling on the calorimeter. As can be seen from Fig. 1, only about 50% of the accelerated current reaches the calorimeter.

Electrons from stripping. To a first approximation the fraction of electrons stripped during the passage of the H⁻ or D⁻ through the accelerator is proportional to the source pressure. Thus if stripping were the cause of the "lost" beam, the transmission should vary strongly with the source filling pressure. Fig. 2 shows the transmission as a function of the source pressure. Within the experimental errors there is no variation with pressure from 0.18 to 0.7 Pa. (Note that the calculated stripping fraction in the acceleration gap at a source filling pressure of 0.3 Pa is ≈3%.)

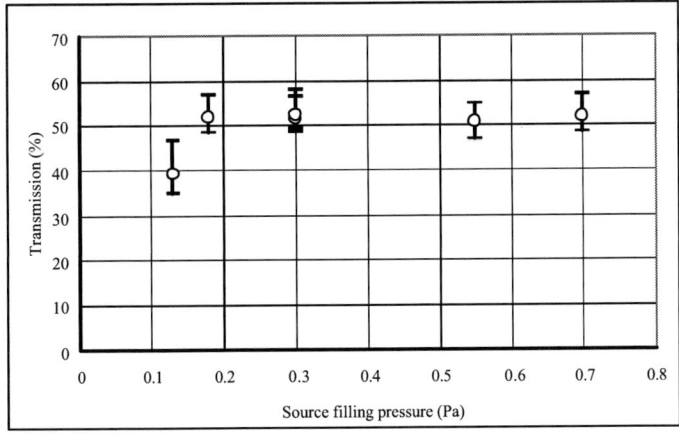

FIGURE 2. Beam transmission as a function of the source filling pressure.

A second possible explanation for the poor beam transmission is that the beam optics is extremely bad. Careful simulations of the beam optics have been carried out with assumed possible variations and errors in extraction and acceleration gaps, grid misalignment and negative ion current density and magnetic field effects. All the simulations predict beams with adequate optics to achieve transmissions of >90%. However it has recently been realised that the acceleration grid could be bowing under the heat load received from intercepted ions and electrons. The circular grid is cooled by water flowing in small (2 mm ID) tubes brazed between the rows of apertures onto its downstream surface. The power to the grid is small, ≈2 kW, and the equilibrium temperature is ≈70 °C. Because it is heated, the grid will try to expand, and, if it is fixed at its circumference, it will bow, or buckle, in order to have the correct, increased, size. If the grid bows the convex surface will have been stretched and concave one compressed. If the forces needed to compress and stretch the material are less than the force due to thermal expansion, the grid will resist bowing. In the case of a temperature increase of 50 °C, grid bowing would lead to a change in the 10 mm acceleration gap of about 3 mm. Now the copper acceleration grid is 3 mm thick over the rectangular aperture array; the thickness increases to 10.5 mm outside the array, and the circular rim is 6.5 mm thick. A photograph of the grid is shown as Fig. 3.

FIGURE 3. Photograph of the acceleration grid

The 3mm thick section will heat first, and it is likely that it will distort as its expansion is resisted by the cooler thicker outer section, then the outer sections will heat by conduction and expand until the expansion is resisted by the contact of the fixing screws with the edges of the holes in the rim. To test this hypothesis the beam transmission was measured as a function of the pulse length, see Fig. 4. The measured data give the average transmission for each pulse, and the instantaneous transmission was derived from a "by eye" fit to those data

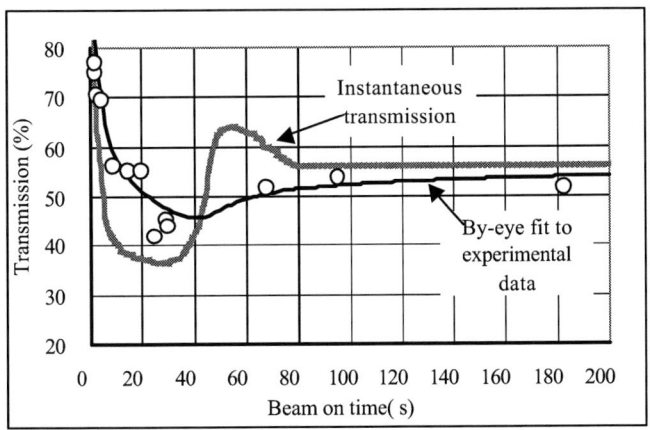

FIGURE 4. Beam transmission as a function of the pulse length. The open circles are the measured values. The instantaneous transmission is derived from the "by-eye" fit through the experimental data.

Fig. 4 shows that for very short pulses, <4 s, the transmission is >70%. The transmission degrades to a minimum at about 25 s, then improves to its long pulse value of ≈55%. This is qualitatively in line with expectations as the inertial time constant for the thin part of the grid is 4.5 s, and that of the rim is 22 s.

EFFECT OF THE PLASMA GRID TEMPERATURE WITH CAESIUM SEEDING

A caesiated source is thought to operate on the principle of the production of negative ions (H⁻ or D⁻) by surface scattering of neutral atoms and positive ions from the PG surface. An essential feature of this hypothesis is the reduction in the work function of the grid surface, due to caesium (Cs) coverage of the surface. In short pulse operation an optimum in the negative ion yield has been found when the PG temperature is increased to between 200 and 300°C [4]. It is reasonable to assume that the Cs coverage on the PG surface is a dynamic balance between evaporation from the grid and the arrival of Cs from the source volume [5]. Heating the PG changes the rate of evaporation loss from the grid but has little effect on the flux to the grid, thus the surface coverage should change. It is assumed that the observed increase of the negative ion yield with the PG temperature occurs because the increased evaporation loss changes the surface coverage towards the optimum value. The increase in negative ion yield with the PG temperature is referred to as the "PG temperature effect".

Two types of long pulse PG's have been designed and fabricated (by JAERI) that have an equilibrium temperature in the desired range, whilst using conventional cooling (water at <10 bar). These are the so-called "frame cooled grid" (FCG, made of Cu/Cr/Zr alloy) and the "actively cooled grid" (ACG, made of molybdenum). Both grids are designed to be heated by the power from the arc discharge arriving at the PG surface, mainly by radiation, and both are designed to have an equilibrium temperature of ≈300 °C with an arc power of ≈45 kW. Both grids perform correctly in that they reach, approximately, the design temperature at the design power flux to them. No difference in source efficiency (with or without Cs seeding of the ion source) was observed with the two different grids, i.e. with different materials.

The PG temperature effect measured with the two grids is similar, ≤40%. The effect of the temperature of the FCG on the negative ion yield from the KAMABOKO III ion source can be seen in Fig. 5. This variation in the negative ion yield is typical for the KAMABOKO III ion source on MANTIS when operating in long pulses. The arc power was 40 kW, and the filling pressure was 0.3 Pa of H_2. The negative ion yield is assumed to be proportional to the accelerated current (I_{drain}) which increases from 0.59 A to 0.73 A, an increase of 24%, with the increase of the temperature of the PG. JAERI have reported PG temperature effects resulting in increases in the negative ion yield of >100 % during short pulse operation, 0.1 s [6], and it had been anticipated that a similar increase would occur during long pulses.

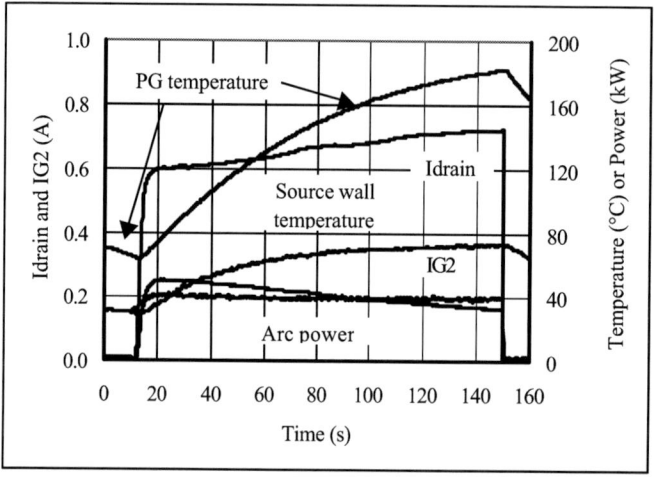

FIGURE 5. The effect of the temperature of the plasma grid (the frame cooled grid) on the negative ion yield from the KAMABOKO III ion source. Note that at the start of this shot the PG and the source walls were still cooling from the previous shot and their temperatures at the start of the shot were 64 °C and 30 °C respectively.

During the long pulse experiments it is found that the equilibrium temperature of the ion source walls is significantly above that of the cooling water (see Fig. 5). This led to the speculation that the reason for the low PG temperature effect in long pulse operation is due to the increased temperature of the walls. As mentioned above, the Cs coverage of the PG surface is a dynamic balance between the arrival of Cs from the source volume and evaporation from the grid surface. Thus if the flux of Cs to the PG were increased by increased evaporation of Cs from the source walls, and then to the PG, the Cs thickness on the PG would be above the optimum with the PG at the design temperature. If this is correct, increasing the PG temperature or decreasing the source wall temperature would give a higher PG temperature effect. As both the source walls and the PG are heated by the discharge, their temperatures are coupled. To de-couple the temperatures the cooling water was removed from the PG and a sequence of ≈30 s arc discharge pulses used to increase the PG temperature. The source wall temperature does not increase significantly in this situation as it cools in between each pulse. The result was that the accelerated negative ion current increased almost linearly with the PG temperature, with a maximum of ≈60% increase in negative ion yield being obtained at the highest PG temperatures reached, see Fig. 6.

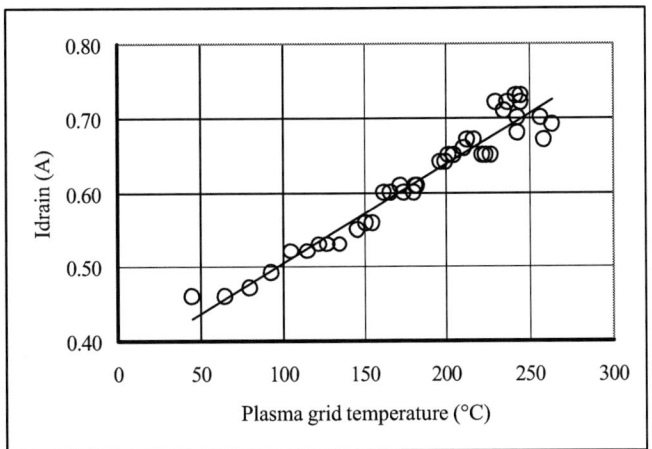

FIGURE 6. Accelerated negative ion current as a function of the plasma grid temperature with "cold" source walls.

SOURCE CHARACTERISTICS

In an attempt to further understand the physics of negative ion formation in the KAMABOKO III ion source the experiments described below have been carried out.

Negative Ion Yield

Here the negative ion yield is assumed to be proportional to the accelerated current, I_{drain}. I_{drain} has been measured as a function of the source gas filling pressure, the arc current and of the anode to cathode voltage, in each case keeping all other parameters constant. All were measured with caesium seeding of the ion source. Unless otherwise stated all are with H_2 at a source filling pressure of 0.3 Pa..

No variation of the H⁻ yield was found with source pressure for source filling pressures of 0.2 to 0.8 Pa, see Fig. 7. The variation of the H yield with arc current is found to be linear see Fig. 8. The variation of the H yield with the anode to cathode voltage is offset linear, see Fig. 9. Here it is to be noted that operation at different anode to cathode voltages has only recently become possible on MANTIS, due to the introduction of a feedback between the filament heating current and the arc current. This feedback system allows the arc current and anode to cathode voltage to be selected independently. In most previous experiments the anode to cathode voltage

9

was 45 to 50 V. The data shown here demonstrate that operating with an anode to cathode voltage of >80 V results in an efficiency increase compared with 45 V of >50%, see also Fig. 10.

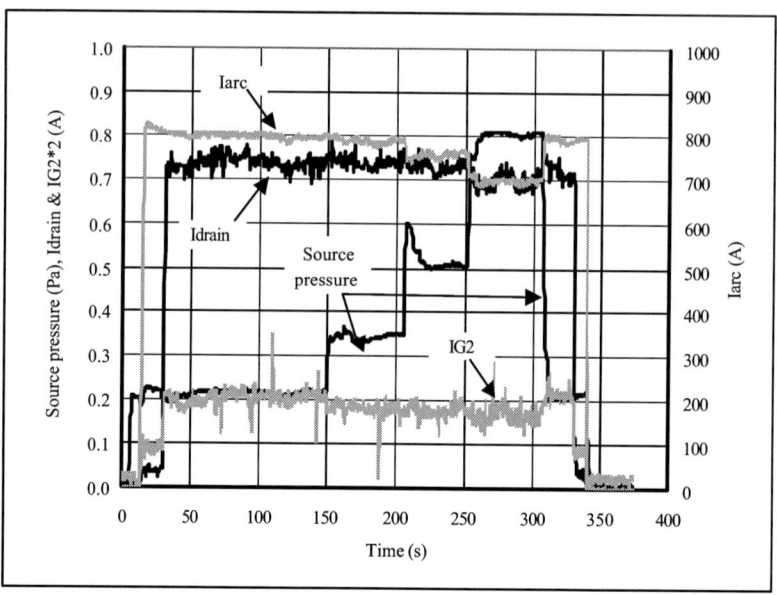

FIGURE 7. Shot demonstrating that the accelerated H⁻ current (assumed equal to I_{drain}) does not vary with the source pressure.

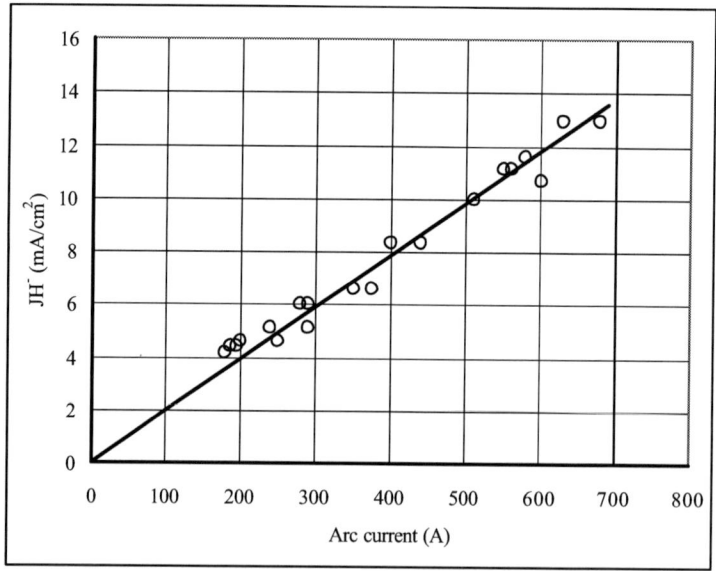

FIGURE 8. JH⁻, the accelerated negative ion density derived from Idrain, as a function of the arc current at constant anode to cathode voltage (75 V).

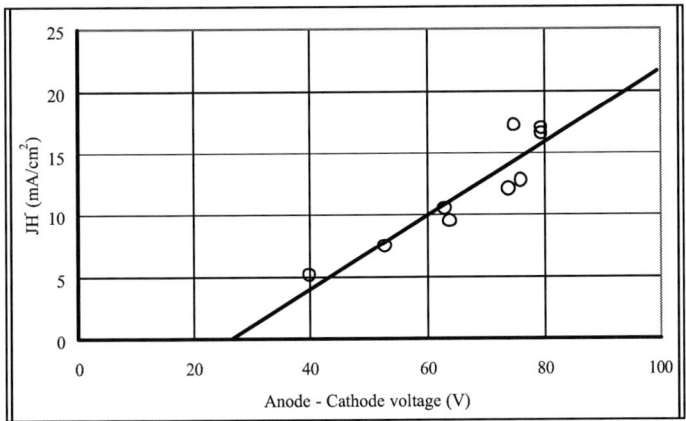

FIGURE 9. JH⁻, the accelerated negative ion density derivedfrom Idrain, as a function of the anode to cathode voltage at constant arc current (600 A).

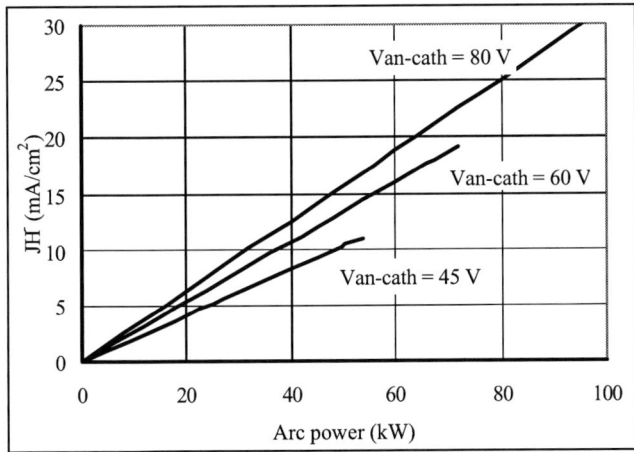

FIGURE. 10 Calculated accelerated negative ion density as a function of arc power for different anode to cathode voltages, with the maximum arc current limited to 1200 A. It is to be noted that the actual negative ion current depends on the operating conditions such as the amount of Cs in the source and the time of operation since the Cs was introduced into the source. The curves shown here are only valid for the operating conditions pertaining when the data of Figs. 8 and 9 were taken.

Plasma Characteristics

The plasma characteristics (electron density and temperature) 17 mm in front of the PG have been measured using a small cylindrical Langmuir probe as a function of the arc current and of the anode to cathode voltage, in each case keeping all other parameters constant. For these measurements there was no caesium seeding of the ion source, but it has been previously found that caesium seeding has no effect on the plasma density or the "bulk" (see below) electron temperature. Judged from the probe characteristics the electron energy distribution seems to be reasonably well described as a bi-Maxwellian with >99% of the electrons having the lower, "bulk", temperature. The variation of the plasma density with arc current is found to be linear, see Fig. 11.

The variation of the plasma density with the anode to cathode voltage is approximately offset linear, but tends to saturate above 80 V, see Fig. 12.

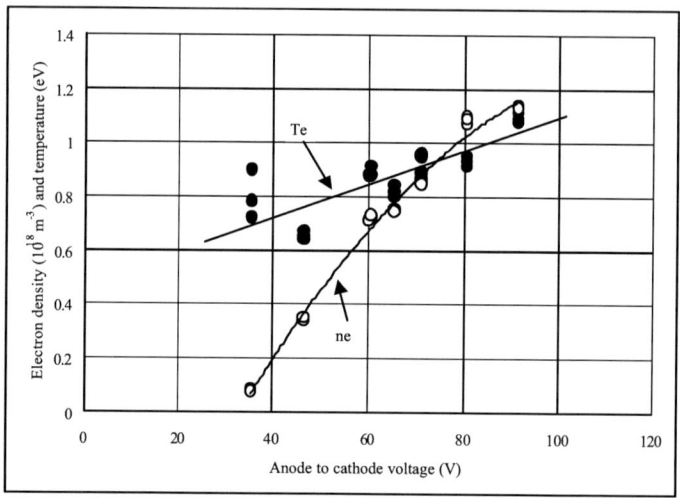

FIGURE 11. The plasma density and bulk electron temperature 17 mm in front of the PG as a function of the arc current at constant anode to cathode voltage (70 V). The probe was at x = 20 mm, y = 0 mm where x is horizontal, y vertical and the origin is the centre of the source in the x-y plane.

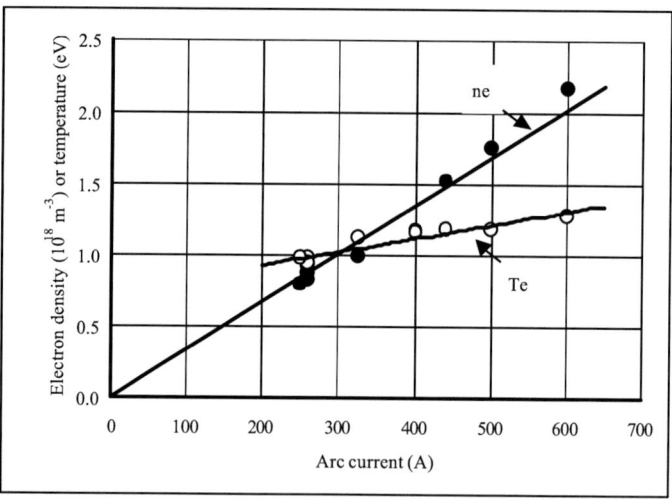

Figure. 12. The plasma density and bulk electron temperature 17 mm in front of the PG as a function of the anode to cathode voltage at constant arc current (260 A). The probe was at the centre of the source in the x-y plane, where x is horizontal and y vertical.

Fractional Dissociation

The fractional dissociation has been measured as a function of the arc current. It is speculated that in a caesiated ion source most of the accelerated H⁻ arises from surface scattering of neutral atoms and positive ions from the PG surface. On MANTIS the magnetic filter in front of the PG has been changed from ≈450 G.CM to ≈900 G.cm. The result was that the electron flux to the PG was reduced by a factor 10 without significantly changing the negative ion yield [2]. (Note that in reference 2 the filter strength, S, is defined as $S = \int_{-\infty}^{\infty} B.d\ell$ where B is the local field strength and ℓ is the distance from the PG. This gives a value of S about twice the definition used in section 1 of this paper.) As any reduction in electron flux implies a similar reduction in the positive ion flux, it can be concluded that the main source of the H⁻ is H atom scattering and not H⁺ scattering. As the H⁻ yield is found to vary linearly with the arc current, if this hypothesis is valid the H atom fraction should increase monotonically with the arc current. The ratio of the Hγ light to that from the d $^3\Pi_u$ to a $^3\Sigma_g^+$ transition (Fulcher spectrum around 600 nm) of H_2 is correlated with the ratio of H atom density to the H_2 molecule density [7]. Using this technique the fractional dissociation (H/H_2) has been measured along a vertical line of sight 17 mm in front of the PG as a function of the arc current, see Fig. 13. The variation is found to be approximately linear with arc current, supporting the aforementioned hypothesis.

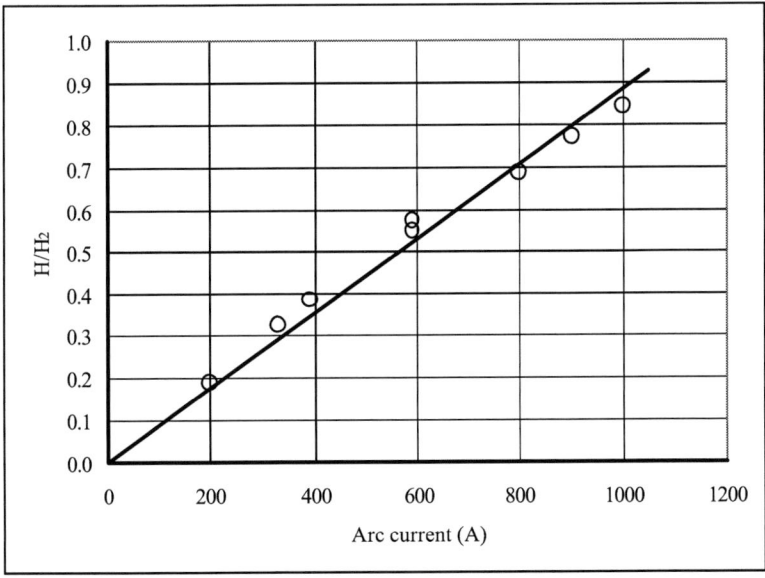

FIGURE 13. The variation of the fractional dissociation as a function of the arc current at constant anode to cathode voltage (51 V).

Gas Temperature

The temperature of the gas in the source has been measured. As the power in the arc discharge is rather high ($\leq$80 kW) it has been speculated that the gas emerging from the ion source might be hot. This would be beneficial as it would result in a lower gas density, hence lower stripping losses, in the accelerator. Adding a small amount of nitrogen to the discharge (10%) the gas temperature can be obtained from the radiation of the N_2^+ molecule (B - X transition, around 391 nm) [7]. The gas temperature is found to be between 2000 and 2500 K at a discharge power of 40 kW and a pressure of 0.3 Pa.

CAESIUM CONSUMPTION

Very high Cs consumption rates have been found during long pulse operation: the amount of Cs "consumed" per aperture in the PG is up to 1500 times that assumed for the ITER source, which is based on extrapolation from short pulse operation. A possible, partial, explanation is that during the operation of long pulses (>100 s), the source walls reach thermal equilibrium at a temperature (typically 60 °C at an arc power of 45 KW) substantially higher than during short pulses ($\approx$20 °C). The increase in the vapour pressure of Cs on the source walls would result in an increase in the Cs flow from the walls into the discharge by up to a factor of 60.

In order to understand better what happens to Cs injected into the ion source, when the ion source was opened after an experimental campaign it was examined carefully. The source itself was covered with what looked like a moist tungsten layer. In order to determine the percentage of Cs left in the source, it was cleaned with water, and the water kept for analysis. Initially the cleaning water was opaque, dark grey, but overnight it became clear with a grey precipitate at the bottom. Chemical analysis of the clear water showed that approx. 4.5 ±0.9 g of Cs was inside the source when it was cleaned. The grey precipitate is presumed to be tungsten (W). As $\approx$5 g had been injected into the source since it was last cleaned, this shows that essentially all the Cs was still present within the source. This was unexpected as the Cs effect has started to disappear and evaporation and loss through the accelerator apertures alone should have significantly reduced the quantity of Cs in the source. It is speculated that the Cs on the walls of the source was either covered by a layer of evaporated W or trapped in a matrix of W on the wall.

If the Cs is "buried", or "blocked", on the wall by W evaporated from the filaments, control of the evaporation of W could prove a key part in the operation of this source for high current density. The W filaments are operated between 2800 and 3000 K in order to obtain the required electron emission current density. At this temperature the evaporation of W from the filaments is significant: it is calculated that the W flux is sufficient to cover all the inner surface of the ion source with a monolayer of W in 125 s of operation. It is proposed to reduce the operating temperature of the W filaments, reducing the evaporated W into the source. This could be achieved by operating at higher anode-cathode voltages and lower emission current, therefore reducing the filament temperature, or by operating with thoriated tungsten filaments, which would allow for operation at 2100 K with the required electron emission density [8].

CONCLUSIONS AND DISCUSSION

The experiments reported here show that:

- The low H⁻ or D⁻ current density measured at the calorimeter on the MANTIS test bed during long pulse operation cannot be explained by lost accelerated electrons arising from either extraction from the ion source or creation by stripping in the accelerator.
- The low H⁻ or D⁻ current density measured at the calorimeter can be explained, at least partly, by poor beam transmission, which may be due to thermal expansion resulting in distortion of the acceleration grid.
- The negative ion yield has been measured as a function of the arc current and anode to cathode voltage. This has demonstrated that the negative ion yield at a given arc power is optimum at anode to cathode voltages of ≈80 V.
- The reduced PG temperature effect measured during long pulse operation could be partly explained by enhanced evaporation of Cs from the source walls at the equilibrium temperature reached during long pulse operation perturbing the dynamic balance between the arrival of Cs from the source walls and the evaporation from the plasma grid.
- The H atom density in front of the PG is measured to vary approximately linearly with the arc current, which supports the hypothesis that the extracted and accelerated H⁻ is created mainly by surface scattering of H as H⁻ from the PG surface.
- The consumption of Cs is many times larger than expected, which may be partly explained by the increased evaporation from the "hot" source walls. However it is found that most of the Cs remains in the source even after it was expected to have been lost from the source. It is speculated that the Cs could be "blocked" on the walls either by burial under layers of tungsten evaporated from the filaments or by being trapped in a loose matrix of tungsten on the source walls.
- The temperature of the gas emerging from the Kamaboko III source is found to be between 2000 and 2500 °C for an arc discharge power of 40 kW and a filling pressure of 0.3 Pa.

REFERENCES

1 T. Inoue, E. Di Pietro, M. Hanada, R. S. Hemsworth, A. Krylov, V. Kulygin, P. Massmann, P. L. Mondino, Y. Okumura, A. Panasenkov, E. Speth and K. Watanabe.
Design of neutral beam system for ITER-FEAT.
Fus. Eng. and Des. 56 – 57, pp 517 – 521, (2001).
2 R. Trainham, C. Jacquot, D. Riz and A. Simonin.
Negative ion sources for neutral beam injection into fusion machines.
Rev. Sci. Inst., 69, 2, pp 926 – 928, (1998).

3   D. Boilson, H. P. L. de Esch, R. S. Hemsworth, M. Kashiwagi, P. Massmann and L. Svensson.
    Long pulse operation of the KAMABOKO III negative ion source.
    Rev. Sci. Inst., 73, 2, pp 1093 – 1095, (2002).
4   Y. Okumura, M. Hanada, T. Inoue, H. Kojima, Y. Matsuda, Y. Ohara, M. Seki and K. Watanabe.
    Caesium mixing in the multi-ampere volume H⁻ ion source.
    5th Int. Symp. on the Production and Neutralisation of Negative Ion Beams, pp 169 - 183, (1990).
5   J. B. Taylor and I. Langmuir.
    The Evaporation of Atoms, Ions and Electrons from Caesium Films on Tungsten.
    Phys. Rev., 44, 6, pp 423 – 458, (1933).
6   M. Hanada, M. Kashiwagi, T. Morishita, M. Taniguchi, Y. Okumura, T. Takayanagi and
    K. Watanabe.
    Development of negative ion sources for the ITER neutral beam injector.
    Fus. Eng. And Des. 56 – 57, pp 505 - 509, (2001).
7   U. Fantz.
    Atomic and molecular emission spectroscopy in low temperature plasmas containing hydrogen and
    deuterium.
    Max-Planck Institute fuer Plasmaphysik, IPP-Report, IPP10/21, (2002).
8   B. Gellert and W. Rohrbach.
    Thermionic emission investigation of materials for directly heated cathodes of electron tubes.
    XVI International Symposium on Discharges and Electrical Insulation in Vacuum, Moscow to St.
    Petersburg, pp.501-504, (1994).

Status of Negative-Ion-Based Neutral Beam Injectors in LHD

Osamu Kaneko, Yasuhiko Takeiri, Yoshihide Oka, Katsuyoshi Tsumori,
Masaki Osakabe, Katsunori Ikeda, Kenichi Nagaoka,
Toshikazu Kawamoto, Tomoki Kondo, and Mamoru Sato

National Institute for Fusion Science
322-6 Oroshi Toki 509-5292 Japan

Abstract. Recent progress in LHD N-NBI system is described. The system was designed to provide 15 MW of port-through beam power by three Beam Lines (BL-1 to BL-3) with 180 keV hydrogen for 10 sec. However, starting from 3.2 MW injection by two beam lines in 1998, their performance has not yet satisfied all the specifications at the same time. This is because several difficulties were found in large negative ion sources while the specifications were determined based on the R&D results of small ion sources. Hence, continuous efforts have been done to improve the performance of negative ion source during the machine maintenance period between the experimental campaigns of LHD. As a result, the injection power has been increased year by year, and in the latest LHD experimental campaign the total port-through power of 13.1 MW was achieved, where R&D on the ion source of BL-1 was most successful. Two improvements have been done on the ion source of BL-1. One is the optimization of magnetic cusp structure by modifying the shape of plasma source, which increased the negative ion current. The other is the adoption of multi-slot aperture for the grounded grid, which reduces the conditioning time of the accelerator dramatically. As a result, the specifications of beam energy and port-through power were satisfied in BL-1 at the same time. The remaining problem is a pulse length which is still limited due to high heat load on the grounded grid.

INTRODUCTION

The neutral beam injector (NBI) for the Large Helical Device (LHD) is the first negative-ion-based NBI (N–NBI) system that was constructed as a main plasma heating facility in helical systems [1]. Because tangential beam injection is necessary and the required beam energy is so high (180 keV for hydrogen) that it is only possible to design a system by using negative ions. Therefore no conventional positive-ion based NBI has been adopted in LHD. Consequently high reliability (as in the P-NBI system) is required as well as the high available power for N-NBI system. It is challenging to meet these requirements because negative ion technology has not been matured to the P-NBI system. Since LHD is designed to confine net current free plasma, more than one beam line is required to balance the direction of beam and to compensate the induced current. Actually an interlock system is constructed to quit the beam injection when the plasma current exceeds an upper limit in LHD. It is a different situation from JT-60U tokamak, which also has introduced an N–NBI facility

CP763, *Production and Neutralization of Negative Ions and Beams*, edited by J. D. Sherman and Y. I. Belchenko
© 2005 American Institute of Physics 0-7354-0248-5/05/$22.50

specified for proof of principle experiments on beam induced current drive [2]. The NBI system in LHD has three beam lines, and each beam line was designed to deliver 180 keV, 5 MW, 10 s neutral beam into LHD by two high efficient cesium seeded negative ion sources with an external magnetic filter. This negative ion source had been developed at NIFS [3]. The construction of the injection system began in 1996 [4], and the injection experiments started in 1998.

Although the high energy NBI heats mostly plasma electrons, it has revealed an excellent heating performance in LHD as in other small helical devices where about a half of beam power directly goes to plasma ions [5]. In LHD, NBI has also shown its potential other than plasma heating, such as the plasma initiation [6] and the control of rotational transform by its inducing current [7]. The plasma production by beam is a unique and a very reliable method in LHD. It is very simple to inject the beam into the thin gas target, and it works even under the low confining magnetic field strength. High beta plasma studies have made a remarkable progress owing to this method [8]. The tangential high energy NBI is now the most useful auxiliary tool to initiate, to heat and to control plasma in LHD.

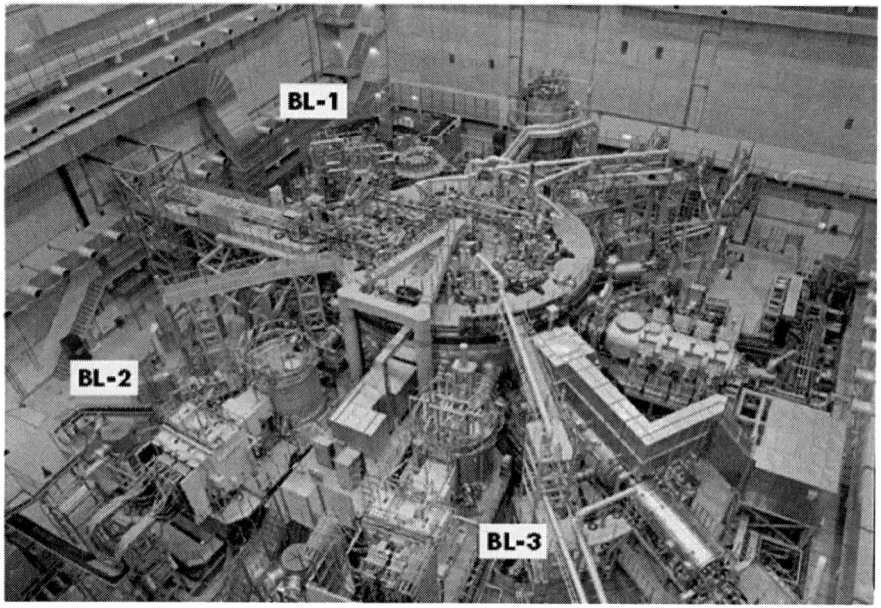

FIGURE 1. A bird's-eye view of LHD experimental hall. Three tangential beam lines are installed

PRESENT STATUS OF LHD N-NBI

The neutral beam injection experiments started in 1998 with 2 MW of port-through power by two ion sources with beam energy of 100 keV. Because the operation of large ion source still had several technical problems, the achieved injection power was far from its design value. Therefore, the efforts to improve the performance of negative ion source have been carried out even after the beam injection experiments started. LHD usually has a six-month-long experimental campaign inevery year, and the improvement of negative ion sources and of other components were carried out during the off-machine time. As a result, the performance of NBI has been improved year by year. The progress of the total injected port-through power and those achieved in each beam line are summarized in Table 1. It can be said that negative ion beam technology has been established at the same level of conventional positive ion technology such as handling 100keV, 10MW beam [9]. A big achievement in the latest experimental campaign is that one of the specifications (port-through power) has been achieved at the beam line #1. Here, the port-through power is evaluated based on the measurement of heat load on the armor plate located in the vacuum vessel of LHD. The measurements are done in the presence of magnetic field of LHD, and therefore both the geometrical loss and the re-ionization loss of the beam in the drift duct are eliminated from the evaluation. We believe that the evaluation of port-through power in LHD is most accurate compared with other NBI systems in the world [10]. The injection power efficiency (the ratio of the injected port-through power to the electric power consumed for ion beam production) is around 0.35. The highest averaged negative ion current density is 30 mA/cm^2 over beam extracting area. Although the specified pulse length is 10 seconds, it is able to be extended at low beam power. Stable long pulse discharges were obtained by NBI in LHD [11]. So far the maximum length of 110 s was achieved, which was limited by insufficient control of plasma grid temperature.

TABLE 1. Progress of Achieved Port-Through Power

Year	Beam Line # 1	Beam Line # 2	Beam Line # 3	Maximum Total Power
1998	2MW / 100keV / 2s	2MW /100keV / 2s 0.6MW / 66keV / 20s	-	3.7 MW
1999	2MW / 133keV / 2s	3.1MW /164keV / 2s 0.5MW / 100keV / 80s	-	4.5 MW
2000	3MW / 152keV / 2s	3.6MW /166keV / 2s 0.8MW / 86keV / 64s	-	5.2 MW
2001	3.5MW / 165keV / 2s 1.1MW / 104keV / 10s	3.6MW /166keV / 2s 0.8MW / 86keV / 64s	3.3MW /165keV / 2s 0.1MW* / 81keV / 110s	9.0 MW
2002	4.4MW / 180keV / 2s	2.9MW /162keV / 2s	3.7MW /165keV / 2s	10.3 MW
2003	5.7MW / 183keV / 1.6s	4.1MW /181keV / 1s	3.9MW /174keV / 1s	13.1 MW

use only one ion source

19

Figure 2 shows a history of injected port-through power for three beam lines in the latest experimental campaign. The beam conditioning precedes the start of LHD experiment, but the beam pulse length is limited by the calorimeter and it is not long enough for plasma experiments. Therefore the long pulse beam conditioning is only possible by using LHD plasma as a target. This is why the port-through power increases gradually with the beam shot number, as it is shown in Fig. 2. It is noted that the beam is injected in every three minutes in LHD, which is possible because the magnetic field of LHD exists in steady state and the duty of plasma shot is determined by the heating devices. This interval (3 min) is desirable for operating negative ion source because the temperature of plasma grid should be kept high for efficient negative ion production, and the arc discharge power is utilized to raise the grid temperature in our ion source. On the other hand, a large amount of shots owing to this high repetitive operation is hard on the cathode lifetime. The filament cathode does not last throughout the experimental campaign due to high current arc discharge of the ion source. The arc chamber must be opened to air for cathode changing during the campaign. This opening affects the high voltage withstanding of grids, and additional beam conditioning is needed. This is the reason why the maximum injection power is usually attained at the last phase of the experimental campaign.

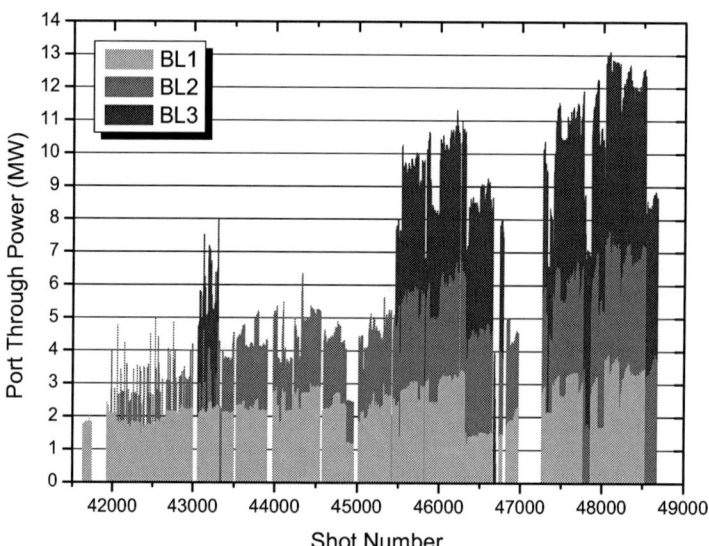

FIGURE 2. History of injected port-through power in the seventh experimental campaign of LHD. The distribution of three beam lines is shown by different gray scaled bars.

Unfortunately, the beam line #1 suffered several troubles in this campaign, and its operation period was shorter compared with other two beam lines. Nevertheless the highest performance was achieved successfully in this beam line, as it can be seen in Fig. 3a. The high performance was obtained because a large modification has been done in the sources of this beam line, as described in the next section. There are two reasons for achieving high power. Usually it takes a long time (many shots) to increase the beam energy by conditioning. The new ion source of beam line #1 has solved this problem by adopting a new accelerator grid system. As it is seen in Fig. 3b, that once BL-1 joined the experiments from shot #46000, the beam energy attained the specified value (180 keV) quickly and stably maintained its level. The new ion source has also succeeded to improve the efficiency of negative ion production, and to get high negative ion current as shown in Fig. 3c.

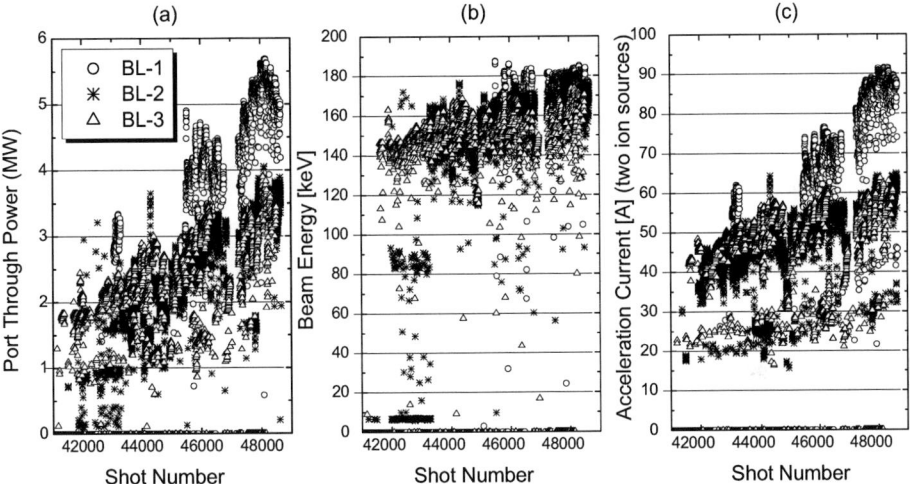

FIGURE 3. History of injected port-through power, beam energy and acceleration current of each beam line. Two ion sources are operated by single acceleration power supply in each beam line.

The practical pulse length is limited by the heat load on the grounded grid which is less than design value of 10 seconds. So far, the extracted electrons are suppressed well by the magnetic filter and by the electron suppression magnet in the extraction grids. The amount of extracted electrons is smaller than that of negative ions, and the heat load on the extraction grid is in the tolerable level. On the other hand, the heat load on the grounded grid is about 15% of acceleration power, which is very high. We limit ed the pulse length by monitoring the temperature increase of grounded grid outlet cooling water. A typical pulse length is 2 s at the maximum beam power 2.5 MW / ion source and correspondingly 10 s at 0.5 MW / ion source. This is shown in Fig. 4.

When the beam power is reduced to 0.5 MW, the cooling capacity becomes higher than input power, and ion source can operates in steady state. The power supplies and the cryo-sorption pumps of NBI can operate 30 minutes continuously at this low beam power. However, two constraints exist: the cooling capacity of the inner wall protector of drift tube is not designed for a such long pulse heat load, and that the cooling of plasma grid is not enough to keep its temperature during a long pulse arc discharge. As a result, the efficiency of negative ion production becomes worse due to excess temperature in plasma grid, and the beam divergence becomes worse, and the temperature of the protector plates becomes high. The optimization of long pulse operation is an important issue in future.

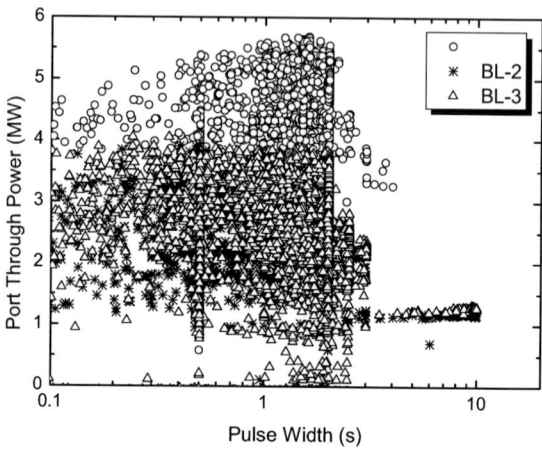

FIGURE 4. Port-through power vs pulse length for each beam line in the last campaign. Typical pulse length is 2 s. The maximum pulse length depends on injection power, and was limited by the heat load on the grounded grid, i.e. by the absorbed energy in the grid during the pulse.

IMPROVEMENT OF NEGATIVE ION SOURCE

High Energy Beam Acceleration

Usually the voltage withstanding of the ion source accelerator increases only after suffering many break downs during beam acceleration. It takes more times when the beam energy becomes high. Although the current density of negative ion source is much lower than that of positive ion source, it still needs many shots for conditioning. One reason is considered that the number of beam extraction apertures is large. Our ion source has a beam extraction area of 25 cm by 125 cm, and has about 900 extraction holes. If the break down occurs due to the interference of the beam with the peripheral of grid hole, more conditioning shots are required when the number of holes increases. The other reason of occurrence of many break downs may be attributed to the spatial non-uniformity of negative ion current density over the grid, because the optimum perveance does not much over all the beam extracting area. Then the two ways to shorten the conditioning process are considered; one is to reduce the area of interference between the beam and the grid, and the other is to improve the spatial uniformity of the negative ion production. The latter is also important to increase the total beam current. In the design of new accelerator of the ion sources of beam line #1, the interference area has been reduced by adopting multi-slot rather than multi-holes for grounded grid (Fig. 5). It was also expected to reduce the heat load on the grounded grid by the beam (ions, and also electrons and neutrals due to low gas pressure between the grids). The result was remarkable. The conditioning time redu-ced dramatically. The specified beam energy (180 keV) was obtained very quickly

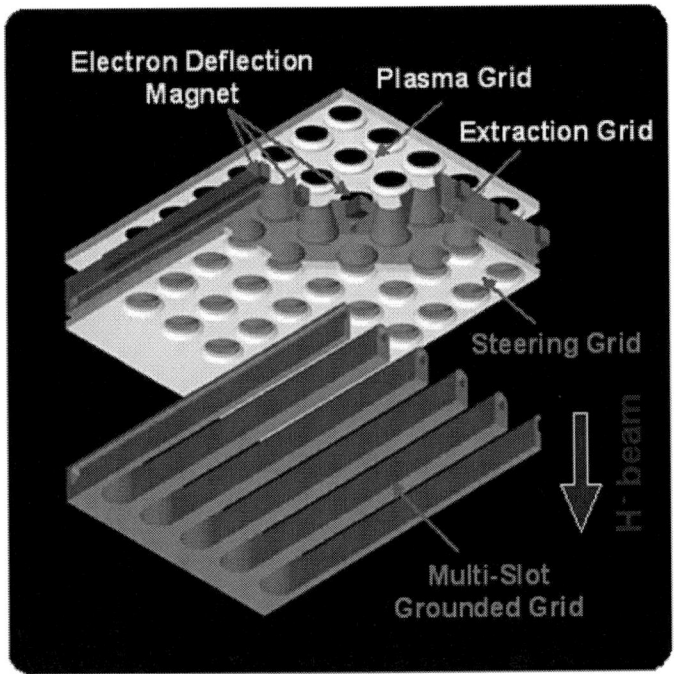

FIGURE 5. A schematic drawing of new grid system with multi-slot grounded grid.

and it was the first achievement of designed value [12]. It is also noted that the heat load on the grounded grid reduced to about a half of that on multi-aperture grid, which is consistent with the increase of transparency of the grid aperture, that is, reduction of the area of interference between the beam and the peripheral of aperture.

The other interesting experimental result was observed recently in the beam line #2. We are carrying out different R&D's in different beam lines. In the beam line #2, new plasma grid was introduced, which was designed to cool the grid more firmly and uniformly by adapting water cooling channels at the periphery of each grid sector. The objective of this grid is to keep the grid temperature at the optimum value during a long pulse operation. When these new plasma grids were installed and started testing, an unexpected result was obtained that the conditioning time was apparently reduced. It is considered that the distributed cooling channel becomes a reservoir of cesium which helps to improve the spatial uniformity of negative ion production. A uniform negative ion density makes the good optical beam conditions over the beam extracting area.

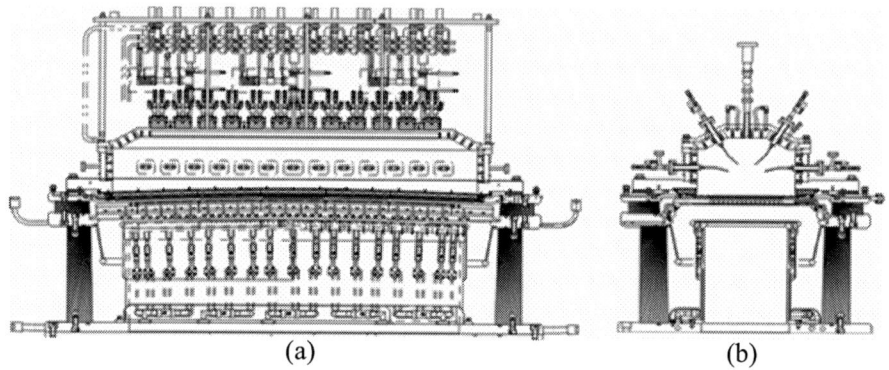

(a)   (b)

FIGURE 6. Cross sectional views of negative ion source of the beam line #1. The size of the arc chamber is 140 cm (L) x 35 cm (W) x 24 cm (H). The filaments at the back plates are spares. The accelerator adopted single stage acceleration, and the grid is divided into five segments along the long direction. Each segment is inclined toward the focal point 13 m downstream of the ion source.

High Efficiency Negative Ion Production

The efficiency of negative ion production depends on the cesium condition in the arc discharge. The amount of cesium and the temperature of plasma grid are the key issues of controlling the efficiency. In our ion sources, the injection of cesium is carried out little by little even during arc discharge. The total consumption of cesium is 9 g for one ion source throughout an experimental campaign where about 20,000 shots are used including conditioning shots. The plasma grid temperature is also important. The optimum temperature is around 250 $^{\circ}$C at the periphery of extraction area. In our ion source, the plasma grids are heated by using the power of arc discharge, and the temperature is kept by insulating the grids from the water cooled supporting structure. The level of thermal insulation is delicate to determine the grid temperature. Usually it is difficult to keep the temperature constant during the beam pulse. The temperature increases during arc discharge. Therefore the temperature should be sufficiently high to produce negative ions during beam extraction, and should be cooled enough between the shots to prevent continuing temperature increases shot by shot. Of course this method is not appropriate to very long pulse operation because the grid temperature continues to increase during arc discharge and exceeds the optimum value. As for increasing the total ion current, the improvement of spatial non-uniformity is another issue. The arc chamber of beam line #1 was designed to fit the shape of its confining magnetic cusp structure to the magnetic filter field as shown in Fig. 6, where the corners of the arc chamber were cut. However the plasma density was still small at both ends of long direction. Two more filament cathodes were added in this direction to adjust the arc current distribution. Then the total current increased about 15%, which directly contributed to increase the injection power.

Beam Divergence

Good optics of negative ion beam can compensate the disadvantages of N-NBI such as its long beam line and needs of large ion source. In our negative ion source, beam divergence is around 10 mrad which assures a good port through efficiency of the beam. As described in the previous section, a new accelerator of the beam line #1 with multi-slot grounded grid has shown an excellent performance from the view point of beam conditioning and the heat load on the grid. However, an issue remains on beam optics.

In the conventional multi-aperture grid system, the holes of grounded grid are displaced systematically for beamlet steering and multi-beamlet focusing, but it is not possible for the combination of multi-aperture multi-slot to focus the beamlets in the direction parallel to the slot. In order to steer the beamlets in this direction, an extra grid was added downstream side of the extraction grid as shown in Fig. 5. The electric potential of this extra grid (steering grid) is the same as that of extraction grid, and its holes are systematically displaced. Although the steering angle becomes more sensitive to the shift of hole at the steering grid compared with that at the grounded grid, i.e. the more accuracy of manufacturing and assembling is required for steering grid, this technique works well.

Another problem still remains. It is well known that the beam divergence depends on the current density, the extraction electric field strength and the acceleration electric field strength. Once the optimum condition between the current density and the extraction field is determined and fixed, the beam divergence depends on the acceleration field and usually an optimum value exists. The problem is that the optimum optical conditions are different between the parallel and perpendicular directions along slot. The electric potential near the slot makes a strong fringe field across the direction of perpendicular to the slot and the electrostatic lens is formed. However, it is almost flat and no electrostatic lens is formed in the direction parallel to the slot. Therefore the dependence on the acceleration field strength is considered to be weak in this direction. This situation is shown in Fig.7, where the horizontal axis shows the ratio of the electric field of acceleration region to that of extraction region. The vertical axis shows the width of the beam profile measured at the calorimeter. The horizontal width indicated in the figure corresponds to the direction parallel to the slot of the grounded grid. It should be noted that the calorimeter is located at a half of the beam focal length, and the measured profile still reflects the size of the ion source. Therefore a large vertical width does not always mean that the beam divergence is large. The problem is that the optimum field ratio is different between vertical and horizontal direction. We chose an intermediate field ratio around 1.8 in the operation, but in this case it is found that the beam divergence in the vertical direction was still large. This caused not only the reduction of port through efficiency but also the damage inside the drift duct due to the increased heat load in the last campaign.

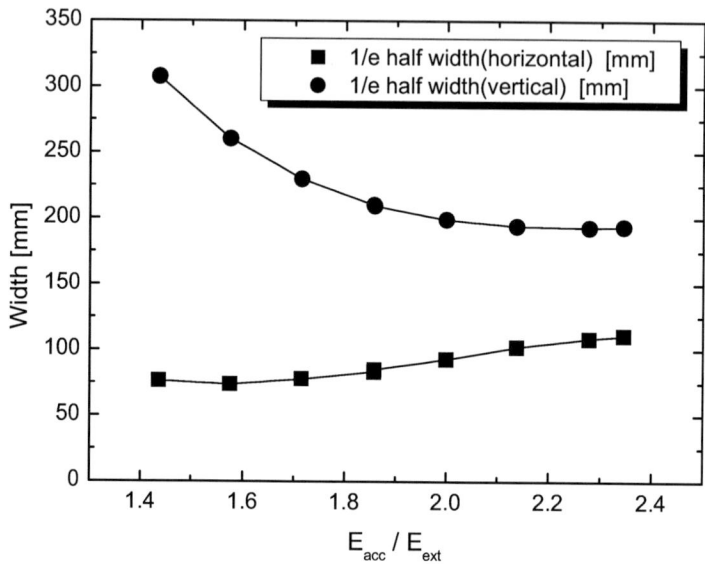

FIGURE 7. Width of measured beam profile on the calorimeter vs the extraction and acceleration electric field ratio. The slots of grounded grid are lined up in the vertical direction.

FOR FURTHER IMPROVEMENT

The efforts on improving the performance of negative ion source still continue for the next experimental campaign which will start in September 2004. The mismatch of optimum electric field ratio owing to the multi-slot structure will be improved by modifying the shape of holes of beam steering grid. The availability of this method was confirmed using a 1/3 scale ion source at the test stand, and it is applied for full size ion sources of the beam line #1. If this method worked well in the next campaign, high power beam would be injected longer than 2 s, because the heat load on the grounded grid is small for the multi-slot structure. After confirming that the beam optics is improved, we will adopt a multi-slot grounded grid for other two beam lines.

A new plasma grid will also be applied for the ion sources of the beam line #2, the cooling of which has been reinforced for long pulse operation. As described in the previous section, more spatially uniform negative ion production is expected on the plasma grid. We hope that the best accelerator will be made by combining the results of these R&D, that is, the plasma grid with cooling channel, the steering grid with deformed aperture, and the multi-slot grounded grid.

REFERENCES

1. T. Mutoh, et al., "Heating Systems of Large Superconducting Helical Device," in *Fusion Technology 1988*, 552 (1989).
2. M. Kuriyama, et al., "Development of Negative-Ion Based NBI System for JT-60U," *J. of Nucl. Science and Technol.*, **35**, 739 (1998).
3. Y. Takeiri, et al., "Negative hydrogen ion source development for Large Helical Device neutral beam injector," *Rev. Sci. Instrum.* **71**, 1225 (2000).
4. O. Kaneko, et al., "Negative-Ion-Based neutral Beam Injector for the Large Helical Device," in *Fusion Energy 1996*, 539 (1997).
5. A. Komori, et al., "Recent results from the Large Helical Device," *Plasma Phys. Control. Fusion* **45**, 671 (2003).
6. O. Kaneko, et al., "Plasma startup by neutral beam injection in the Large Helical Device," *Nucl. Fusion* **39**, 1087 (1999).
7. N. Ohyabu, et al., "Influence of Beam Flow on the Electron Transport in Low Density LHD Discharges," in *Proc. of 30th EPS Conf. on Controlled Fusion and Plasma Physics*, St.Petersburg (2003).
8. S. Sakakibara, et al., "Effect of MHD activities on pressure profile in high-beta plasmas of LHD," *Plasma Phys. Control. Fusion* **44**, A271 (2002).
9. O. Kaneko, et al., "Engineering prospects of negative-ion-based neutral beam injection system from high power operation for the large helical device," *Nucl. Fusion* **43**, 692 (2003).
10. M. Osakabe, et al., "In situ calibration of neutral beam port-through power and estimation of neutral beam deposition on LHD," *Rev. Sci. Instrum.*, **72**, 590 (2001).
11. Y. Takeiri, et al., "Plasma characteristics of long-pulse discharges heated by neutral beam injection in the Large Helical Device," *Plasma Phys. Control. Fusion*, **42**, 147 (2000).
12. K. Tsumori, et.al., "High power beam injection using an improved negative ion source for the large helical device," *Rev. Sci. Instrum.* **75**, 1847 (2004).

Experimental Results with the New ITER-like 1 MV SINGAP Accelerator

L. Svensson, D. Boilson*, H. P. L. de Esch, R. S. Hemsworth and P. Massmann

Association EURATOM-CEA, CEA/DSM/DRFC, CEA -Cadarache,
13108 ST PAUL-LEZ-DURANCE (France)
*Association EURATOM-DCU , PRL/NCPST, Glasnevin, Dublin 13, Ireland

Abstract. A new "ITER-like" accelerator, which is a scaled down version of the ITER SINGAP (SINgle GAP, SINGle APerture) accelerator, has been built and installed on the Cadarache 1 MV test bed. The objective is to demonstrate reliable D^- beam acceleration as close as possible to 1 MeV with a current density $\bar{j} \approx 200\,A/m^2$ with the beam optics required for ITER, i.e. a beamlet divergence of ≤ 7 mrad and beamlet steering within ± 2 mrad of that specified. High voltage hold off tests have been performed and 940 kV has been held without breakdowns. The first beams up to 850 keV (D^-, 15 A/m^2) have been obtained after 4 weeks of experiments and the highest current density that has been obtained so far is 85 A/m^2 (D^-, 580 keV).

INTRODUCTION

The European concept for a 1 MeV, 40 A negative ion based accelerator for the neutral beam system on ITER, the SINgle GAP, SINgle Aperture (SINGAP), is an attractive alternative to the ITER reference design, the so-called Multi-Aperture, Multi-Grid accelerator. A prototype SINGAP accelerator has been used for several years and produced D beams of 910 keV, 60 A/m^2 [1]. The measured beam profiles on the target agreed well with those predicted by calculations [2]. However with the prototype accelerator it was not possible to produce beams with the optical quality required for ITER [2], i.e. a beamlet divergence of ≤ 7 mrad and beamlet aiming within ± 2 mrad of that specified. Therefore a new accelerator has been designed and built in order to demonstrate that the beam optics required for ITER can be achieved with a SINGAP accelerator.

THE SINGAP TESTBED

The Cadarache 1 MV negative ion beam facility is capable of accelerating 100 mA of H or D^- up to 1 MeV. The negative ions are first accelerated in the pre-accelerator to energies of 10-50 keV and thereafter up to 1 MeV in the post-accelerator. The ion source is at ground potential and the negative ion beam is extracted and accelerated up to high positive potential and stopped on the calorimeter which is at the same high

CP763, *Production and Neutralization of Negative Ions and Beams*, edited by J. D. Sherman and Y. I. Belchenko
© 2005 American Institute of Physics 0-7354-0248-5/05/$22.50

potential. The calorimeter is made of Mitsubishi MFC1-A graphite which has a much higher thermal conductivity in the beam direction than in the orthogonal directions. The beam footprint is measured at the rear of the target with an IR camera.

THE ACCELERATOR

The pre-accelerator consists of a plasma grid, an extraction grid and a pre-acceleration grid. Each grid and its associated water tubes are embedded in a circular stainless steel (SS) grid support plate. These plates are mounted on alumina post insulators from a common SS base plate. The extraction grid and the pre-acceleration grid have aperture patterns of 5 x 5 with a horizontal and vertical pitch of 20 mm. A 20 mm high "kerb" made of stainless steel is fitted at the exit of the pre-accelerator. The kerb modifies the electrostatic potential such that the outer beamlets are deflected towards the beam centre whilst traversing the post-acceleration gap. A cross section of the "ITER-like" accelerator with the ion source can be seen in Fig. 1.

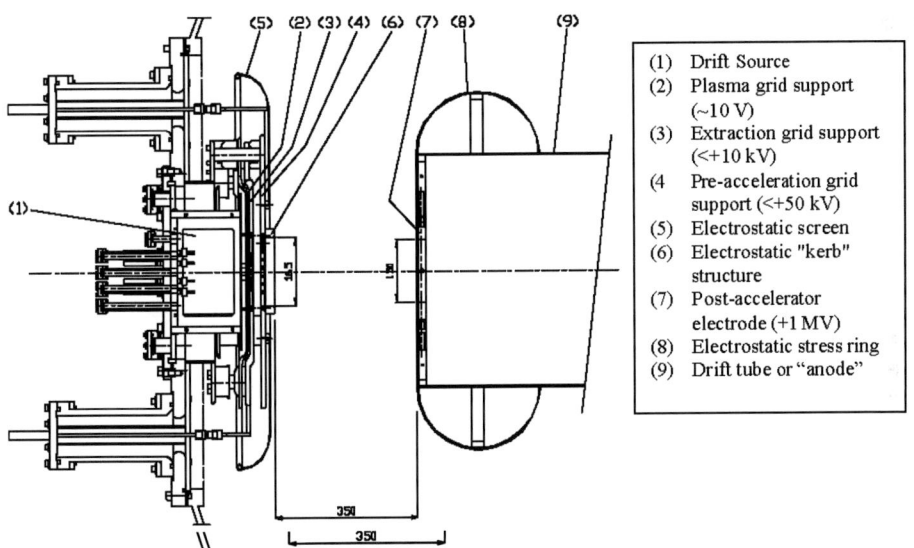

(1) Drift Source
(2) Plasma grid support (~10 V)
(3) Extraction grid support (<+10 kV)
(4 Pre-acceleration grid support (<+50 kV)
(5) Electrostatic screen
(6) Electrostatic "kerb" structure
(7) Post-accelerator electrode (+1 MV)
(8) Electrostatic stress ring
(9) Drift tube or "anode"

FIGURE 1. Vertical section of the "ITER-like" SINGAP beam source.

The Cadarache 1 MV power supply has a current limit of 100 mA, which limits the numbers of apertures on the plasma grid to three when $200 \ A/m^2$ beams are to be produced. The plasma grid has two thermocoax heater elements embedded in the source side of the grid. This enables heating of the plasma grid to ≈300 °C for efficient negative ion production with Cs seeding of the source [3]. The extraction grid and the pre-acceleration grid are both water cooled through horizontal channels between the aperture rows and incorporate CoSm magnets for electron suppression and/or ion trajectory correction. A schematic of the grids with their water channels and the magnets are shown in Figs. 2 and 3. The beamlets formed in the pre-accelerator are accelerated to ≤1 MeV in one step across the main acceleration gap of 350 mm. The post-accelerator electrode has only one large square opening, 150 x 150 mm², and is made of OFHC copper. It can be displaced vertically and horizontally, thus providing aperture offset beam steering to simulate the vertical steering (±0.55°) required on ITER or just for correction of misalignment.

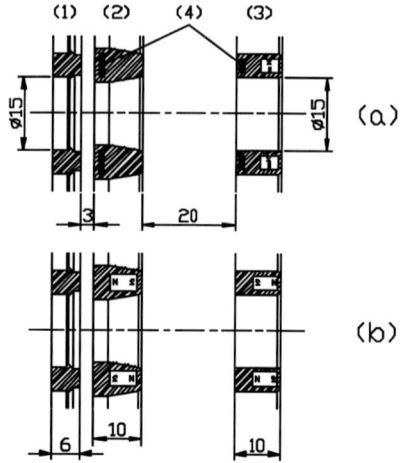

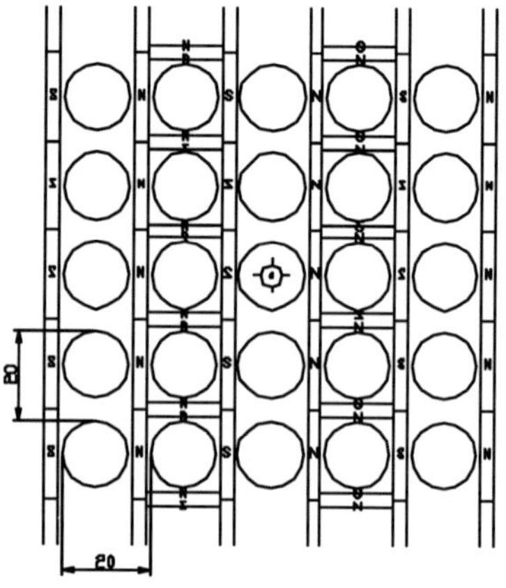

FIGURE 2. Pre-accelerator aperture geometry: (a) vertical (b) horizontal cut.
(1) plasma grid, (2) extraction grid
(3) pre-acceleration grid, (4) water channels

FIGURE 3. Magnet configuration in the pre-acceleration grid viewed from downstream

THE FIRST EXPERIMENTAL RESULTS AND COMPARISON WITH SIMULATIONS

Voltage Holding

A 110 kΩ high power resistor has been installed in series with the 1 MV power supply, replacing the 1 MΩ used previously. The new resistor allow operation with pulse lengths of several minutes but in practise the pulse length is restricted by the existing timing system, to 40 s. It is found that with the extended pulse length the operating time required to condition the accelerator is drastically reduced.

On application of high voltage to the system with no gas flowing and no discharge in the ion source the so called "dark current" appears, the cause of which is not known. This dark current flows from the ground potential structure (mainly the vacuum vessel walls) to the structure at high voltage. The long pulse lengths result in substantial heating of the drift tube and its support structure by the dark current (see Fig. 1) and some hot spots have been observed. Helium is introduced into the vacuum tank to give a pressure of ≈0.03 Pa in order to suppress dark currents [4]. This is higher than the predicted pressure around the accelerator of the ITER injector

(0.02 Pa) [5], but increasing the pressure in the ITER injector to 0.03 Pa (if it is necessary) should pose no serious problems. Breakdown free HV pulses up to 940 kV were achieved after only 160 minutes of accumulated voltage on-time. Higher voltages have not been attempted in order to minimise the risk of breakdowns at higher voltages damaging the 1 MV power supply.

Beam Optics Simulations

The first comparisons between simulations and experiments have been done with the 3 apertures in the plasma grid. Shot 7545 was chosen for the simulation because the experimental data show the three beamlets clearly resolved, which allows a maximum of details to be deduced and compared with the simulation. This shot had 1.8 s of 28 A/m^2 D$^-$ beams, 13 mA in total, as determined from the energy deposited onto the calorimeter. Taking the calculated stripping losses (22%) into account, the extracted current density from the source was 36 A/m^2. The extraction voltage was 2.5 kV, the pre-acceleration voltage 18 kV and the post-acceleration voltage 625 kV. The source pressure was 0.4 Pa, the plasma grid was at 225 °C and the source had been caesiated.

The simulation procedure has been described in detail in [2] and [6]. An addition to the previously described procedure is that the calculated temperature profiles are now corrected for the time and temperature dependent 3D heat diffusion occurring during the transit from the exposed front towards the rear face of the carbon target where the temperature distribution is measured experimentally.

For the conditions pertaining to shot 7545 it is calculated that the pre-accelerated beamlets are 2.8 mm in radius and convergent by 25 mrad. The calculated deflection for the beamlets on the right in Fig. 4 is almost zero because the deflections by the field from the magnetic filter of the ion source and the fields from the magnets embedded in the grids nearly cancel. The beamlet on the left, however, is deflected 12 mrad downwards and 11 mrad to the right (before post-acceleration). The calculated beamlet divergence is around 2.5 mrad. The horizontal direction angle is around 1 mrad, the vertical direction angles are larger due to the magnetic fields. The power density on the carbon target, 3.1 m downstream of the accelerator is calculated assuming Gaussian beamlets with the starting position, width, divergence and steering direction as described above. Due to the low divergence, the calculated power density is very high, up to 2 kW/cm^2, with a rather narrow profile. Fig. 4 gives the calculated profiles at t = 1.0 s and t=2.8 s. In the simulation the beam was on from t= 0 to t = 1.8 s as in the experiment. The profile is taken 1 s after the beam is off (at 2.8 s) in order to allow heat to diffuse from the front to the back of the target. Due to lateral diffusion, the power density in Fig. 4 is lower than the input power density.

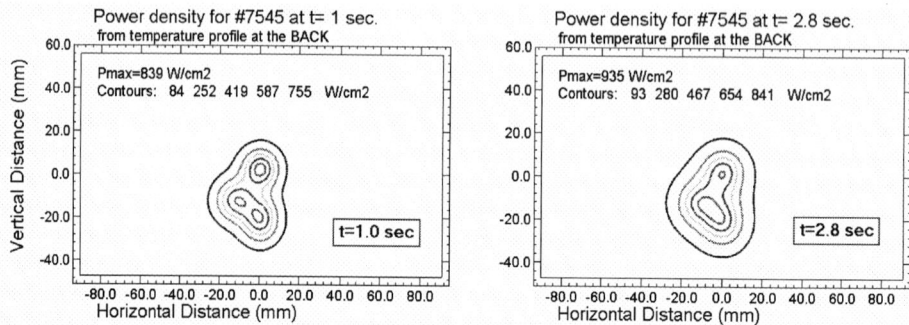

FIGURE 4. Calculated power density profiles at the back of the carbon target. The beam was on from t = 0 to t = 1.8 s. The power per beamlet is 2.7 kW.

Experimental Data

The measured data are shown in Fig. 5. The right picture is derived from the standard Agema 782 IR camera system and taken at 2.8 s, which is 1.0 s after the beam was turned off. The left picture is from a higher resolution Flir 550 IR camera system (on loan from UKAEA, Culham, UK) taken at 1.0 s into the beam shows clearly the separation of the 3 beamlets.

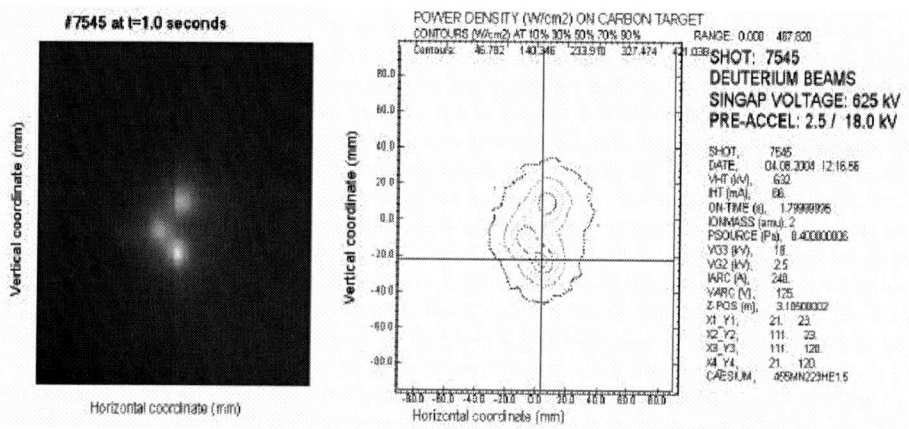

FIGURE 5. Infrared data from the back of the carbon target for shot 7545 (2.8 mA/cm^2 D$^-$). LHS: from the Flir 550 camera, taken during the beams at 1.0 s. RHS: from the Agema 782 camera, taken 1.0 s after the 1.8 s beam.

From Fig. 5 we can deduce that:

- The two beamlets on the right are vertically 30 mm apart (the calculation gives 23 mm).
- The lower two beamlets are horizontally 11 mm apart (the calculation gives 12 mm).
- The maximum power density is only half the calculated value in Fig. 4.
- The power density profile is wider than calculated.

The reasons for the differences between the calculated and apparent beam optics is not understood and it is the object of the ongoing studies.

With the experimental profile information available, we have tried to determine the actual beam optics. If we assume that the starting positions of the beamlets are correct and adjust the steering angles to match the measured positions on the target, the beamlet positions at the target will be correct, but the current density will still be too high. The measured power density indicates that the beamlet optics are worse than those calculated; either the beamlet divergence is higher than calculated, or the beamlet profile not the assumed simple Gaussian. Simply degrading the beamlet divergence to match the peak power density results in calculated profiles that are too narrow at the edge and too wide in the centre (smearing out the individual beamlets). A reasonable match can be found if it is assumed that the beamlet profile is bi-Gaussian with 60% with a divergence of ≈3 mrad and 40% with a 7 mrad divergence. Fig. 6 shows the calculated (solid) and measured (dotted) power density profiles overlaid.

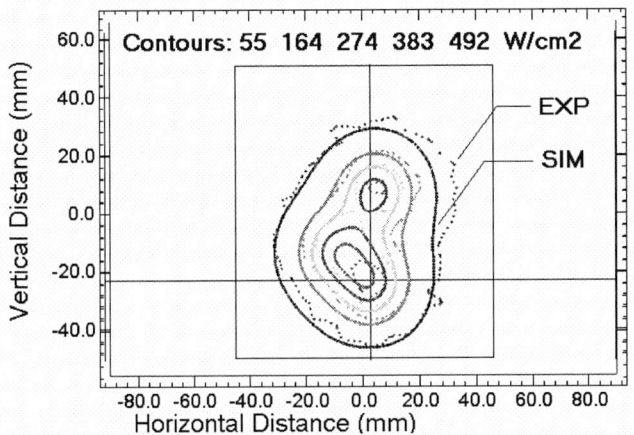

FIGURE 6. Overlay of a calculated profile (including halo and heat diffusion) and the measured profile of shot 7545, 1 s after the shot.

Higher Current Density Experiments

There has been a limited number of shots done so far with the new accelerator (all with D⁻). Before introducing caesium into the source a beam of 850 keV was obtained with a current density of 15 A/m^2. With caesium seeding of the source, beams with a current density of 85 A/m^2 have been obtained at an energy of 580 keV. The analysis of these data is on-going.

CONCLUSIONS AND FUTURE WORK

HV conditioning pulses have demonstrated that the ITER-like accelerator can hold 930 kV without breakdowns. D^- beams have been produced at 850 keV with a current density of 15 A/m^2 at 580 keV, and with a current density of 85 A/m^2. The power is measured calorimetrically on the graphite target.

The first experiments have so far confirmed some aspects of the design of the new ITER-like accelerator, but not all. In particular the experiment data suggest that the beamlets have a bi-Gaussian power density distribution (60% with a divergence of $\approx$3 mrad and and 40% with a 7 mrad) as opposed to the single Gaussian with 2.5 mrad divergence of the simulation. The positions of the beamlets relative to each other are correct (within 1 mrad), except the central beamlet, which is almost 3 mrad too high. The reasons for these differences are not yet understood. Further experiments and simulations will be carried out in an attempt to understand the differences between the calculated and experimental beam profiles and the accelerated current density and the beam energy will be increased to as close as possible to 200 A/m^2 and 1 MeV.

REFERENCES

1. L. Svensson, D. Boilson, H. P. L. de Esch, R. S. Hemsworth, A. Krylov and P. Massmann 22nd SOFT, Helsinki, 2002
2. H. P. L. de Esch, R. S. Hemsworth and P. Massmann, *Updated physics design ITER-SINGAP accelerator*, submitted to Fusion Engineering and Design.
3. Y. Okumura, 18th International Conference on Fusion Energy, Sorrento, Italy, 4-10 October 2000.
4. P. Massmann, D. Boilson, H. P. L. de Esch, R. S. Hemsworth and L. Svensson, 20th ISDEIV, Tours, 2002
5. A. Krylov, R. S. Hemsworth, Gas losses and related beam losses in the ITER NBI, submitted to Fusion Engineering and Design.
6. H. P. L. de Esch, D Boilson, R S Hemsworth, P Massmann and LSvensson, Proc. 9th International Symposium on the Production and Neutralization of Negative Ions and Beams, Ed. Martin P Stockli, Gif-sur-Yvette, France 2002. AIP conference proceedings volume 639, pages 184-196.

Correction of Beam Distortion in Negative Hydrogen Ion Source with Multi-Slot Grounded Grid

Katsuyoshi Tsumori, Osamu Kaneko, Yasuhiko Takeiri, Yoshihide Oka,
Masaki Osakabe, Katsunori Ikeda, Kenichi Nagaoka,
Toshikazu Kawamoto, Eiji Asano, Mamoru Sato and Tomoki Kondo

National Institute for Fusion Science
322-6 Oroshi Toki 509-5292 Japan

Junko Watanabe, Shiro Asano and Yasuo Suzuki

TOSHIBA Co. Power and Industrial Systems R&D Center
4-1 Ukishima Kawasaki-ku Kawasaki Kanagawa, 210-0862 Japan

Abstract. The new beam accelerator with multi-slot grounded grid (MSGG) has been developed to increase the port-through power into large helical device (LHD). Using the accelerator, the maximum power of 5.7 MW was achieved at the beam energy of 186 keV in the beam injection to LHD plasma last year. Although the port-through power increased compared with conventional accelerators with multi-hole grounded grid (MHGG), the accelerator with the MSGG includes a disadvantage of bi-focal condition in parallel and perpendicular direction to the long side of the slots. When the beam width in one of those directions gets narrower, the width in another direction becomes wider. This disadvantage includes the loss of beam port-through power and induces internal damages in neutral beam line. In order to reduce the disadvantage, an experiment has been done using a small-scaled negative ion source with racetrack-shaped apertures for the steering grid installed at beam upstream of the MSGG. By applying the racetrack apertures to the accelerator, it is observed that the beam widths in the parallel and perpendicular directions to the slot long side have almost the same focal condition to obtain minimal beam widths.

INTRODUCTION

Higher energy is required for neutral beam injection (NBI) to obtain deep deposition profiles as required by the increasing plasma target sizes [1-4]. Hydrogen negative ions (H-) have high neutralization efficiency in the energy range more than 100 keV [5], and negative-ion-based NBIs are indispensable to large scaled plasma confinement devices. In this decade, the H- ion based NBIs have become one of the established devices in the nuclear fusion researches. In order to enhance the H- current density, Cs vapor is seeded in the ion sources. There are two differences in current H-ion sources for NBIs comparing to positive ion sources. They are the electrode heat loads deposited by stripped electrons and neutralized hydrogen atom beam, and the

CP763, *Production and Neutralization of Negative Ions and Beams*, edited by J. D. Sherman and Y. I. Belchenko
© 2005 American Institute of Physics 0-7354-0248-5/05/$22.50

optimization for production rate of H⁻ ions related to Cs condition inside the arc chamber. High port-through efficiency, which is the same as the positive ion sources, is required to increase the injection power.

Cesium seeded H⁻ ion sources have advantage to obtain not only high H⁻ yield but also low beam divergences. In case of plasma confinement devices consisting of superconducting coils, such as the Large Helical Device (LHD) [2], the beam injection ports are narrow and the drift tubes are long because of the thermal insulating structure for the coils. Low divergence character of neutral beams based on Cs seeded negative ion sources matched well to the condition of injection ports and the tubes. Beam parameters required for LHD-NBI are 180 keV for maximum injection energy, 5 MW of the maximum port-through power for 10 sec of pulse duration. The beam divergent angle is less than 10 mrad.

Beam injections have started with beam line #1 (BL1) and #2 (BL2) since the 1998 fiscal year, and a new beam line (BL3) has been constructed in the 2001 fiscal year, prior to the 5th LHD experimental campaign. All beam lines produce the tangential injection and each beam line consists of two ion sources. The sources installed have a set of electrode grids with multi-circular-hole apertures. The maximum port-through powers were limited by the voltage breakdowns between the grids. Since the 2002 fiscal year, the grounded grid (GG) for the beam accelerator of the ion sources of BL1 was replaced from previous multi-hole grounded grid (MHGG) to multi-slot grounded grid (MSGG) in order to improve the injection power. By introducing the accelerator with MSGG, the beam energy increased up to 180 keV, and the port-through power of 4.4 MW has been achieved simultaneously [6,7]. Although the beam energy has reached the design value of LHD-NBI, the power and pulse duration have not been simultaneously satisfied. Improvements for the ion source has been continued to increase the injection power.

We describe in the following sections the scheme of equipping negative ion accelerator with MSGG, the achieved injection status, the problem on the beamlet profiles for MSGG accelerator, and its solution by modifying the aperture shapes of the grid located upstream of MSGG.

NEGATIVE ION SOURCE WITH MULTI-SLOT GRID

The overall structure of negative ion source with MSGG is explained elsewhere [7]. The source is a multi-cusp ion source with the external magnetic filter. Cesium vapor is dosed to the arc chamber from its back plate. The new ion source has two parts different from the previous source. One of the differences is the cross-sectional shape of the arc chamber, whose back-plate corners were cut to reduce the linkages of the filter magnets lines with the cusp magnets on the back plate. Electron emission from filaments induces an uncontrollable localized discharge and abnormal arcing in the case of intersectional linkage of the magnetic lines. Cross-sectional view of the arc chamber is shown in Fig. 1(a). Hydrogen negative ions produced in the arc chamber are extracted and accelerated as multi-beamlets by the accelerator shown on the lower side of Fig. 1(a). The accelerator grids are enclosed by dotted line in Fig. 1(a). Another difference is the composition of the accelerator grid set, whose partial cut-view is

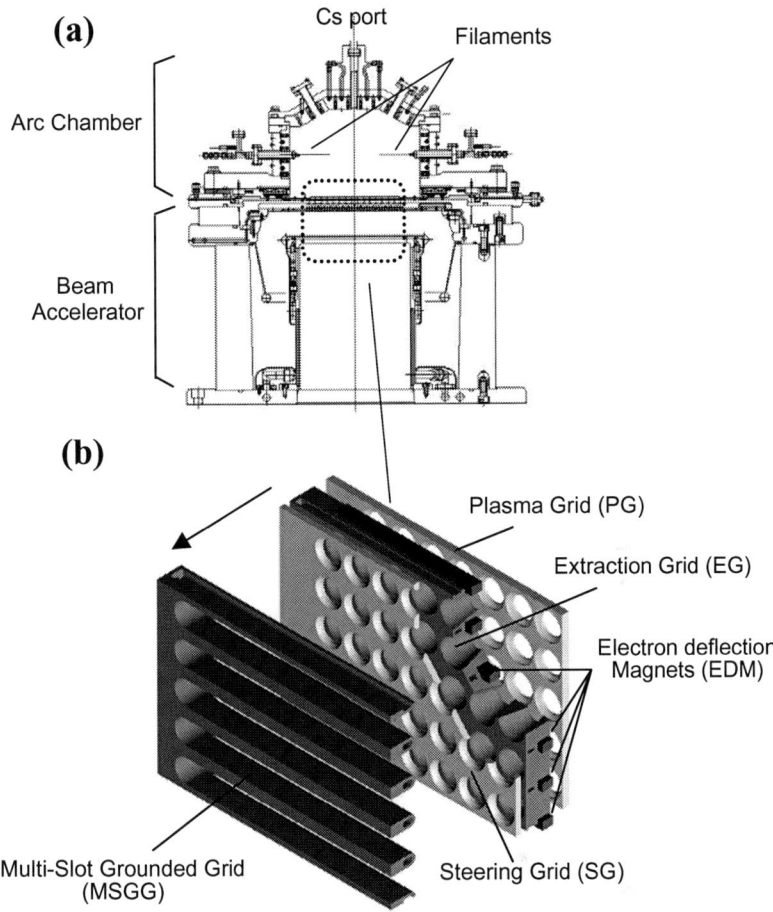

FIGURE 1. (a) Cross-sectional view of the BL1 ion source. The short-side cross section is shown. (b) Layout of the accelerator with multi-slot grounded grid. Accelerator parts ringed in Fig. 1(a) by dotted line is shown.

illustrated in Fig. 1(b). The accelerator consists of plasma grid (PG), extraction grid (EG), steering grid (SG) and MSGG. To sweep electrons extracted from the arc chamber, electron deflection magnets (EDM) were installed inside the EG. These magnets also deflect the H⁻ beamlets. The trajectories of deflected H⁻ beamlets were corrected by displacing the aperture axes of MHGG as compared with that in the previous accelerators, with PG, EG and MHGG. In case of MSGG accelerator, it is not possible to steer the H⁻ beamlets in the direction of slot long side. The new SG is added to steer the deflected beamlets. PG, EG and SG grids have circular apertures.

In the single stage accelerators, where the main acceleration voltage is applied to a single gap, the beam heat load onto grounded grids is more serious than that on the other grids. The heat load is carried by stripped electrons, neutralized atoms H⁰ and by the unfocused part of H⁻ beams. The unfocused H⁻ beam can be decreased by improving the beam focusing design for electrodes. On the other hand, it is impossible to eliminate the stripping process at all because the finite gas pressure is necessary for the ion source operation. For gas pressure reducing at the acceleration gap, the wide evacuation windows had been done at the retainer walls of grounded grid. In that case, however, the back streaming ions from beam plasmas were accelerated back onto the walls of acceleration insulator. Frequent voltage breakdowns occurred by gas emission form the insulator material, and the surface of the insulator was finally carbonized.

In accelerators with high transparent grounded grid, such as SINGAP [8], the gas pressure in acceleration gap is reduced, and beam heat load can be simultaneously decreased due to small beam interception. A transparency of single slot of the MSGG corresponds to transparency of a row in the previous MHGG, and the transparency of MSGG is about twice as large as that of previous MHGG. The area difference between MHGG and MSGG is shown in Fig. 2. Gray area C in the left figure indicates the additional part of MHGG comparing to MSGG. This additional part is far from the water-cooling channel. Heat removal from the C points is worse than from the other parts, and surface temperature at this area is higher during beam exposure. The temperature rise can induce gas emission and together with sputtering, may cause voltage break down. In contrast, this area is absent in MSGG and only the direct cooled area is remained around the cooling channels. The top double-side arrows in Fig. 2 show the direction of H$^-$ beamlet in the magnetic field. Electrostatic correction

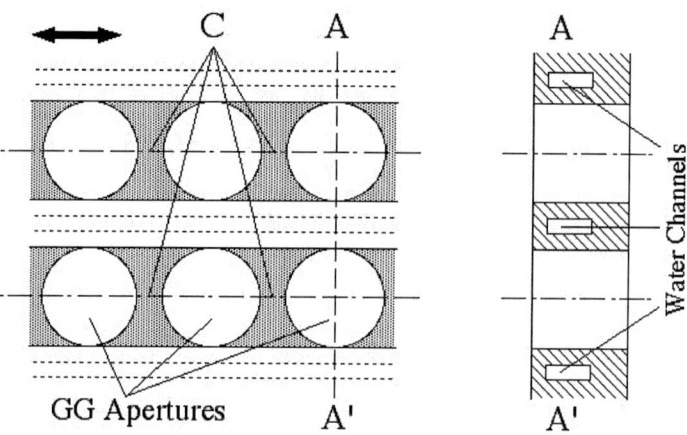

FIGURE 2. Area difference of MHGG and MSGG. Gray area in the left figure is removed in MSGG. The water-cooling channels are indicated as dotted lines in the left figure, and A-A' cross section is shown in the right figure.

of deflected beam trajectory is done by the displacement of the SG apertures with respect to the EG apertures. The beam aberration is expected due to the mismatching with such correction. The degree of the aberration is larger in the direction of the displacement. The long side of slot in MSGG corresponds to the direction of magnetic deflection, and amount of colliding H^0 beams is expected to be lower in MSGG than that in MHGG.

The plasma and steering grids are made of molybdenum. The PG temperature is raised by the radiation and flux of hot particles from arc plasmas. The H⁻ current is indirectly adjusted by controlling the PG temperature and Cs seed to the arc chamber. Higher intensity of back streaming was expected in the accelerator with MSGG. To decrease the sputtering of SG by back streaming ions, it is made of molybdenum. The rest of the grids are made of oxygen-free-copper and those grids are cooled by water. The extraction voltage (V_{ext}) and acceleration voltage (V_{acc}) are applied to the PG-EG and SG-MSGG gaps respectively; while the EG and SG are electrically connected.

INJECTION BY USING THE ACCELERATOR WITH MULTI-SLOT GROUNDED GRID

By replacing the grounded grid from MHGG to MSGG, the maximum beam energy increased from 165 keV to 180 keV. The 180 keV energy is the design value of LHD-NBI, but it had never been obtained before in LHD ion sources with MHGG. With the beam energy rise, the maximum port-through power reached 4.4 MW from 3.5 MW. Although the port-through power is increased by adopting the accelerator with MSGG, power growth degradation appeared at energies higher than 160 keV. The port-through power obtained as a function of beam energy is shown in Fig. 3. The port-through power values were evaluated by using a calorimeter array installed at the armor plate on the LHD wall located at opposite side of BL1 [9]

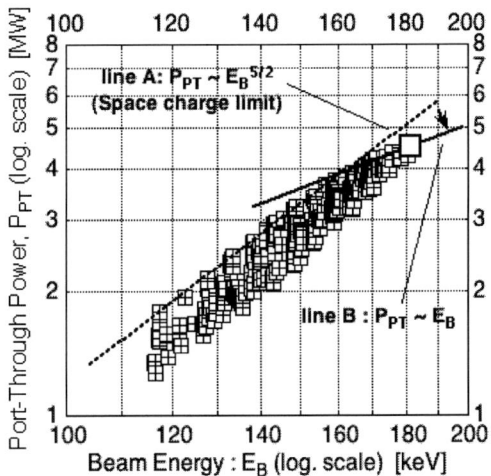

FIGURE 3. Port-through power (P_{inj}) as a function of beam energy (E_B) in the 2002 fiscal year. The line A indicates the condition of space charge limit and is proportional to the 5/2 powers of beam energy. The line B shows current saturation line obtained, which is linear to the beam energy.

The line A power is proportional to the 5/2 powers of beam energy (E_B). In this condition, the accelerated H⁻ current follows the Child-Langmuir's law. The line B, on other hand, is linear proportional to E_B, and shows the degradation of the port-through power at energies higher than 160 keV. It seems beam focusing was not optimal for port through beams. Varying the focal condition, however, did not increase beam power transmission. It was suggested that extracted H⁻ current was saturated due to insufficiency of the input arc power and low Cs effect. The feature was not changed by varying the input arc power. Assuming the surface production of H⁻ ions as the dominant process, extracted H⁻ current relates to area-averaged work function of PG. In practice, the H⁻ production is adjusted by controlling the Cs seed to arc chamber and PG temperature. First of all the Cs vapor pressure was changed by duration of Cs seed, but that influence to the H⁻ current was very small. The temperature of PG is adjusted by changing input arc power and arc duration keeping the H⁻ beam perveance value constant. In the 2002 FY, the maximum PG temperature was about 210 °C. It was below the optimal value of 250 °C.

Heat is indirectly removed from PG by cooling the PG frame connected with PG by thermal insulators. The thermal insulators were replaced to enhance H⁻ current, and the maximum PG temperature was increased to 240 °C. Fig. 4 shows the averaged PG temperatures as a function of input arc power before and after insulator replacement. The temperature average was taken over the ten PGs used, i.e. for two ion sources times five plasma grids per each source. An input arc power in Fig. 4 is summarized for two ion sources.

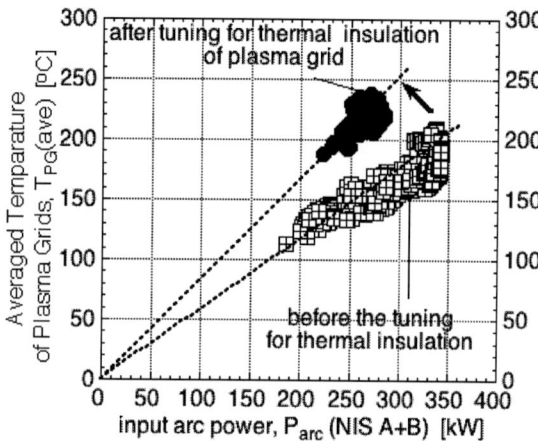

FIGURE 4. Averaged PG temperature before (□) and after (●) adjusting the thermal insulation. The temperature is obtained by averaging over 10 PGs of two ion sources.

The arc-plasma distribution was improved by increasing the number of the filaments. The distribution of PG temperature over all the area became more homogenous by additional filaments. The port-through powers in both cases are compared in Fig. 5. As shown in this figure, the power growth degradation shown in the 2002 FY disappeared. The maximum port-through power consequently reached 5.7 MW at the energy of 186 keV [10]. The pulse duration was 1.6 sec. The beam power and energy exceeded the design value of LHD-NBI.

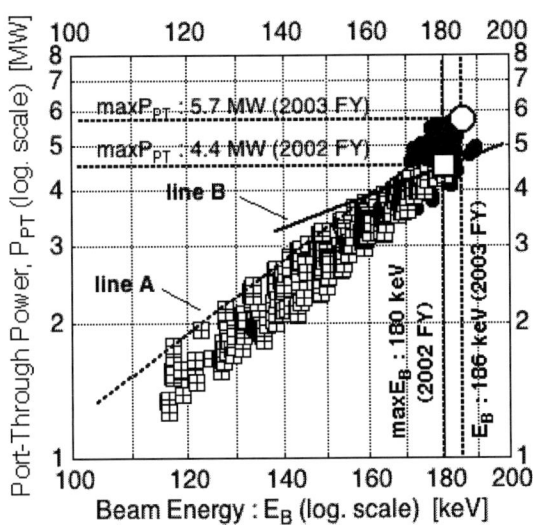

FIGURE 5. Port-through power (P_{inj}) as a function of beam energy (E_B) in the 2003 FY (●) and in the 2002 FY (□). The line A indicates the condition of space charge limit and the line B shows current saturation line. By tuning H⁻ production, the port-through power was enhanced from line B to line A. The maximum power increased from 4.4 MW at the beam energy of 180 keV to 5.7 MW at 186 keV.

PROBLEM ON ACCELERATION WITH MULTI-SLOT GROUNDED GRID

The accelerator with MSGG delivers more intense port-through powers because of its high transparency for neutral gas and the high acceptance to beam heat load. The accelerator, however, includes a problem on beam focusing. The problem was observed by monitoring beam profiles at the movable calorimeter array installed in the neutral beam dump. The beam profiles are monitored for each ion source by moving the array. The distance between the array and grounded grid is 8.6 m apart from the grounded grid. Although the neutral beam focal point is located 13 m downstream from the grid, the focal point is considered to be close enough to the calorimeter position for meaningful measurements. Figure 6 indicates the schematic view of BL1 for LHD.

A voltage ratio R_v (= V_{acc}/V_{ext}), where the V_{acc} and V_{ext} denote acceleration and extraction voltages, is the main parameter of the accelerator focal characteristics. Fig. 7 shows the change of beam widths by scanning the R_v ratio value. Beam width along the slot is shown in Fig. 7 by squares, perpendicular to slot – by circles. The extraction voltage is fixed at 7.5 keV in that case. Although the ratio is a crude parameter to argue the exact beam focusing characteristics, we adopted this parameter to obtain the problem scope. The beam profile at the calorimeter array is fit by Gaussian, and the

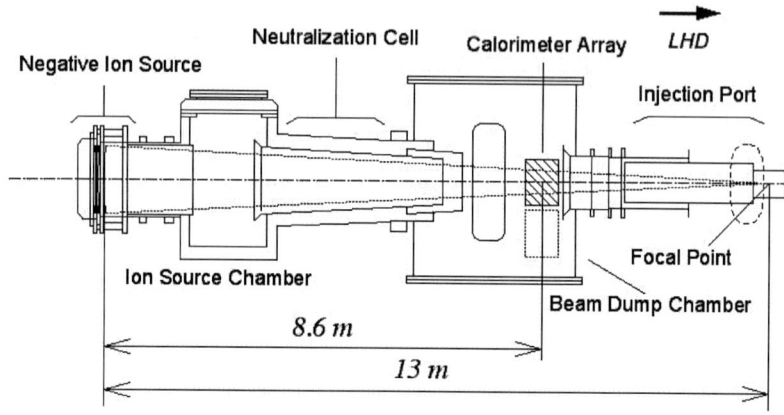

FIGURE 6. Schematic view of BL1 for LHD. The positions of ion source, calorimeter array and beam focal point are indicated. Ring step position is marked by the dotted oval line.

e-folding half widths are indicated in the figure. The minimum widths in the direction parallel and perpendicular to the slot were achieved at different voltage ratios of 14.3 and 21.6, respectively. This difference in the focal conditions is originated by the asymmetry of slot aperture and it was not observed in case of MHGG. The periodicity of the slots is almost the same as the periodicity of MHGG apertures in the corresponding direction (perpendicular to slot). The focal characteristic in this direction, therefore, is not so different from that in the case of MHGG. The minimal beam

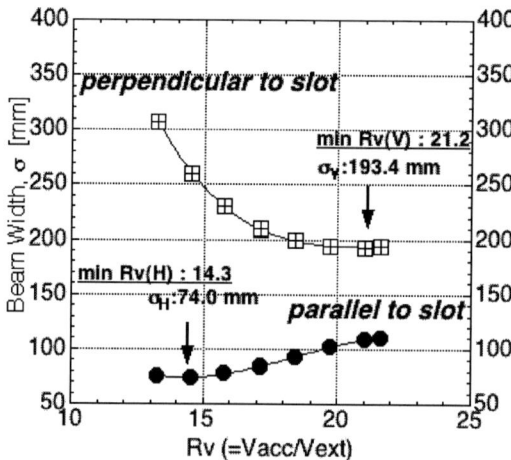

FIGURE 7. Bi-focal condition in the accelerator with multi-slot grounded grid. The beam widths are half of e-folding lengths of beam profile in the direction of parallel (●) and perpendicular (□) to slot. The profile is measured at calorimeter array in BL1.

42

width in MHGG case was obtained at the voltage ratio of about 20.5. On the other hand, the optimal ratio has much smaller value for the direction parallel to slot.

The beam profiles were also monitored by calorimeter array installed at the LHD wall opposite to BL1 [9]. The voltage ratio in practical operation is optimized to obtain the largest port-through powers. The optimal ratio is about 16.5-18.5, and elongation of beam profile at the armor plate is monitored in this ratio range. Direction of the elongation corresponds to the direction vertical to the beam line. The beam divergence in the directions parallel and perpendicular to the slot are evaluated to be about 8 mrad and 15 mrad.

Due to the elongated beam profile, the tails of neutral beams collides with the beam line inner walls, especially with the narrowest part of the injection port, shown at the right side of Fig.6. The port is composed of two drift tubes with different diameters, connected each other by an intermediate ring shaped step (see Fig.8). The ring step is almost vertical to beam axes. To prevent the damage by neutral beams, the injection port is covered by molybdenum armor plates. A photograph of the ring step is shown in Fig 8, and its position is marked by the dotted oval line in Fig. 6. The damaged armor segments are shown by four circles in the photograph. Thickness of the armors is 40 mm, and they are water cooled on the backside. Nevertheless, the armor plates were melted and slipped off their regular positions. Neutral beams collided directly on the outer shell of injection port. Due to thermal expansion of the outer port tube, air leakages occurred at the port flange. Finally, the port-cooling channel was torn and water leakage occurred from the channel too. The ring step is located near the neutral beam focal point. Heat load carried by beam tails, therefore, is severe near the ring. The mentioned problem was happened for the port-through power of more than 5 MW. The new ring armors with the slope of about 30° to beam axis were installed and the cooling system was enhanced.

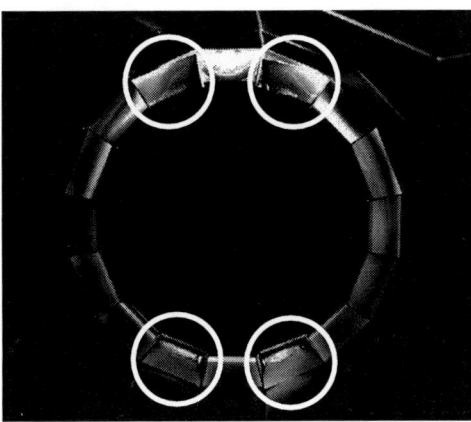

FIGURE 8. Photograph of internal structure of the drift tube. The damaged armor plates are indicated by the areas surrounded with white circles.

CORRECTION OF BEAM PROFILE BY MODIFYING
APERTURE-SHAPE OF STEERING GRID

The bi-focal character of the accelerator with MSGG originates from mismatching of equipotential surfaces near the exit of SG and entrance of MSGG. The curvature of equipotential surfaces has an axial symmetry near the exit of circular SG aperture and the mirror symmetry near the entrance of MSGG. Each H⁻ beamlet converges axi-symmetrically during propagation through the SG aperture and then is expanded near MSGG in the direction perpendicular to the slot. In contrast, much weaker divergence acts to the beamlet in the direction parallel to the slot long axis, because the curvature of the potential surface is smaller than that in the other directions. The combination of SG circular apertures and MSGG is similar to a set of axial symmetric convex lens and one directional concave lens. It is impossible to remove the aberration in this lens system completely. To reduce the distortion of beamlet profile, a compensation method was applied by modifying the shape of SG aperture from circular hole to the racetrack one. The beamlet moves through axisymmetrical apertures of PG and EG, and then through racetrack aperture of SG and finally through slot of MSGG. By changing gradually the aperture symmetry from axial to mirror one, it is possible to avoid the undesirable expansion of the beamlet profile and to reduce the profile distortion.

A smaller negative ion source is used to investigate the correction of beam profile at the test stand. The source has a similar shape of arc chamber as the full-scale ion source for BL1. The smaller source consists of PG, EG, SG and MSGG as well as LHD-source, and the only difference is the SG-MSGG gap distance. The polarities of electron deflection magnets (EDM), which are installed inside EG, change alternatively row by row. Extracted H⁻ beamlets are deflected by the magnetic field too and the directions of deflection indicated by the arrows at the right in Fig. 9.

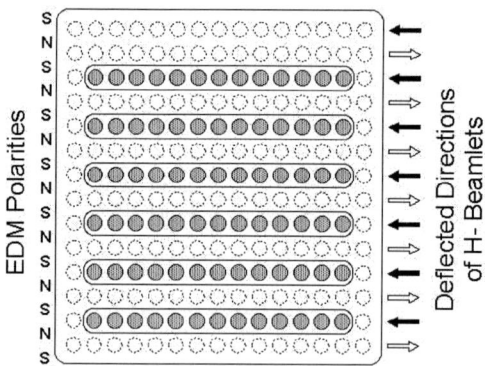

FIGURE 9. Mask for plasma grid. Unmasked PG apertures are indicated by solid gray circles. Polarities of electron deflection magnets installed inside the extraction grid are indicated on the left. The mask opens every even row of aperture, and all the beamlets extracted are deflected in the same direction.

To remove the different directions of deflected H⁻ beams, PG was covered by a mask with opened slots for every even row of aperture holes. Two types of SG with circular and racetrack apertures were prepared for the experiment. The diameter of the circular aperture is 13 mm, and the dimension of the racetrack aperture is 10 x 13 mm. The long axes of the racetrack apertures are oriented along the slot. Near the exit of SG, the curvature of equipotential surface in the racetrack short direction (width) is larger than that in the racetrack long direction. The beamlet passing the racetrack aperture, therefore, is compressed more in the racetrack short direction. This compression could compensate the elongating effect due to the equipotential curvature of the MSGG entrance.

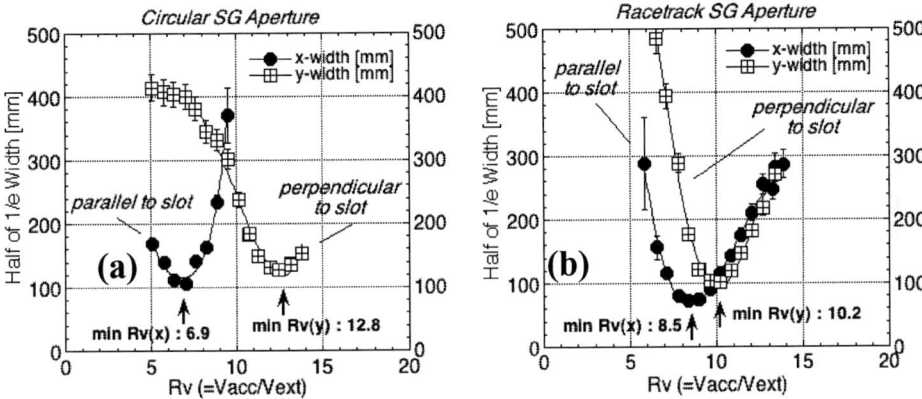

FIGURE 10. The e-folding half widths for accelerators consisting of the steering grids with circular apertures (left) and racetrack apertures (right). The same multi-slot grounded grids were used. The data is obtained with smaller negative ion source at test stand. In both figures, the solid circles and crossed squares shows the widths parallel and perpendicular to slot.

The beam profiles are monitored by one of the two calorimeter arrays installed at the beam dump chamber of the test stand. The distance between the calorimeter array and MSGG of the smaller source is 11.4 m. The e-folding half widths in the long and shot axes directions of GG slot are shown in Fig. 10(a) and 10(b), whose SG aperture are circular or racetrack shaped, respectively. In the case of Fig. 10(a), different optimal voltage ratios are necessary to minimize beam widths in directions parallel and perpendicular to the slot, as it was observed for the full-scale ion source for LHD-NBI. The ratios approach each other in case of SG with racetrack apertures. Although the beam widths become narrower in racetrack case than in circular aperture case, this is considered caused by some non-uniformity of Cs condition on PG surface.

The coincidence of focal conditions in the both directions of slot reduces the beam profile elongation, and the damage of drift tube will be eliminated. The port-through power will also increase by reduction of the beam loss onto the port.

REFERENCES

1. Ushigusa and the JT-60 Team, Proceedings of 16th IAEA Fusion Energy Conference, Montreal, Canada, 1996, (unpublished), IAEA-CN-64/01-3.
2. Motojima, O., et al., *Physics of Plasma* **6**, 1843-1850 (1999).
3. Okumura, Y,. et al., *Review of Scientific Instruments*, **71**, 1219-1224 (2000).
4. Inoue, T., et al., Proceedings of the 20th IAEA Fusion energy Conference, Vilamoura, Portugal, (to be published).
5. Berkner, K.H., et al., *Nuclear Fusion*, **15**, 249-254 (1975).
6. Kaneko, O., et al, *Nuclear Fusion* **43**, 692-699 (2003).
7. Tsumori, K., et al, *Review of Scientific Instruments* **75**, No.5 1847-1850 (2004).
8. de Esch., H. P. L, et al., *Review of Scientific Instruments* **73**, 1045-1047, (2002) .
9. Osakabe, M., et al., *Review of Scientific Instruments* **72**, 590-593, (2001).
10. Tsumori, K., et al., Proceedings of the 20th IAEA Fusion energy Conference, Vilamoura, Portugal, (to be published).

Plasma Injection From Several Cesiated Hollow Cathodes into Large H- Ion Sources

Y. Oka, Yu. Belchenko*, V. Davydenko* K. Ikeda, O. Kaneko,
K. Nagaoka, M. Osakabe, Y. Takeiri, K. Tsumori, E. Asano,
T. Kawamoto, T. Kondo, and M. Sato

National Institute for Fusion Science, Toki-city, 509-5292, Japan
*Budker Institute of Nuclear Physics, Novosibirsk, 630090, Russia

Abstract. Compact cesiated hollow cathodes (CHC) has been developed and studied in two large multicusp H- ion sources at NIFS, aiming the long lifetime cathodes for plasma production in the neutral beam injectors. Results on hollow cathodes and power supplies improvement are described. Basic characteristics of multicusp source discharge, driven by several CHCs were studied. Key issues about simultaneous CHCs operation in high current discharge mode are discussed. Each of advanced CHC delivered stable and quiet income with current up to 70A for 5sec to multicusp source discharge. The hollow cathode operation with no Cs feed was obtained by using 30%Xe+70%H$_2$ gas mixture.

INTRODUCTION

Neutral beam injectors based on acceleration of negative ions has been reliably applied for plasma heating in NIFS and JAERI. At present all large multicusp plasma sources utilize multi-filament driven arc discharges for the production of dense plasma with H$_2$/D$_2$. Use of filament makes the discharge operation simple and reliable, and keeps an optimal temperature of the plasma grid for surface production with Cs seed. However, lifetime of the filament is restricted due to evaporation/sputtering [1]. Also the tungsten evaporation from filaments accelerates the pollution of seeded Cs.

To develop the long lifetime plasma source, we started to establish the system of compact cesiated-hollow cathode which could be effective for the multicusp plasma sources of Neutral Beam Injectors (NBI) of Large Helical Device (LHD). Hollow cathode is considered to be a reliable cathode and can produce plasma injection for a long time. Cesium seed to CHC could provide H- ion production in ion source plasma grid as well. In contrast to previous experiments with a single high current hollow cathode, we have suggested to use the set of compact CHC for production of uniform plasma in the large area multicusp source. CHCs which were investigated and fabricated in BINP [2] were newly fitted to large multicusp H- ion sources on the LHD-NBI test stand [3]. Our purpose was to investigate the applicability of new CHC for LHD-NBI and to make the necessary R&D. This study is mainly concentrated on viability of the combined operation of several CHCs and establishing the necessary systems.

CP763, *Production and Neutralization of Negative Ions and Beams,* edited by J. D. Sherman and Y. I. Belchenko
© 2005 American Institute of Physics 0-7354-0248-5/05/$22.50

SINGLE CHC IN 1/6TH SCALE ION SOURCE

The ion source used consists of 1/6th scale multicusp plasma source (i.e., the dimension of 35 x 35 x ~18cm^3) with external magnetic filter and a small accelerator with single hole of 5mm in diameter. Fig. 1 shows a schematic drawing of 1/6th scale H- ion source.

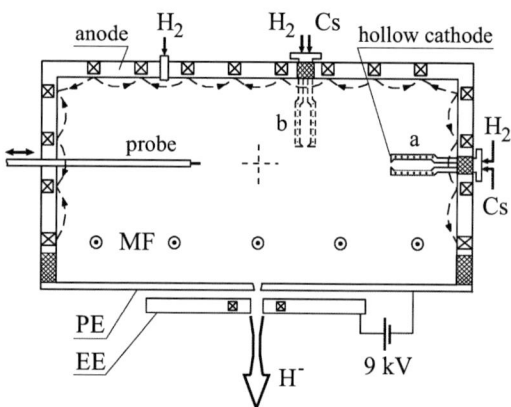

FIGURE 1. 1/6th scale H- ion source with Cesiated-Hollow Cathode on sidewall (a) and on back (b).

Structurally CHC was miniaturized to fit the filament feedthrough port of the source. CHC was made of Mo cylinder and was equipped with the ohmic heater. H$_2$ gas (or H$_2$ + Ar mixture) and Cs vapour were introduced into the cathode hollow (Fig.1). CHC body and Cs oven were preliminary heated. Cs coverage on hot Mo internal surface provided high electron emission current density. Additionally H$_2$ gas was introduced into multicusp plasma source.

The detailed results of this experiment were reported in the ref. 3. CHC provided a high multicusp discharge current up to 60 A in the 10 sec mode and up to 30 A in cw mode. The discharge voltage was varied in the range 40 – 90 V by the Cs feed control. Discharge current attainable under the indicated voltage range was proportional to the H$_2$ flow rate into the CHC and to the area of CHC plasma emission aperture. Plasma density measured by Langmuir probe decreased with the distance from the CHC. The uniform plasma profile is expected for the distance of ~15cm between the CHCs (which is similar to the separation of the filaments in the multicusp source). Negative ion current of ~0.2mA was detected with 9 kV extraction voltage. This current value was similar to that obtained from the filament driven 1/6th source at 6 kW discharge power. Cs consumption was about 50mg/hr for the pulsed operation mode used.

STUDY OF CHC OPERATION WITH TWO CATHODES

Two CHCs were attached to the sidewalls of the 1/6th source (Photo of Fig.2). Two sets of CHC power supplies and two gas systems were used for gas feed to each CHC for studying simultaneous cathode operation with independent bias.

FIGURE 2. Photo of two CHCs attached to filament feedthrough ports of 1/6th scale source.

There were several improvements made after the first experiment with the single CHC. The special oven with the safe Cs pellets was built into each CHC to supply Cs independently (cylindrical volume attached to CHC in Fig.2). Ceramic insulator between CHC housing and filament feedthrough port was applied to operate at high oven temperature (~300 °C). Two Piezo gas valves were used for H_2 feed and for independent control of hydrogen pressures in two CHCs.

Arc power supplies for each CHC consisted of high voltage (up to 400 V) dc power supply (PS1) to start the discharge and of high power (up to 130V x 70A, 5sec) supply (PS2) for discharge support. PS1 and PS2 were connected in parallel to CHC circuit. Arc power supplies were regulated by feedback to provide the constant current (c-c) mode. The other power supplies were used for the Piezo valve control, for the heating of CHC body and of the pellet oven. Power supplies were controlled remotely from the test stand console.

The continuous Cs discharge with the voltage of about 5-25V and current 0.5-1 A was initially ignited. The discharge was changed to high current H_2 + Cs mode when the PS2 power supply and hydrogen were pulsed to CHC.

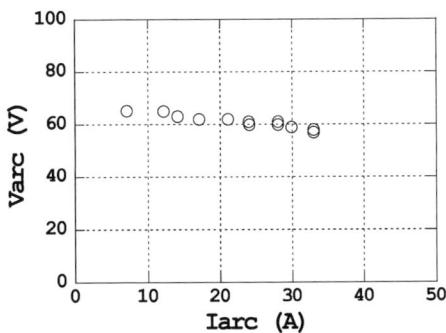

FIGURE 3. Discharge voltage vs discharge current with H_2 + Cs mode for 5s pulse.

49

In H_2 + Cs mode, discharge current was up to 35 A for 5 sec with the discharge voltage ranging from 50 V to 70 V (Fig.3). Thermal capacity of small cathode insert forming the CHC emission aperture limited the pulse duration. Maximum 50 A of discharge current was obtained for 1 sec with the small Mo insert. Essentially higher discharge power (50A for 5sec) was obtained with CHC equipped with the enlarged insert, as tested on BINP test stand. Arcing and aperture melting were recorded at current-duration value >300A·s in the case of enlarged insert.

Simultaneous operation of two CHCs was first demonstrated in the 1/6th scale source at the decreased discharge current. Discharge current pulses for two simultaneously operating CHCs in the multicusp source are shown in Fig.4. CHC #1 delivers 17 A discharge current at the fixed level of c-c control, while CHC #2 produces the smaller current of 8 A with the current oscillations (see Fig.4). The low current of CHC #2 was caused by CHC emission aperture melting by the previous high current pulse with the abnormal several tens seconds duration of the pulse.

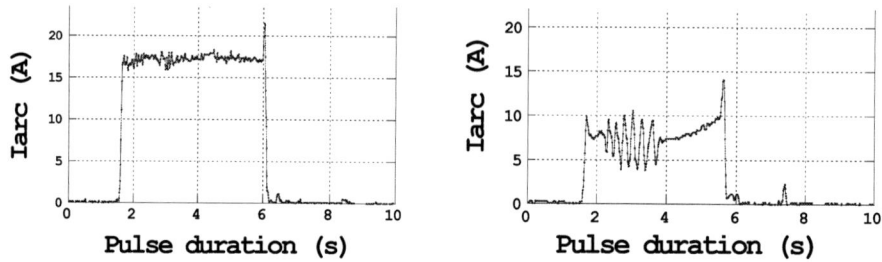

FIGURE 4. Pulses of discharge current for two CHCs simultaneous operation in the multicusp source.

CHC OPERATION IN 1/3RD SCALE SOURCE

Work of three CHCs was tested in the experiment at the larger, 1/3rd scale H- ion source. The plasma chamber of this source has the dimensions of 38 x 62 x ~20 cm³. Plasma grid, extraction and acceleration grids of the source have the multi-aperture structure with openings, drilled over the plasma chamber central area of about 25 x 25cm². Source and accelerator were pumped with the cryopump. Layout of 1/3rd scale source with the CHCs, attached to the sidewalls, is shown in Fig.5.

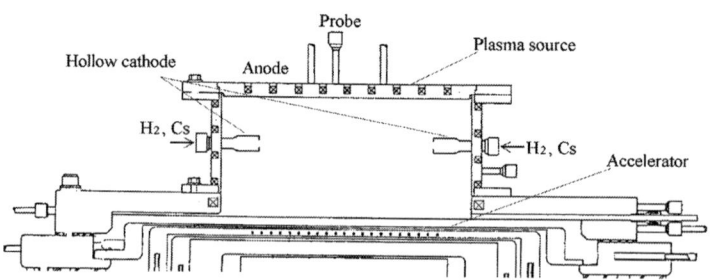

FIGURE 5. 1/3rd scale H- ion source with compact CHCs on sidewalls.

New CHC was designed by Yu. Belchenko for this experiment. For prevention of CHC aperture arcing and melting, the CHC body was isolated from the outer electrode having emission aperture (Fig.6). This outer electrode was connected to the plasma chamber anode with a series resistor for the discharge ignition. CHC was miniaturized to fit the small filament feedthrough port of the source, as it is shown in Fig.7. A mixture of H_2 and Ar gases was used for CHC feed. These changes allowed operation at lower Cs feed to CHC. Argon addition to hydrogen in the range 0.5-30% was tested in cathode cesium mode. Gas mixture was introduced into every CHC with the independent Piezo valve. High current arc power supply with the constant-voltage mode instead of previous c-c-mode was used for the discharge feedback control to reduce the open loop voltage, and its switch was doubled to stop the pulse reliably.

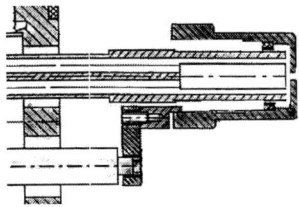

FIGURE 6. Drawing of CHC with the outer electrode.

Operation of single CHC and simultaneous work of two and of three CHCs with discharge current up to 70 A, 5 s per cathode was sequentially tested. Discharge current in range 50~70A and discharge voltage in range 20~70V were independently controlled for each cathode by power supply, gas and cesium seed. Discharge voltage and current has the quiet waveforms for ~5 s pulse, as it is shown in Fig.8 for the simultaneous work of CHC #1, 2, 4 with the 10% Ar and 90% H_2 gas mixture feed. Gas pressure in multicusp plasma chamber was measured with the baratron gauge. It was controlled with range 0.5-4mT during the gas feed to three CHCs. No erosion and arcing of CHCs was recorded after 6 day/480 shots operation. One of CHC (# 3) was broken at the start-up due to operational error with cesium pellets overheating.

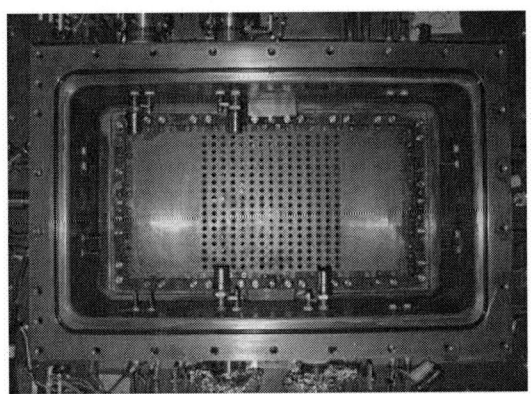

FIGURE 7. Photo of four CHCs in 1/3rd scale H- ion source.

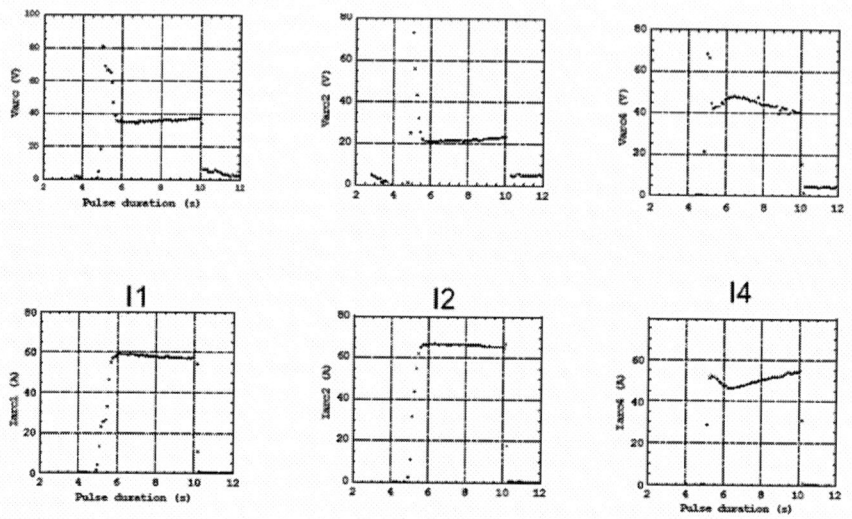

FIGURE 8. Waveforms of the discharge voltage (top) and discharge current (bottom) for simultaneous CHC #1,2 and 4 operation. A mixture of 10% Ar and 90% H_2 was fed.

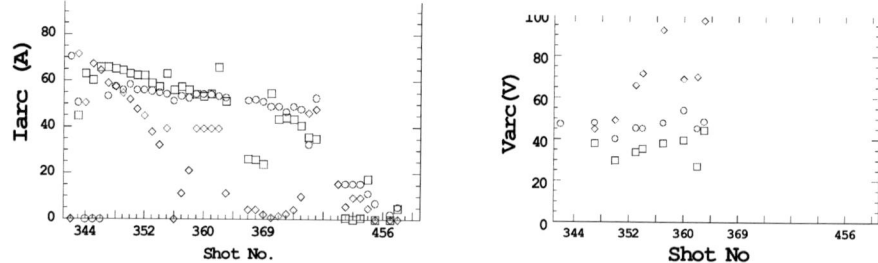

FIGURE 9. Hystory of CHC simultaneous operation. Circles- cathode #1, squares- cathode #2, diamonds- cathode #4. CHC currents (left) and voltages (right) were purposely spread for plasma study.

Hystory of CHC No.1, 2, 4 simultaneous operations are shown in Fig. 9. CHC discharge currents were purposely spread for the plasma study. After shot No. 364, discharge current was diminished for trial of H-extraction/acceleration with the 3 CHC on. This first extraction trial failed due to multi-aperture accelerator cesium contamination, which was produced during the preceding CHC study and during the mistake with the Cs oven overheat.

ARGON AND XENON GAS FEED

Single hollow cathode operation without Cs seed and without cathode preliminary heating was tested in $1/3^{rd}$ scale source too. 100% Ar, 100% Xe, or mixture of 30% Xe + 70% H_2 gases were introduced into hollow cathode in this case. The pressure of Xe in the plasma chamber was varied in the range 0.6 – 1.1 mTor, and of Ar – in the range 0.6 – 0.9 mTor. Stable discharge current up to ~70 A were obtained for 5 s pulse every 120 s (Fig.10). Discharge ignition was strongly dependent on kind of gases. Impedance of the Xe discharge has a lower value, than that for Ar discharge.

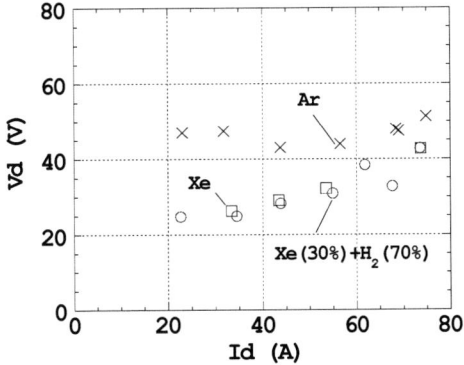

FIGURE 10. Discharge voltage vs discharge current for hollow cathode with rare gas or gas mixture feed. No Cs seed and no cathode preliminary heating was applied. Ar (crosses), Xe (squares) or 30% Xe + 70% H_2 gas mixture (circles) were introduced.

SUMMARY

New system of plasma injection to the LHD-NBI $1/3^{rd}$ scale multicusp plasma source with several hot cesiated hollow cathodes was designed, produced and studied. CHC with the isolated outer electrode operates reliably and delivered quiet and stable discharge current up to 70 A per unit for 5-10 s pulse. System of three CHC and HC with no Cs was successfully tested. Arc power supply with c-v mode control made the operation of CHC and of electronics more reliable. It is important to note, that cesium is blocked inside the CHC hydrogen-argon discharge. The use of dc cesiated CHC can simplify the multi-aperture extraction. Plasma parameters and H- production from the multi-aperture source with HC will be studied in the near future to compare with those of filament driven H- ion source.

ACKNOWLEDGMENTS

The authors would like to acknowledge the NIFS and BINP authorities for continuous support of the collaboration work.

REFERENCES

1. Y. Oka, Y. Takeiri, K. Tsumori, M. Osakabe, O. Kaneko, K. Ikeda, M. Hamabe, E. Asano, T. Kawamoto, and L.Grisham, *Rev. Sci. Instrum.*, **73**, 1054 (2002)
2. Yu. I. Belchenko and A. S. Kupriyanov, *Rev. Sci. Instrum.*, **69**, 929 (1998).
3. Yu. I. Belchenko, Y. Oka, M. Hamabe, O. Kaneko, Λ. Krivenko, Y. Takeiri, K.Tsumori, M. Osakabe, K. Ikeda, E. Asano, and T. Kawamoto, *Rev. Sci. Instrum.* **73**, 940 (2002).

FUNDAMENTAL PROCESSES

Isotope Effect of H⁻/D⁻ Volume Production in Low-Pressure H₂/D₂ Plasmas - Negative Ion Densities versus Plasma Parameters

Osamu Fukumasa and Shigefumi Mori

Department of Electrical and Electronic Engineering, Faculty of Engineering,
Yamaguchi University, Tokiwadai 2-16-1, Ube, 755-8611, Japan

Abstract. Isotope effect on H⁻/D⁻ volume production is studied in a rectangular arc chamber. Axial distributions of H⁻/D⁻ ion densities in the source are measured directly using a laser photodetachment method. Relationship between H⁻/D⁻ production and plasma parameter control with using a magnetic filter (MF) is discussed. Furthermore, relative intensities of extracted negative ion currents are discussed compared with the negative ion densities in the source. Production and control of D₂ plasmas are well realized with the MF including good combination between the filament position and field intensity of the MF. Extracted H⁻ and D⁻ currents depend directly on negative ion densities in the source.

INTRODUCTION

Sources of H⁻ and D⁻ negative ions are required for generation of efficient neutral beams with energies in excess of 150 keV. The magnetically filtered multicusp ion source has been shown to be a promising source of high-quality multiampere H⁻ ions. In pure hydrogen (H₂) discharge plasmas, most of the H⁻ ions are generated by the dissociative attachment of slow plasma electrons e_s (electron temperature $T_e \sim 1$ eV) to highly vibrationally excited hydrogen molecules H₂ (v") (effective vibrational level v" $\geq$ 5-6). These H₂(v") are mainly produced by collisional excitation of fast electrons e_f with energies in excess of 15-20 eV. Namely, H⁻ ions are produced by the following two step process [1, 2]:

$$H_2 (X^1\Sigma_g, v" = 0) + e_f \rightarrow H_2^*(B^1\Sigma_u, C^1\Pi_u) + e_f', \quad (1a)$$
$$H_2^*(B^1\Sigma_u, C^1\Pi_u) \rightarrow H_2 (X^1\Sigma_g, v") + h\nu, \quad (1b)$$
$$H_2 (v") + e_s \rightarrow H^- + H. \quad (2)$$

Production process of D⁻ ions is believed to be the same as that of H⁻ ions. To develop efficient D⁻ ion sources, namely to extract D⁻ ions with high current density, it is important to clarify production and control of deuterium (D₂) plasmas, and to understand difference in the two step process of negative ion production between H₂ plasmas and D₂ plasmas. Cesium seeding into this source is used to enhance negative ion currents and to reduce extracted electron currents. However, there are few studies on optimization of volume-produced D⁻ ions. Then, we focus on understanding the

CP763, *Production and Neutralization of Negative Ions and Beams*, edited by J. D. Sherman and Y. I. Belchenko
© 2005 American Institute of Physics 0-7354-0248-5/05/$22.50

negative ion production mechanisms in the "volume" ion source where negative ions are produced in low-pressure pure H_2 or D_2 discharge plasmas.

For this purpose, we are interested in estimating densities of highly vibrationally excited molecules and negative ions in the source. The production process of $H_2(v")/D_2(v")$ is discussed [3] by observing the photon emission, i.e. VUV emission associated with the process (1b) [4, 5]. To clarify the relationship between plasma parameters and volume production of negative ions, H^- or D^- ions in the source are measured [6, 7] by the laser photodetachment method [11].

In this paper, plasma parameter control by varying the magnetic field intensity of the MF is presented [8, 9]. Influence of these plasma parameter distributions on H^-/D^- production is discussed with using estimated rate coefficients and collision frequencies based on measured plasma parameters [9, 10]. Estimating negative ion densities in the source with the use of laser photodetachment technique, we discuss the relationship between negative ions in the source and extracted negative ion currents [8].

EXPERIMENTAL SETUP

Figure 1 shows a schematic diagram of the ion source [6-10]. The rectangular arc chamber is 25 cm × 25 cm in cross section and 19 cm in height. Four tungsten filaments with 0.7 mm in diameter and 20 cm in length are installed in the source region from side walls of the chamber. The line cusp magnetic field is produced by permanent magnets which surrounded the chamber. The external magnetic filter (MF) is composed of a pair of permanent magnets in front of the plasma grid (PG), and the MF separates the extraction region from the source region. PG potential is kept earth potential throughout the present experiments both for H_2 and D_2 plasmas.

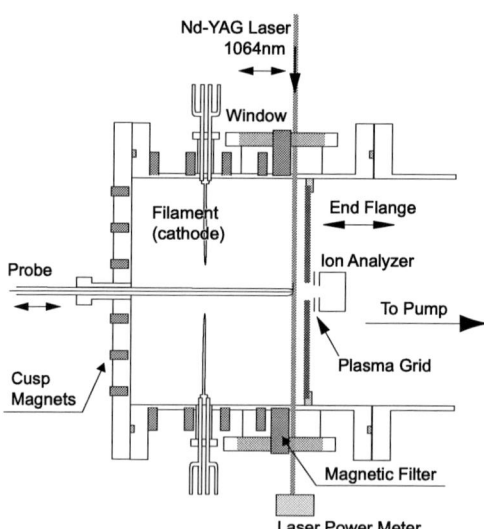

FIGURE 1. Schematic diagram of the ion source. The probe, the laser path, and power meter used in photodetachment experiments are also shown.

Plasma parameters are measured by Langmuir probes. A magnetic deflection type ion analyzer is also used for relative measurements of the extracted H⁻ or D⁻ current. H⁻ or D⁻ densities in the source are measured by the laser photodetachment method [11]. A light pulse from a Nd:YAG laser (wavelength 1064 nm, duration of laser pulse 9 ns, repetition 10 Hz) is introduced from the side wall window of the chamber and passes through the source plasmas. The laser light axis can move across the MF.

EXPERIMENTAL RESULTS AND DISCUSSION

On H⁻/D⁻ volume production, desired condition for plasma parameters is as follows: T_e in the extraction region should be reduced below 1 eV while keeping n_e higher. To realize this condition, namely to enhance H⁻/D⁻ production by dissociative attachment and to reduce H⁻/D⁻ destruction by electron detachment including collisions with energetic electrons, the MF is used. In this purpose, plasma parameter control is studied by varying the intensity of the MF.

Figures 2 and 3 show axial distributions of plasma parameters (n_e and T_e) in H_2 and D_2 plasmas, respectively. By varying the intensity of the MF, axial distributions of n_e and T_e in both H_2 and D_2 plasmas are strongly changed in the downstream region (from $z = 8$ to -2 cm) [8-10]. Production and control of D_2 plasmas are almost the same characteristics as that of H_2 plasmas.

In Figs. 2 and 3, for the MF with 150 G ($B_{MF} = 150$ G), not only T_e, but also n_e are decreased far from the MF, i.e. $z = 8$ cm, in the source region. On the other hand, decreases in n_e and T_e are shifted to downstream region for the case of 80 G and 60 G. In this source, due to the external MF, width of the half-maximum of magnetic field intensity is wider (about 16cm in this case) than the case of rod filter [5]. Namely, varying the intensity of the magnetic filter also indicates varying the strength of magnetic field distribution in both source and extraction region. Thus, the external MF has the merit of gradual control of plasma parameters done precisely with keeping n_e high in the extraction region.

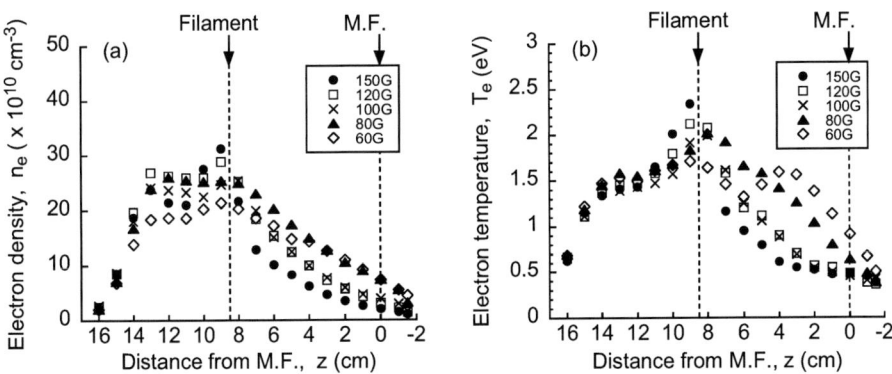

FIGURE 2. Axial distributions of plasma parameters (a) n_e and (b) T_e in H_2 plasmas. Experimental conditions are as follows: $V_d = 70$ V, $I_d = 5$ A, $p(H_2) = 1.5$ mTorr. Parameter is the magnetic field intensity of the MF.

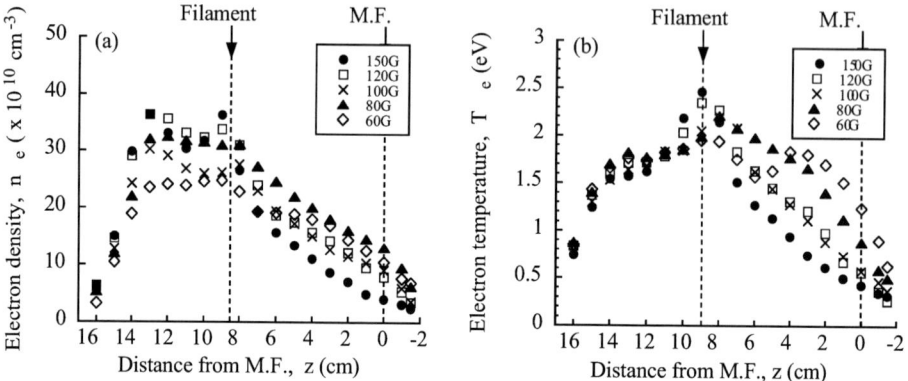

Figure 3. Axial distributions of plasma parameters (a) n_e and (b) T_e in D_2 plasmas. Experimental conditions are as follows: $V_d = 70$ V, $I_d = 5$ A, $p(D_2) = 1.5$ mTorr. Parameter is the magnetic field intensity of the MF.

When $B_{MF} = 60$ G, n_e is slightly higher than that for the case of 80 G. T_e in H_2 plasma is equal to or lower than 1 eV, but T_e in D_2 plasma is above 1eV in the extraction region. Then, plasma conditions are good for H⁻ production, but not good for D⁻ production. When $B_{MF} = 80$ G, values of n_e and T_e in D_2 plasmas are higher than ones in H_2 plasmas. T_e in the extraction region is decreased below 1 eV in both H_2 and D_2 plasmas. Plasma conditions are good for H⁻ and D⁻ production. The stronger MF field is required for control of T_e in D_2 plasmas. Therefore, plasma productions of H_2 and D_2 plasmas are different from each other. Thus isotope effect of plasma production is observed.

As shown in Figs. 2 and 3, plasma parameters in the extraction region depend strongly on the MF intensity. Fig. 4 shows pressure dependence of extracted negative

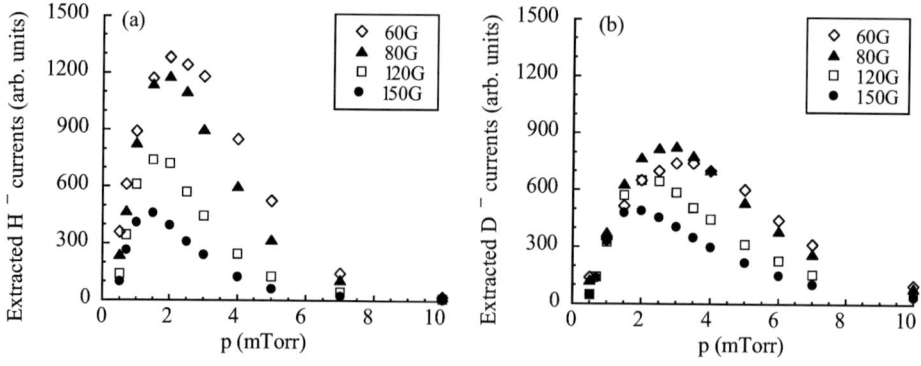

FIGURE 4. Pressure dependence of extracted (a) H- and (b) D- currents. Experimental conditions are as follows: $V_d = 70$ V, $I_d = 5$ A, and $V_{ex} = 1.5$ kV. Parameter is the magnetic field intensity of the magnetic filter.

ion currents from (a) H_2 and (b) D_2 plasmas [8]. The negative ion currents are also strongly dependent on the MF intensity. In both cases, there are some optimum pressures. With increasing gas pressure, negative ion currents (i.e. the H^- current, I_{H^-} and the D^- current, I_{D^-}) increase in their magnitude, reach the maximum value, and then, decrease. Decreasing MF intensity, the optimum pressure p_{opt} shifts to higher pressure. For D^- production, p_{opt} is from 2 to 3.5 mTorr. On the other hand, for H^- production, p_{opt} is from 1.5 to 2 mTorr. Optimum pressure in D_2 plasmas is slightly higher than one in H_2 plasmas.

H^- density distributions across the MF are measured and its dependence on plasma parameters are studied. Figure 5 shows axial distributions of H^- ion densities, where B_{MF} = 150 G and 80 G, respectively [9, 10]. Plasma parameters corresponding to these H^- ion densities are shown in Fig. 2. Spatial distributions of H^- densities are varied by changing plasma parameters. When B_{MF} = 150 G, H^- density distribution decreases toward the extraction hole (i.e. z = -2 cm). On the other hand, when B_{MF} =80 G, H^- density distribution remains nearly constant value although n_e decreases toward to the extraction hole. In front of the extraction hole (i.e. plots at z = -1.5 cm), H^- density with 80 G is higher than that with 150 G by a factor about 2. As is shown in Fig. 4, extracted H^- currents from the source are also the same ratio. Extracted H^- currents depend on H^- densities in front of the extraction hole.

Relationship between these plasma parameter distributions and H^-/D^- production is not well clarified. The variations of H^- and D^- production due to changes in plasma parameter distributions are discussed by taking into account main collision processes for production and destruction. Dissociative attachment (DA: $H_2(v") + e \rightarrow H^- + H$) is main process for H^- production and electron detachment (ED: $H^- + e \rightarrow H + 2e$) is main process for H^- destruction. These two processes are also applicable to the main

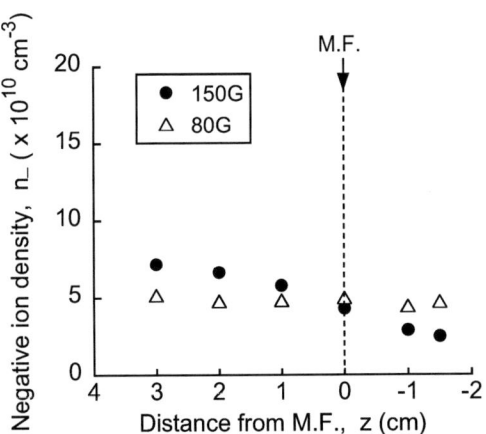

FIGURE 5. Axial distributions of H- ion densities. Experimental conditions are as follows: V_d = 70 V, I_d = 5A, $p(H_2)$ =1.5 mTorr. Parameter is the magnetic field intensity of the MF. Corresponding plasma parameters are shown in Fig. 2 (with B_{MF} = 150 G and 80G).

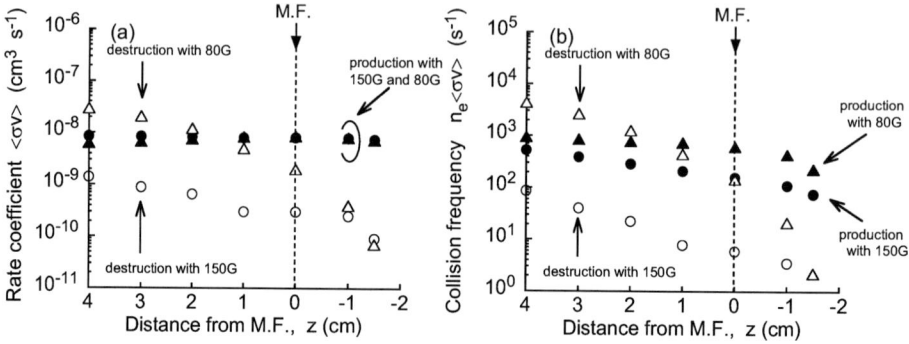

FIGURE 6. Axial distributions of (a) rate coefficient and (b) collision frequency estimated by measured T_e and n_e in H_2 plasmas (closed circle; H- production with 150G, closed triangle; H-production with 80G, open circle; H- destruction with 150G, open triangle; H- destruction with 80G). Corresponding plasma parameters are shown in Fig. 2.

process in D_2 plasmas for D⁻ production and destruction. In the following discussion, it is assumed that only $H_2(v''=8)$ or $D_2(v''=12)$ is present. $H_2(v''=8)$ and $D_2(v''=12)$ have almost the same internal energy [12]. The values of rate coefficient, $<\sigma v>_{DA}$ for DA and $<\sigma v>_{ED}$ for ED, and collision frequency, $n_e<\sigma v>_{DA}$ and $n_e<\sigma v>_{ED}$ are estimated by using measured values of T_e and n_e shown in Figs. 2 and 3.

Figure 6 shows axial distributions of rate coefficients and collision frequencies of DA and ED processes, respectively [9, 10]. With changing T_e distributions, as shown in Fig. 6(a), distributions of ED processes are changed strongly while DA processes keep nearly the same value. It is found that T_e control by varying the intensity of the MF reduces the ED process remarkably. As shown in Fig. 6(b), by taking into account both T_e and n_e changes, the difference between DA with 150 G and with 80 G is caused by n_e in this region (n_e with 80 G is higher than that with 150 G). The distribution patterns of H⁻ densities are mainly determined by ED process in range of T_e = 0.5 to 1.5 eV. H⁻ density with 150 G in Fig. 5 is scarcely decreased by ED process because T_e keeps sufficiently below 1 eV (almost 0.5 eV). The pattern of H⁻ density distribution with 150 G is nearly the same as that of n_e distribution. On the other hand, H⁻ density with 80 G is decreased strongly by ED process because T_e is nearly equal to or above 1 eV in the upstream region from the MF. Thus, the pattern of H⁻ density distribution with 80G is different from that of n_e distribution.

D⁻ density distributions are compared to H⁻ density distributions in the same discharge condition. Figure 7 shows axial distributions of negative ion densities, where B_{MF} = 80 G, $p(H_2) = p(D_2)$ = 1.5 mTorr, respectively [9, 10]. Axial distribution of D⁻ density is lower than that of H⁻ density. According to the plasma conditions shown in Figs. 2 and 3, T_e in D_2 plasma is higher than that in H_2 plasma. With discussion described above, influence of D⁻ destruction by ED process on D⁻ density is higher although n_e in D_2 plasma is also higher. As shown in Fig. 4, extracted D⁻ current is also lower than H⁻ current, and the ratio of H⁻ to D⁻ current is almost the same as the ratio of H⁻ to D⁻ density in front of the extraction hole. Therefore, extracted D⁻ current

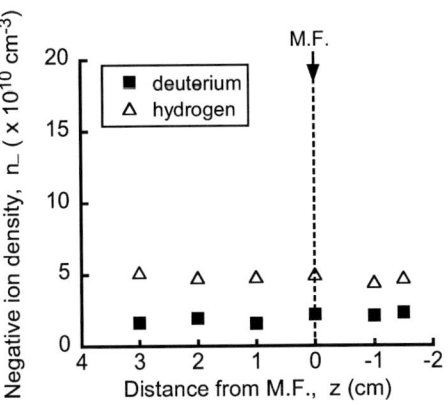

FIGURE 7. Axial distributions of H⁻ and D⁻ ion densities. Experimental conditions are as follows: V_d = 70 V, I_d = 5 A, $p(H_2$ or $D_2)$ = 1.5 mTorr and B_{MF} = 80 G. Corresponding plasma parameters are shown in Fig. 2 (for H_2 plasma) and Fig. 3 (for D_2 plasma).

is mainly determined by D- density in front of the extraction hole. Detailed discussions in relation to negative ions in the source and extracted negative ion currents will be done below.

The relationship between the behavior of negative ions (corresponding to negative ion densities) in the source and the extracted negative ion currents are not well studied. Figure 8 shows axial distributions of negative ion densities in the source, where the field intensity of the MF is 80 G, $p(H_2)$= 1.5 mTorr and $p(D_2)$= 3 mTorr, respectively [8]. These two different pressure conditions correspond to the results in Fig. 4. As shown in Fig. 8, the negative ion densities in front of the extraction hole in D_2 plasmas nearly equal to that in H_2 plasmas. On the other hand, according to the results in Fig.7,

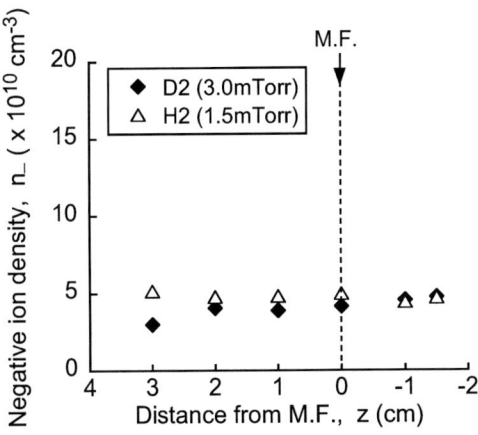

FIGURE 8. Axial distributions of H⁻ and D⁻ ion densities. Experimental conditions are as follows: V_d= 70 V, I_d = 5 A, $p(H_2)$ = 1.5 mTorr, $p(D_2)$ = 3 mTorr, B_{MF} = 80 G and V_{ex} = 1.5 kV.

I_{D^-} at 3 mTorr with 80G is lower than I_{H^-} at 1.5 mTorr, where the same extraction voltage V_{ex} is applied for H- extraction and D- extraction, respectively. Considering the factor of $\sqrt{2}$ due to mass difference, $\sqrt{2}$ times I_{D^-} is nearly equal to I_{H^-}. Then D- ion density in the source are expected to be the same as H- ion density. Results in Fig. 8 support this.

SUMMARY

Production and control of plasma parameters in H_2 and D_2 plasmas are performed by varying the intensity of the MF. The values of T_e and n_e in D_2 plasmas are slightly higher than ones in H_2 plasmas. T_e in D_2 plasmas cannot be decreased and is kept above 1 eV in the extraction region with the same MF intensity for optimizing H_2 plasmas. The stronger MF field is required for control of T_e in D_2 plasmas. Therefore, plasma production between H_2 and D_2 plasmas is different from each other. Namely isotope effect of plasma production is observed. H⁻ and D⁻ densities have different spatial distributions corresponding to those different plasma parameters. Extracted H⁻ and D⁻ currents are mainly determined by H⁻ and D⁻ densities in front of the extraction hole, respectively. According to the discussions based on estimated rate coefficient and collision frequency of main collision processes, it is reconfirmed that T_e in the extraction region should be reduced below 1 eV with n_e keeping higher for enhancement of H⁻ and D⁻ production. For further studying enhancement of D⁻ production, simultaneous measurements of VUV emission and negative ion density in the source is necessary.

ACKNOWLEDGEMENTS

The authors would like to thank Y. Tauchi (Yamaguchi University) for his support in the experiments and Prof. H. Naitou (Yamaguchi University) for his variable discussions and encouragement. The authors also would like to thank Prof. Y. Takeiri and Dr. K. Tsumori (National Institute for Fusion Science) for their discussions. A part of this work was supported by the Grant-in-Aid for Scientific Research from Ministry of Education, Culture, Sports, Science and Technology, Japan. This work was also carried out as the collaboration research program (the LHD project) of National Institute for Fusion Science.

REFERENCES

1. Hiskes, J. R., and Karo, A. M., *J. Appl. Phys.* **56**, 1927 (1984).
2. Fukumasa, O., *J. Phys.* **D22**, 1668 (1989).
3. Fukumasa, O., Tauchi, Y., Yabuki, Y., Mori, S., and Takeiri, Y., *Proceedings of the 9th International Symposium on the Production and Neutralization of Negative Ions and Beams* (2002) pp.28-34.
4. Graham, W. G., *J. Phys.* **D17**, 2225 (1984).
5. Fukumasa, O., Mizuki, N., and Niitani, E., Rev. Sci. Instrum. **69**, 995 (1998).
6. Mori, S., Tauchi, Y., Fukumasa, O., Hamabe, M. and Takeiri, Y., *Abstracts of the 30th IEEE International Conference on Plasma Science* (2003) p.209.

7. Fukumasa, O., *Proceedings of the 26th International Conference on Phenomena in Ionized Gases* (2003) **1**, pp.13-14.
8. Fukumasa, O., Mori, S., Nakada, N., Tauchi, Y., Hamabe, M., Tsumori, K. and Takeiri, Y., *Contrib. Plasma Phys.* **44**, 516 (2004).
9. Mori, S., Tauchi, Y., Fukumasa, O., Hamabe, M., Tsumori, K. and Takeiri, Y., *Proceedings of the Novel Materials Processing by Advanced Electromagnetic Energy Sources 2004* (2004) (to be published).
10. Mori, S. and Fukumasa, O., *Abstracts of 7th Asia Pacific Conference on Plasma Science and Technology and 17th Symposium on Plasma Science for Materials* (2004) p.451.
11. Bacal, M. and Hamilton, G. W., *Phys. Rev. Lett.* **42**, 1538 (1979).
12. Wadehra, J. M., *Appl. Phys. Lett.* **35**, 917 (1979).

Twenty Five Years of Vibrational Kinetics and Negative Ion Production in H₂ Plasmas: Modelling Aspects

M.Capitelli[1,2], O.De Pascale[2], P.Diomede[1], A.Gicquel[3], C.Gorse[1], K.Hassouni[3], S.Longo[1], D.Pagano[1]

[1]*Department of Chemistry, University of Bari, Bari, Italy*
[2]*CSCP-IMIP(CNR), Sezione Terriroriale di Bari, Bari, Italy*
[3]*LIMHP-CNRS, Univerité Paris Nord, Villetaneuse, France*

Abstract. Different approaches to study vibrational kinetics coupled to electron one for modeling different kinds of negative ion sources are presented. In particular two types of sources are investigated. The first one is a classical negative ion source in which the plasma is generated by thermoemitted electrons; in the second one, electrons already present in the mixture are accelerated by an RF field to sufficiently high energy to ionize the gas molecules. For the first kind of ion source a new computational scheme is presented to couple heavy particle and electron kinetics. Moreover models developed for an RF inductive discharge and for a parallel plate discharge are described.

INTRODUCTION

The development of high current negative ion sources for generating intense neutral beams represents an important topic in the thermonuclear research field. Negative hydrogen ions are very important for the generation of neutral beams for heating magnetically confined fusion plasmas. Negative ion based neutral beam injector (N-NBI) is required because negative ion beams present a higher neutralization efficiency respect to positive ion ones at high energies [1]. Different kinds of negative ion sources are studied both from theoretical and experimental point of view, in order to optimize negative ion production. At present the volume plasma sources are widely used for H⁻ generation, even though the addition of alkaline metals increases the concentration of negative ions.

VIBRATIONAL KINETICS IN H₂ PLASMAS

Vibrational kinetics of molecular hydrogen under plasma conditions started his long way many years ago in an attempt to understand alternative mechanisms in the dissociation of H₂ under electrical discharges. Molinari and coworkers in an attempt to rationalize their experimental dissociation rates introduced the concept of a vibrational temperature higher than the translational one able to increase the dissociation of H₂ by heavy particle collisions [2]. This mechanism could prevail over the dissociation

CP763, *Production and Neutralization of Negative Ions and Beams*, edited by J. D. Sherman and Y. I. Belchenko
© 2005 American Institute of Physics 0-7354-0248-5/05/$22.50

mechanism by electron impact in the cases in which this well known mechanism failed to describe the dissociation process. Capitelli *et al.* [3] building up a vibrational kinetic model mutuated by the laser community first tried to implement the Molinari's ideas. The first model was very simple including the pumping of vibrational quanta through the e-V (electron-vibration energy exchange) processes and the redistribution of vibrational quanta on the vibrational ladder by V-V (vibration-vibration) and V-T (vibration-translation) energy transfer processes. V-V up pumping mechanism or Treanor's mechanism was able to bring the vibrational quanta up to the dissociation limit leading to dissociation rates higher than the dissociation rate induced by electron impact. This mechanism was called [3] a laser type mechanism or pure vibrational mechanism (PVM). This kind of name was given because the vibrational distribution of molecular hydrogen presented a long plateau typically met in CO lasers. The enthusiasm in the new mechanism was strongly damped when becomes clear that atomic hydrogen should be considered a killer of vibrational level populations. V-T deactivation of vibrationally excited hydrogen molecules, not considered in Ref.3, presented indeed large rates which strongly destroy the vibrational quanta introduced by e-V processes. In a subsequent work the V-T term due to atomic hydrogen was inserted in the vibrational master equation having as result a strong decrease of the vibrational plateau and a consequent loss of importance of the pure vibrational mechanism in the dissociation process [4]. Only at very high electron density i.e. at high vibrational quanta pumping rates the pure vibrational mechanism could be effective in dissociating H_2 molecules. On the other hand large plateaux again arose as result of the recombination process which selectively pump vibrational quanta on the top of vibrational ladder.

Despite the strong deactivating effect of atomic hydrogen on the vibrational distributions of H_2 the research continued. In particular the authors of Ref.5 developed a joint vibrational dissociation mechanism (JVD) which included both the pure vibrational mechanism (PVM) and the electron impact dissociation (DEM) mechanism in the same model. Moreover the DEM model considered the dissociation transitions by electron impact from each vibrational level of vibrational ladder. The same authors realized the possibility of non-Maxwellian distribution functions for the electrons so that a Boltzmann equation for the electron energy distribution function (eedf) was solved to get the actual eedf. Soon after the importance of second kind collisions involving vibrationally excited H_2 molecules and electrons in affecting eedf was realized as well as the dependence of eedf on the presence of atomic hydrogen. For long time only superelastic vibrational collisions were considered in the kinetics; recently however two groups of research emphasized the role of metastable electronic states in forming structures in the eedf of H_2 plasmas.

At the same time a large effort was done to calculate electron impact dissociation and ionization cross sections involving each vibrational level of the H_2 vibrational manifold [6], a problem which is still under study today [7].

Three independent codes were built up in Europe mainly in Bari (Gorse [8]), Lisboa (Lourciro and Ferreira [9]) and Saint Petersburg (Baksht [10]). The research in the field shows a sharp increase when Bacal and Hamilton [11] discovered the presence of large concentrations of negative ions (H$^-$) in multicusp magnetic plasmas. It was soon realized that the mechanism at the basis of the creation of negative ions in this kind of research was dissociative attachment from vibrationally excited molecules [12]. The relevant cross sections in fact were shown to dramatically depend on the vibrational quantum number. The development of negative ion sources for fusion applications gives new impetus to the research in the field of vibrational kinetics in H_2 plasmas from both experimental and theoretical points of view. Different codes (Bari-Gorse [13], Palaiseau-Bacal [14], Livermore-Hiskes [15], Japan-Fukumasa [16]) were built up to describe the complex phenomenology occurring in the plasma. In particular Hiskes [17] discovered a new elementary process for exciting the whole vibrational manifold of H_2 the so called E-V process which consists in the excitation of singlet electronically excited states of H_2 followed by radiative decay on the ground state. Corresponding cross sections first calculated by Hiskes have been then refined by Celiberto et al. [18].

Different diagnostics were used to monitor the quantities entering in the model i.e. concentration of H, H⁻, H_n^+ species as well as electron number density and electron temperatures. At the same time sophisticated experiments were dedicated to measure the vibrational distribution (Essen-Dobele [19]), the eedf (Hopkins-Graham-Dublin [20]) and the ratio between cold and hot atomic hydrogen (Sultan-Orsay [21]). Atomic and molecular physics methods were also implemented to shed light on different elementary processes important to understand the physics of these discharges. In particular Laganà et al. [22] and Esposito et al. [22] presented complete sets of H-H_2(v) rate coefficients, while Billing and Cacciatore [23] start their pioneristic work on the interaction of vibrationally excited states with copper surfaces and on the recombination of atomic hydrogen on the same surface. The last process indeed is important also for pumping vibrational energy in the molecules formed during heterogeneous recombination and finally desorbed by the surface. Atomic hydrogen changes his role from being the killer of vibrationally excited molecules to being a source of them. On the other hand the experimental works of Hall et al. [24] and of Eenshuistra et al. [25] at the end of 1980s confirmed the production of vibrationally excited H_2 molecules during atom recombination thus starting a topics of large actuality at the present [23].

It should be noted that the synergy between atomic/molecular physics and vibrational plasma kinetics has contributed to the advancement of negative ion production reactors. We believe indeed, that without the work of Wadehra and Bardsley [12] on the dependence of dissociative attachment cross sections on the vibrational quantum number and the work of Hiskes on the pumping of vibrational states by high energy electrons through the E-V process the concept of hybrid reactor for the formation of negative ions could not be probably developed. Nowadays another elementary process that one involving dissociative attachment on Rydberg states could push the negative ion community towards the development of other kind of reactors [26]-[29].

Negative ion sources are still a fascinating topic in which vibrational kinetics constitutes one of the most important aspects.

At the beginning of 90s the H_2 plasma community was attracted to other kind of reactors which are used in material science. In particular the microwave discharges used for the production of diamond films and parallel plate RF discharges for microelectronics acquired a noticeable technological importance. Also in this case two sophisticated codes were developed in Villatencuse [30] and Bari [31] for finding the optimum conditions in these reactors. At the same time a complete kinetic model was developed by Matveyev *et al.* [32] for describing high energy H_2 plasma expansion.

In the last 5 years a new impetus on vibrational kinetics arose for handling problems met by fusion people in the divertor plasma [33]. In particular the increase of ionic recombination assisted by vibrational kinetics at the edge of the divertor as well as the use of Monte Carlo simulation for H^- ion and neutral transport are topics of current interest [34][35].

In particular our team [36]-[40] is trying to improve the numerous input data entering in the kinetics describing the negative ion production in multipole, microwave and parallel plate reactors. Details of this modeling will be reported in the present lecture.

NEGATIVE ION SOURCES

Different methods can be used to generate plasmas. In this section we consider two different kinds of plasma sources that can be used to generate negative hydrogen ions.

In multicusp ion sources the plasma is produced by high energy electrons, emitted by hot filaments and accelerated by the negative voltage between the filaments and the source wall. The flux of accelerated electrons impinges on the H_2 target and gives rise to molecular and atomic reactions such as ionization, excitation and dissociation. In particular, negative ion production in the plasma volume is driven by the dissociative attachment of low energy electrons to highly vibrationally excited molecules. On the other hand high energy electrons cause H^- destruction through electron detachment. This production mechanism evidences the necessity of dividing the source into two regions where electron energy distributions are optimized respectively for the plasma excitation and the dissociative attachment process. This separation is reached in multicusp ion sources by two kind of "filters": the first one acts separating spatially electrons with different energies by means of a magnetic filter; the second one, realized pulsing the discharge (temporal filter), creates different electron energy distributions at different times.

Multicusp ion sources present some technical limitations due to the damage of the emitting filaments and their evaporation that causes also a contamination and then a variation in the operating conditions. These problems are absent in a RF discharge: because of their low mass, electrons already present in the gas can be easily accelerated by the electric field to energies which are sufficient to ionize a gas molecule, then generating the plasma.

PLASMA KINETIC MODELING

In order to model kinetically a plasma and in particular a negative ion source we need to solve the vibrational kinetics of $H_2(v)$, the dissociation kinetics of $H_2(v)$, the kinetics of electronically excited states of H_2 and H, the ion kinetics.

The time evolution of the heavy particle densities can be described by a set of nonlinear differential equations that reads as:

$$\left(\frac{dN_v}{dt}\right) = \left(\frac{dN_v}{dt}\right)_{e-V} + \left(\frac{dN_v}{dt}\right)_{E-V} + \left(\frac{dN_v}{dt}\right)_{V-V} + \left(\frac{dN_v}{dt}\right)_{V-T} +$$

$$\left(\frac{dN_v}{dt}\right)_{e-D} + \left(\frac{dN_v}{dt}\right)_{e-I} + \left(\frac{dN_v}{dt}\right)_{e-da} + \left(\frac{dN_v}{dt}\right)_{e-E} + \left(\frac{dN_v}{dt}\right)_{wall}$$

(1)

where each term on the right hand side represents the gain or the loss for the v-th species due to a specific reaction and is given by:

$$\pm\left(\frac{dN_v}{dt}\right)_{TOT} = \sum_{i=1}^{reaz}\left(k_i \cdot \prod_{j=1}^{react} N_j\right)$$

(2)

where k_i is the rate coefficient for the i-th reaction and the product runs over the reactants for each reaction.

A complete description of the plasma kinetics of negative ion sources requires a self-consistent coupling between the heavy particle kinetics and the electron one. This coupling occurs as heavy particle density evolution depends on the rate coefficients of electron processes,

$$k_i^e = \int_{\varepsilon_{th}}^{\infty}\sigma_i(\varepsilon)v(\varepsilon)n(\varepsilon)d\varepsilon$$

(3)

which are themselves linked to the EEDF variations, however EEDF behaviour is strongly connected to the heavy particle distribution.

The EEDF is governed by the electron Boltzmann equation, which describes the time evolution of the electrons with energy between ε and $\varepsilon + d\varepsilon$. It takes into account of different terms involving the flux of electrons in the energy space due to various contributions. Some of the terms appearing in the Boltzmann equation are characteristic of the system we are investigating.

In the following paragraphs we describe three different models developed independently, but which apply the same state-to-state philosophy to describe the heavy particle kinetics.

Zero-dimensional Model of Multicusp Ion Sources

In multicusp ion sources the Boltzmann equation can be written as [13]:

$$\frac{\partial n(\varepsilon,t)}{\partial t} = -\left(\frac{\partial J_{el}}{\partial \varepsilon}\right)_{e-M} - \left(\frac{\partial J_{el}}{\partial \varepsilon}\right)_{e-e} + In + Ion + Sup + S - L$$

(4)

where $-(\partial J_{el}/\partial \varepsilon)_{e-M}$ accounts for the flux of electrons along the energy axis due to elastic collisions, $-(\partial J_{el}/\partial \varepsilon)_{e-e}$ for that due to electron-electron Coulomb collisions, and the other terms corresponds to inelastic (In), ionizing (Ion) and superelastic (Sup) collisions and electron losses due to recombination. The term S characterizes the system under consideration and corresponds to the injection of electron through the hot filaments and reads as:

$$S = \frac{I}{Ve\Delta\varepsilon} \qquad (5)$$

where I is the current of the injected electrons, V is the plasma volume and $\Delta\varepsilon$ is the energy spreading of the injected electrons.

In previous works the coupling between heavy particle kinetics and electron kinetics was realized by solving at the same time step on the one hand the master equations describing the temporal evolution of the heavy particle density and on the other one the Boltzmann equation for the electron energy distribution function.

In this work we introduce a new approach [40] to couple heavy particle and electron kinetics, which shows a higher self-consistent character. In our scheme the electron energy range is discretized and represented by a set of intervals. At each interval, characterized by its mean energy, a different "kind" of electron is associated, then these "representative electrons" behave as the vibrationally or electronically excited levels of the hydrogen molecule or atom.

$$n(\varepsilon,t) \approx n(\varepsilon_i,t) \qquad \text{for} \qquad \varepsilon_i - \frac{1}{2}\Delta\varepsilon_i \leq \varepsilon \leq \varepsilon_i + \frac{1}{2}\Delta\varepsilon_i \qquad (6)$$

It means that also for electrons we can write a state-to-state set of kinetic equations whose rate coefficients can be written in the following way:

$$k_i^e = \sigma_i(\varepsilon)v(\varepsilon) \qquad (7)$$

which represents the rate coefficient for the excitation of such a species promoted by an electron with velocity:

$$v(\varepsilon) = \sqrt{\frac{2\varepsilon}{m_e}} \qquad (8)$$

These rate coefficients are no more global and referred to the overall electron energy distribution function but to an electron with a specific energy.

The described approach is easily applicable to all electronic processes, but it can be applied also to describe the flux of electrons along the energy axis due to elastic and electron-electron Coulomb collisions. Indeed, according to Rockwood [41] and Elliot *et al.* [42], Boltzmann equation can be reformulated and terms due to these collisions can be rewritten and interpreted as the rate at which electrons are promoted or demoted between adjacent energy intervals.

The numerical scheme resulting from this approach has been used to build a zero-dimensional model of multicusp ion sources in the driver region. The negative ion concentration is described as a function of some discharge parameters (pressure, discharge current) together with electron and heavy particle distributions.

Figure 1a shows the vibrational distribution of H_2 as a function of the vibrational energy for different pressures: increasing the filling pressure, the number density of the vibrational levels increases; low vibrational levels are stronger affected by the increasing pressure. A pressure variation reflects on the plasma potential that governs the electron wall loss: the plasma potential decreases if the discharge current and the voltage current are kept constant. As a consequence the number of low energy electrons increases because they are not lost on the walls (Figure 1b).

71

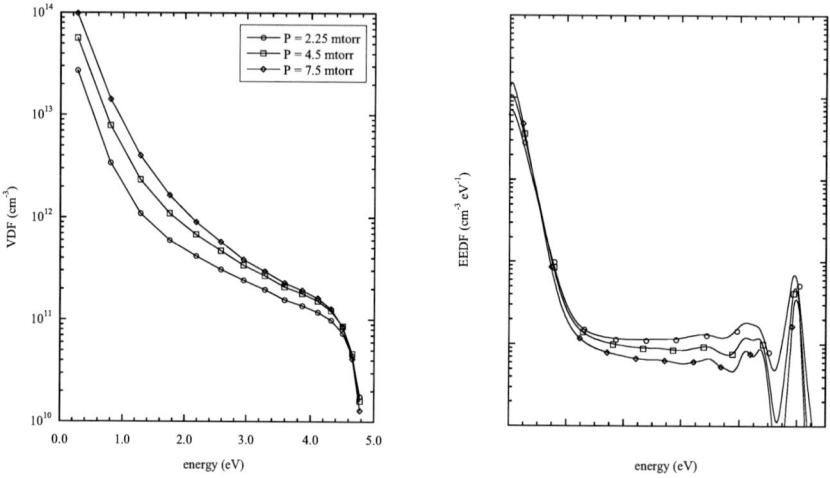

FIGURE 1. VDF (a) and EEDF(b) for different pressures (T_g=500 K; discharge current=10 A; discharge voltage=100 V).

A variation of the discharge current does not cause a significant variation of the vibrational distribution of molecular hydrogen (Figure 2a). On the other hand the electron energy distribution function is shifted at higher number density, reflecting the greater number of electrons injected from the filaments (Figure 2b).

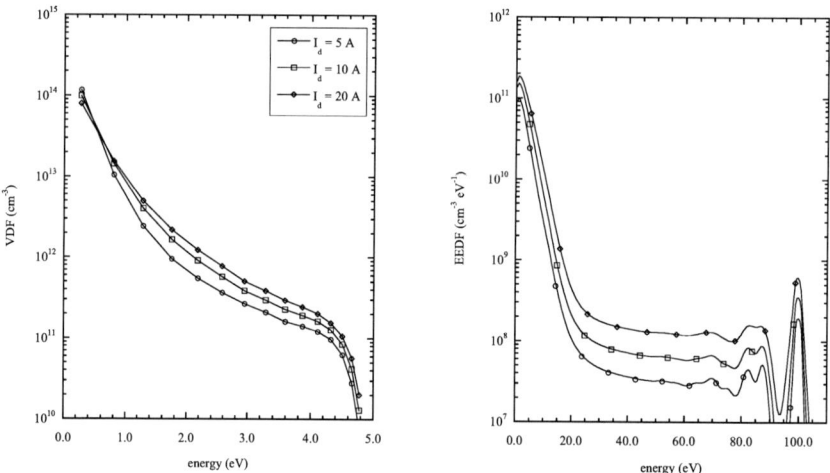

FIGURE 2. VDF (a) and EEDF (b) for different discharge current (T_g=500 K; pressure=7.5 mtorr; discharge voltage=100 V).

RF Ion Sources: Microwave discharge

Boltzmann equation assumes a slightly different form when we consider an RF discharge. In this case the source term S is lost, while another one appears, which describes the flux of electrons in the energy space due to the electric field (see Ref.41). Then Boltzmann equation reads as [38][39]:

$$\frac{\partial n(\varepsilon,t)}{\partial t} = -\left(\frac{\partial J_{el}}{\partial \varepsilon}\right)_{field} - \left(\frac{\partial J_{el}}{\partial \varepsilon}\right)_{e-M} - \left(\frac{\partial J_{el}}{\partial \varepsilon}\right)_{e-e} + In + Ion + Sup - L \qquad (9)$$

The integration of the Boltzmann equation requires the determination of the RF electric field amplitude. This amplitude, or the corresponding root mean square (rms) value, can be determined from the absorbed RF power density that is a model input parameter known from the experiment. For this purpose, the following additional algebraic equation that expresses the dependence between the electric field and the absorbed RF power density (MWPD) was coupled to the electron Boltzmann and species balance equations:

$$E_{rms} = \left(MWPD \frac{m_e}{n_e e^2}\right)^{1/2} \left(\int_\varepsilon \frac{\nu}{\nu^2 + \omega^2} f(\varepsilon)d\varepsilon\right)^{1/2} \qquad (10)$$

where ω is the angular frequency of the RF field, m_e, e, n_e, ε and $f(\varepsilon)$ are the mass, the charge, the density, the energy and the distribution function of electrons, $\nu(\varepsilon)$ is the electron-heavy particle collision frequency. The model includes a total energy equation that yields the gas temperature that governs the rate constants of the collisions that involve heavy species and a quasi-homogeneous plasma transport model for the estimation of species and energy losses at the plasma reactor wall.

The coupling between all the sub-models is realized through a detailed radiative and collisional model [30].

We analyzed and interpreted vibrational and experimental temperatures of molecular hydrogen obtained by Coherent Anti-Stokes Raman Spectroscopy (CARS) in radiofrequency inductive plasmas in the following discharge conditions: (a) pressure=1torr, injected power=0.5W, plasma length=27cm, radius=1.27cm, wall temperature=370K; (b) pressure=6torr, injected power=2.0W, plasma length=27cm, radius=1.27cm, wall temperature=550K. We report comparisons between theoretical and experimental vibrational temperatures defined on the basis of the v=1 level to v=0 level population ratio as a function of the recombination factor (Figure 3). Theoretical vibrational temperatures increase with the increase of γ_H. The difference in the γ_H values can be partially explained by the different experimental wall temperatures. The variation of the recombination factor reflects obviously on the other features of an RF discharge. In particular increasing γ_H leads to the increase of both the vibrational temperature and the density associated with the plateau of the vibrational distribution (Figure 4). On the other hand the increase of γ_H results in warming up the EEDF (Figure 5) with a subsequent increase of the electron temperature (Figure 6).

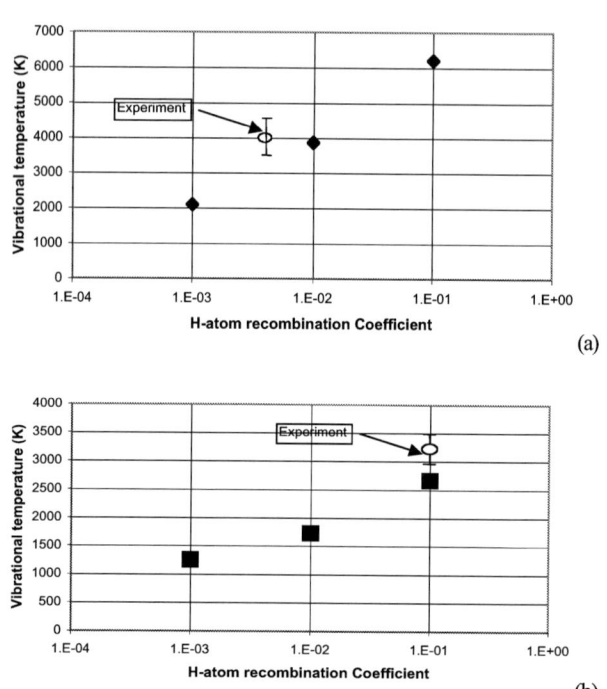

(a)

(b)

FIGURE 3. Theoretical vibrational temperature as a function of the atom recombination coefficient. (a) pressure=1torr, injected power=0.5W, wall temperature=370K; (b) pressure=6torr, injected power=2.0W, wall temperature=550K.

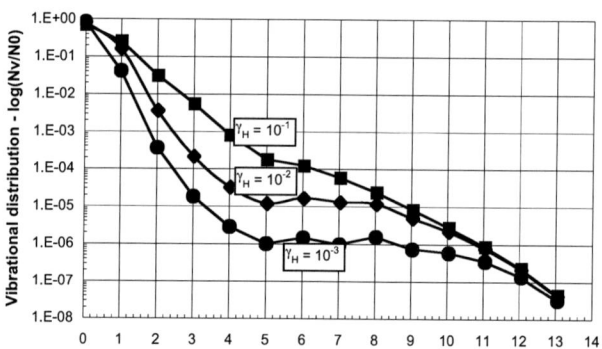

FIGURE 4. Vibrational distribution functions for different γ_H values.

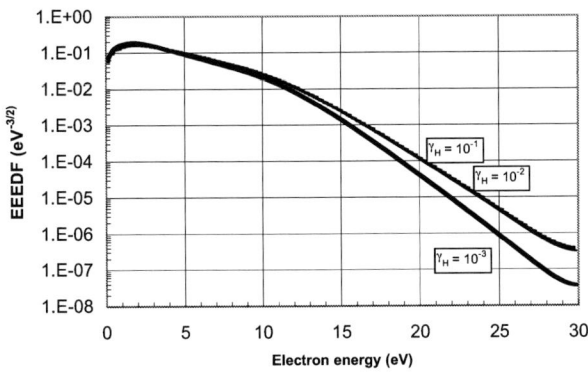

FIGURE 5. Electron energy distribution function for different γ_H values.

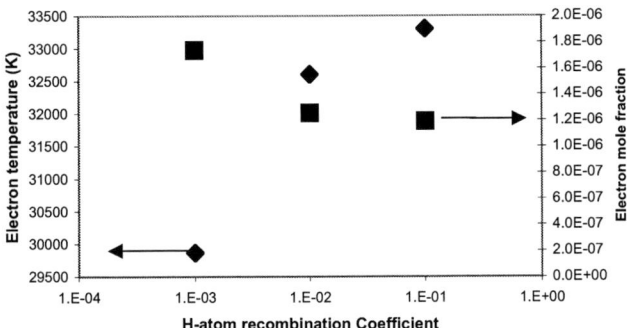

FIGURE 6. Total electron density and electron temperature as a function of γ_H.

1D(r)2D(v) PIC/MCC Model of RF Parallel Plate Discharges

A 1D(r)2D(v) fully self-consistent particle/continuum model has been developed to study capacitively coupled RF discharge plasmas in hydrogen [37]. The code includes a state-to-state reaction diffusion model.

Also in this case it is necessary to realize a self-consistent coupling of the electron transport with the chemical kinetics, i.e. to solve at the same time the electron transport and chemical kinetics problems taking into account their reciprocal connection.

During the calculations, the densities of different species will be updated by solving appropriate equations. The approach is different for charged and neutral particles.

To solve the problem we use a Particle in Cell/Monte Carlo method for the transport equation and a grid-discretized relaxation technique for the reaction-diffusion part. In the PIC/MCC, applied to electrons and four ionic species (H_3^+, H_2^+, H^+ and H^-), the Newton equation for a large ensemble of mathematical point particles is solved taking into account the local electric field as it results from local interpolation within a cell of a mathematical mesh.

The problem can be formalized as follows:

$$\left(\frac{\partial}{\partial t} + v_x \frac{\partial}{\partial x} - \frac{q_s}{m_s} \frac{\partial \varphi(x,t)}{\partial x} \frac{\partial}{\partial v_x}\right) f_s(x,\mathbf{v},t) = C_s(\{F_c\}) \tag{11}$$

$$\frac{\partial^2 \varphi(x,t)}{\partial x^2} = -\frac{1}{\varepsilon_0} \sum_s q_s \int d^3 v f_s(x,\mathbf{v},t) \tag{12}$$

$$-D_c \frac{\partial^2 n_c(x)}{\partial x^2} = \sum_r (v'_{rc} - v_{rc}) k_r (\langle f_e \rangle_t) \prod_{c'} n_{c'}^{v_{rc}} \tag{13}$$

where f_s and F_c are the kinetic distribution functions for the s-th charged species and the c-th neutral species respectively, q_s and m_s are the s-th species electric charge and mass, φ is the electric potential, n_c is the number density of the c-th neutral component, D_c is its diffusion coefficient, k and v are, respectively, the rate coefficient and the molecularity of the c-th species in the r-th elementary process.

In eq. (11) C_s is the Boltzmann collision integral for charged/neutral particle collisions:

$$C_s(\{F_c\}) = -f_s(v) \int d^3 v' p_{v \to v'} + \int d^3 v' p_{v' \to v} f_s(v')$$
$$p_{v' \to v} = \int d^3 w d^3 w' |v' - w'| \sum_k \sigma_k(v',w',v,w) F_{c(k)}(r,w') \tag{14}$$

where k is an index addressing a specific collision process, σ_k and $c(k)$ are the differential cross section and the neutral collision partner of the k-th process.

The PIC method delivers a solution of the Vlasov-Poisson plasma problem in the following form

$$f_s(\mathbf{r},\mathbf{v},t) = \frac{W_s}{N_s} \sum_{p=1}^{N} S(\mathbf{r} - \mathbf{r}_p) \delta(\mathbf{v} - \mathbf{v}_p) \tag{15}$$

W_s is the ratio between real and simulated particles and S is the particle shape factor which describes the way particles are assigned to the mesh.

The density for the neutral species are obtained by finding a stationary solution for the a set of non-linear equations equal to that given in equation (1) except for a diffusion term that in one dimension reads as:

$$\left(\frac{dN_v}{dt}\right)_{diffus} = D_v \frac{\partial^2}{\partial x^2} N_v \tag{16}$$

After any calculation step of the motion equations, the electric charge in any cell of the mesh is determined from the number of electrons and ions found in the cell itself, according to their statistical weight. Known the electric charge density, the electric potential and field are determined by solving the Poisson equation on the mesh.

The electron-molecule process rates are calculated as a function of the position taking into account the translational non equilibrium of electrons and the vibrational non equilibrium of molecules.

As regard the inclusion of the Boltzmann collision term C_s, as it has been demonstrated [43], a stochastic calculation of C_s in the von Neumann sense delivers directly an improved version of the null-collision Monte Carlo method including the thermal distribution of neutrals.

Recombination processes cannot fit the basic PIC/MCC formalism since they involve two charged particles. These processes are treated as a combination of two first order ones, each including one of the two particle species involved in the process.

The model has been applied to a discharge in pure hydrogen, produced by into a parallel-plate high frequency reactor. One of the plates (x=0) (the so-called "grounded electrode") is constantly kept at zero voltage, while the other one (the "powered electrode") is assumed to be driven by an external generator to an oscillating voltage.

We present some results calculated considering the following physical conditions: gas temperature=300K, voltage amplitude=300V, gas pressure=1torr, discharge frequency=13.56MHz, discharge gap=0.02m, DC voltage (bias)=0V. In Figure 7 plasma phase quantities in typical conditions for the probability parameters (γ_V=0.05, γ_H=0.2) are shown.

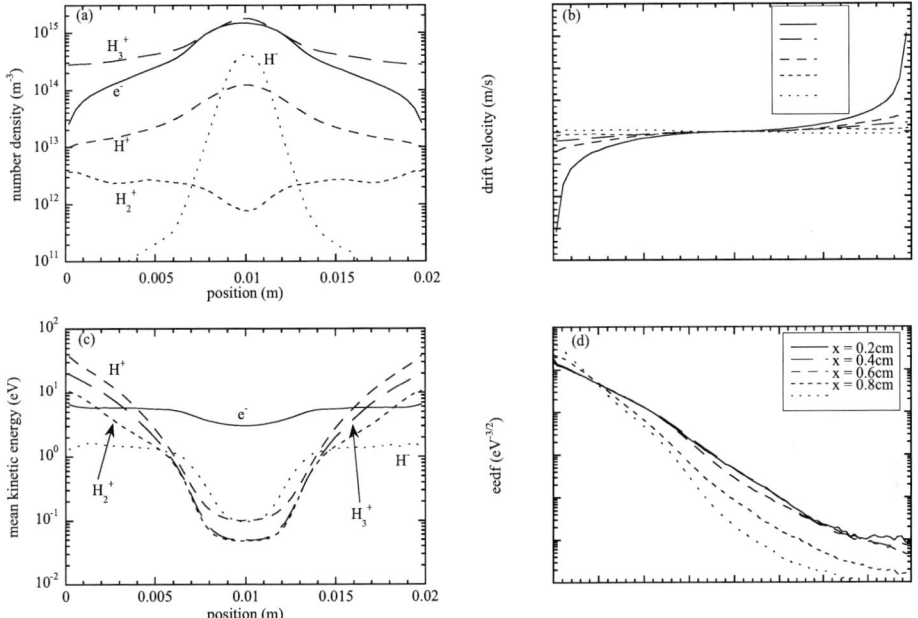

FIGURE 7. Plasma phase quantities in typical conditions for the probability parameters (γ_V=0.05, γ_H=0.2): (a) number density, (b) drift velocity, (c) mean kinetic energy of charged species as a function of position and (d) EEDF as a function of energy.

In particular: (a) number density, (b) drift velocity, (c) mean kinetic energy of charged species as a function of position and (d) electron energy distribution function as a function of energy. From (a) it is evident a slight electronegative behaviour in the centre of the discharge as shown by the separation of the positive ion and electron density in the bulk plasma. The EEDF as shown in (d) significantly deviates from the Maxwell-Boltzmann law showing the necessity of including a kinetic level description of the electron transport.

Table 1 reports the results obtained for H⁻ ions, H atom and electron density varying the vibrational level on which H atoms recombine at the reactor walls, according to the reaction

$$H \rightarrow \tfrac{1}{2} H_2(v)$$

It can be noticed that the level of recombination has a significant effect on H⁻ density, while the other quantities are not considerably affected.

TABLE 1. Influence of the vibrational level for H-atom wall recombination.

	$n_{H^-} \, / \, \mathrm{m}^{-3}$	$n_H \, / \, \mathrm{m}^{-3}$	$n_e \, / \, \mathrm{m}^{-3}$
$v=0$	3×1014		
$v=7$	5.72×1014	7.4×1018	1.18×1015
$v=14$		7.4×1018	1.18×1015

CONCLUSIONS

In conclusion we can say that vibrational kinetics in H_2 plasmas is a subject in continuous evolution having a strong impact in different pure and applied research fields. In particular the future ITER project will use negative ion sources for heating the tokamak. This point is generating a renewing interest in the vibrational kinetics of H_2/D_2 plasmas as can be appreciated by the large European activity in the field by fusion and cold plasma communities.

We presented different approaches to study vibrational kinetics coupled to electron one for modelling different kinds of negative ion sources. In particular we improved the numerous input data describing the negative ion production in multipole, microwave and parallel plate reactors.

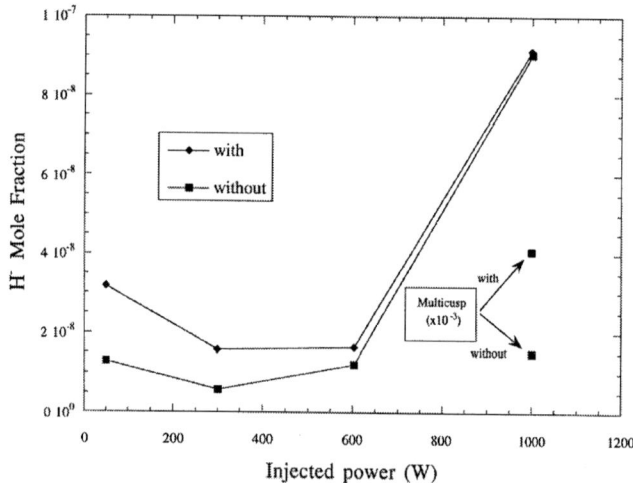

FIGURE 8. H⁻ concentration as a function of the injected power calculated including (with) and neglecting (without) dissociative attachment from Rydberg states [29].

However, a complete understanding of the production mechanism of negative ions should require also the inclusion in the model of Rydberg excited states of atomic hydrogen, as it has been shown by Hassouni *et al.* [29].

Moreover, the role of alkaline metals atoms (such as Cs) in enhancing the negative ions production will be investigated.

ACKNOWLEDGEMENT

This work has been partially supported by ASI (I/R/055/02) and MIUR (cof.2003 Project No. 2003037912_010).

REFERENCES

1. F.Melchert, M.Benner, S.Kruender, E.Salzborn, *Nucl. Instr. and Meth. in Phys. Res. B*, **99**, 98 (1995)
2. P.Capezzuto, F.Cramarossa, R.d'Agostino, E.Molinari, *J. Phys. Chem.*, **79**, 1487 (1975)
3. M.Capitelli, M.Dilonardo, E.Molinari, *Chem. Phys.*, **20**, 417 (1977)
4. M.Capitelli, M.Dilonardo, *Chem. Phys.*, **24**, 417 (1977)
5. M.Cacciatore, M.Capitelli, M.Dilonardo, *Chem. Phys.*, **34**, 193 (1978)
6. M.Cacciatore, M. Capitelli, C.Gorse, *J. Phys .D: Appl. Phys.*, **13**, 575 (1980)
7. R.Celiberto et al., *Atomic Data and Nuclear Data Tables,* **77**, 161 (2001)
8. C.Gorse, M.Capitelli, J.Bretagne, M.Bacal, *Chem Phys.*, **93**, 1 (1985)
9. J.Loureiro, C.M.Ferreira, *J. Phys. D: Appl. Phys.*, **22**, 1680 (1989)
10. F.G.Baksht, G.A.Djuzhev, L.I.Elizarov, V.G.Ivanov, A.A.Kostin, S.M.Shkolnik, *Plasma Sources Sci. Technol.*, **3**, 88 (1994); F.G.Baksht, V.G.Ivanov, M.Bacal, *Plasma Sources Sci. Technol.*, **7**, 431 (1998)
11. M.Bacal, G.W.Hamilton, *Phys. Rev. Lett.*, **42**, 1538 (1979)
12. J.N.Bardsley, J.M.Wadehra, *Phys. Rev. A*, **20**, 1398 (1979)
13. J.Bretagne, G.Delouya, C.Gorse, M.Capitelli, M.Bacal, *J. Phys. D: Appl. Phys.*, **18**, 811 (1985); *J. Phys. D: Appl. Phys.*, **19**, 1197 (1986)
14. M.Bacal, A.M.Bruneteau, W.G.Graham, G.W.Hamilton, N.Nachman, *J. Appl. Phys.*, **52**, 1247 (1981)
15. J.R.Hiskes, A.M.Karo, *J. Appl. Phys.*, **56**, 1927 (1984); J.R.Hiskes, A.M.Karo, P.A.Willmann, *J. Appl. Phys.*, **58**, 1759 (1985)
16. O.Fukumasa, *J.Phys.D: Appl. Phys.*, **22**, 1668 (1989)
17. J.R.Hiskes, *J. Appl. Phys.*, **51**, 4592 (1980)
18. R.Celiberto, M.Capitelli, U.Lamanna, *Chem. Phys.*, **183**, 101 (1994)
19. T.Mosbach, H.M.Katsch, H.F.Dobele, *Phys. Rev. Lett.*, **85**, 3420 (2000)
20. M.B.Hopkins, W.G.Graham, *Rev. Sci. Instrum.*, **57**, 2210 (1986); *J.Phys.D: Appl. Phys.*, **20**, 838 (1987)
21. G.Baravian, J.Jolly, P.Persuy, G.Sultan, *Chem. Phys. Lett.*, **159**, 361 (1989)
22. E.Garcia, A.Laganà, *Chem.Phys. Lett.*, **123**, 365 (1986); F.Esposito et al., *Chem. Phys. Lett.*, **303**, 636 (1999)
23. G.D.Billing, *Pure and Appl. Chem.*, **68**, 1075 (1996); M.Rutigliano, M.Cacciatore, G.D.Billing, *Chem. Phys. Lett.*, **340**, 13 (2001)
24. R.Hall et al., *Phys. Rev. Lett.*, **60**, 337 (1988)
25. P.J.Eenshuistra et al., *Phys. Rev. Lett.*, **60**, 341 (1988)
26. L.A.Pinnaduwage, W.X.Ding, D.L.McCorcle, S.H.Lin, A.M.Mebel, A.Garscadden, *J.Appl. Phys.*, **85**, 7064 (1999)
27. A.Garscadden, R. Napgal, *Plasma Sources Sci. and Tech.*, **4**, 268(1995).
28. J.Hiskes, *Appl. Phys. Lett.*, **69**,755 (1996)
29. K.Hassouni, A.Gicquel, M.Capitelli, *Chem.Phys.Lett.*, **290**, 502(1998)

30. K.Hassouni, A.Gicquel, J.Loureiro, M.Capitelli, *Plasma Sources Sci. and Technol.*, **8**, 494 (1999)
31. S. Longo, *Plasma Sources Sci. Technol.*, **9**, 468(2000).
32. V.P.Silakov et al., *J. Phys. D*, **29**, 2111 (1996)
33. S.I.Krasheninnikov, *Physica Scripta* T96 (2002) 7; A.Yu Pigarov, *Physica Scripta* T96,16(2002).
34. A.Hatayama et al., *Rev. Sci. Instrum.*, **73**, 910 (2002)
35. T.Sakurabayashi et al., *Rev. Sci. Instrum.*, **73**,1048 (2002)
36. S. Longo, M. Capitelli, P. Diomede, Particle Models of Discharge Plasmas in Molecular Gases, *Lecture Notes in Computer Science* 3039, pp. 580-587, 2004.
37. P. Diomede, S. Longo, M. Capitelli, "Capacitively coupled radio frequency discharge plasmas in Hydrogen: particle modeling and negative ion kinetics", RGD24 Conference, Bari 2004
38. M.Capitelli, O.De Pascale, V.Shakatov, K.Hassouni, G.Lombardi and A.Gicquel, "Non-Equilibrium Vibrational Kinetics in Radiofrequency H2 plasmas: a Comparison Between Theoretical and Experimental Results", RGD24 Conference, Bari 2004
39. V.Shakatov, M.Capitelli, O.De Pascale, K.Hassouni, G.Lombardi and A.Gicquel, *Physics of Plasmas* (2004) in press
40. D.Pagano, C.Gorse, M.Capitelli, F.Carbutti, "Hydrogen plasmas for negative ion production", RGD24 Conference, Bari 2004
41. S.D.Rockwood, *Phys. Rev. A*, **8**, 2348 (1973)
42. C.J.Elliott, A.E.Greene, *J. Appl. Phys.*, **47**, 2946 (1976)
43. S. Longo, P. Diomede, *Eur. Phys. J. AP*, in press

Progress in Elementary Processes for Negative Ion Source Modeling

M. Capitelli[†*], M. Cacciatore[†], R. Celiberto[**], F. Esposito[†], A. Laricchiuta[†]
and M. Rutigliano[†]

Dipartimento di Chimica Università di Bari, Italy
†IMIP, Sezione di Bari, CNR, Italy
***Dipartimento di Ingegneria Civile ed Ambientale, Politecnico di Bari, Italy*

Abstract. Cross sections for elementary processes relevant in the modeling of negative ion sources are reviewed, with particular attention to the dependence on the vibrational excitation of the target molecule. Electron impact induced excitation and dissociation processes in H_2 molecule and its isotopic variants are considered. Ro-vibrationally resolved cross sections for dissociation and de-excitation occurring in atom-molecule collision processes are presented. Finally the theoretical determination of hydrogen recombination probability on graphite surface is also considered.

Keywords: elementary processes, excited states, electron-molecule collisions, atom-molecule collisions, heterogeneous recombination probability

INTRODUCTION

A great effort has been devoted in the last decade to the theoretical characterization of negative ion sources, emphasizing the role of non-equilibrium in enhancing the negative ion production [1, 2].

Dissociative attachment is recognized as the main channel for H^- production

$$H_2(X^1\Sigma_g^+, v_i) + e \rightarrow H_2^- \rightarrow H + H^- \tag{1}$$

The efficiency of this process is greatly enhanced by vibrational excitation of molecular target [3, 4], therefore, in order to optimize the source operating conditions, the knowledge of all the elementary processes affecting the vibrational distribution, destroying excited states or populating vibrational levels, is of paramount importance. Furthermore extraordinarily high rate of H^- formation has been observed, in particular experimental conditions [12], opening question about the relevance of alternative mechanisms involving highly excited Rydberg states. In this frame processes involving vibrationally as well as electronically excited H_2 molecule are characterized through the corresponding state-resolved cross sections, being these data relevant for both kinetic modeling of negative ion sources [5] and experimental diagnostic methods [6].

Electron-molecule collision processes have been extensively studied [7] due to their efficiency in affecting the H_2 distribution on the internal degrees of freedom. Also heavy collisions act in the plasma promoting dissociation and energy redistribution in the vibrational ladder. Finally the catalytic activity of source's walls in recombination processes determine a nascent vibrational distribution with a strong non-Boltzmann

CP763, *Production and Neutralization of Negative Ions and Beams*, edited by J. D. Sherman and Y. I. Belchenko
© 2005 American Institute of Physics 0-7354-0248-5/05/$22.50

character. In this paper a review of the work done on elementary processes by the plasma chemistry group of Bari is presented.

ELECTRON-MOLECULE COLLISION PROCESSES

Electronic Excitation

Allowed transitions to the singlet terms of the H_2 spectrum promoted by electron impact have been widely investigated, representing the first step of the indirect mechanism of vibrational excitation, the so-called *E-V process*

$$H_2(X^1\Sigma_g^+, v_i) + e \to H_2^*(\text{excited singlet states}) \to H_2(X^1\Sigma_g^+, v_f) + e + h\nu. \quad (2)$$

As sketched the fate of excited states is to radiatively decay back to the ground electronic state populating preferentially the high vibrational levels (v_f).

The principal excitations in this excitation-radiative sequence proceed through the lowest singlets $B^1\Sigma_u^+$ and $C^1\Pi_u$ [10]

$$H_2(X^1\Sigma_g^+, v_i) + e \to H_2(B^1\Sigma_u^+) + e \quad (3)$$

$$H_2(X^1\Sigma_g^+, v_i) + e \to H_2(C^1\Pi_u) + e \quad (4)$$

Total cross sections (summed over final bound and continuum vibrational states) for the processes (3),(4), calculated in the framework of the semiclassical impact-parameter method, have been reported in refs. [8, 9, 7] for H_2, D_2, T_2 and DT molecules, as function of both the collision energy and the initial vibrational quantum number. These results allow an interesting study on the existence of isotopic effect. In Fig. 1a the dependence of total cross section on the initial vibrational quantum number for the $X \to B$ excitation process is presented, at a fixed incident energy, for all isotopic variants of hydrogen molecule. The observed shift of cross sections to higher values of v_i, in passing from H_2 to T_2 molecule, seems to correspond to the increase of molecular mass. However this isotopic effect is only apparent and the cross sections collapse if plotted as a function of the vibrational eigenvalues (Fig. 1b), suggesting a dependence of excitation cross section on the vertical transition energy.

A not negligible contribution in process (2) is also represented by radiative cascade processes from higher singlet states [10]. Excitations to the B', B'', D, D' states, usually referred as low-lying Rydberg states, have been also studied [11]

$$H_2(X^1\Sigma_g^+, v_i) + e \to H_2(B'^1\Sigma_u^+) + e \quad (5)$$

$$H_2(X^1\Sigma_g^+, v_i) + e \to H_2(B''^1\Sigma_u^+) + e \quad (6)$$

$$H_2(X^1\Sigma_g^+, v_i) + e \to H_2(D^1\Pi_u) + e \quad (7)$$

$$H_2(X^1\Sigma_g^+, v_i) + e \to H_2(D'^1\Pi_u) + e \quad (8)$$

Total cross sections for processes (5)-(8) are presented in Figs. 2a-d, respectively.

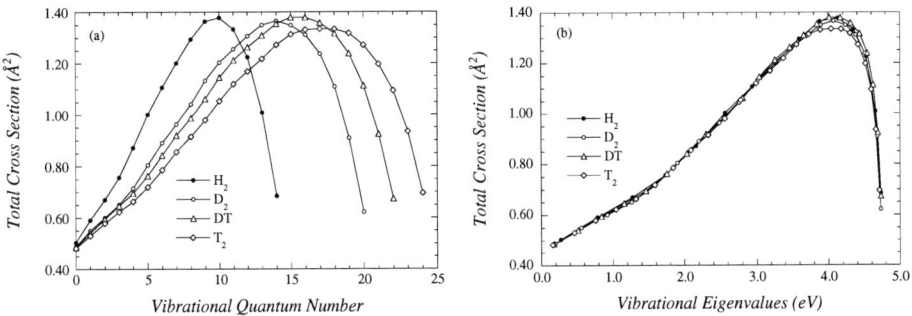

FIGURE 1. Cross section for the transition $(X^1\Sigma_g^+, v_i) \rightarrow (B^1\Sigma_u^+)$ for different isotopes, at a fixed energy of 40 eV, (a) as a function of initial vibrational quantum number; (b) as a function of vibrational eigenvalues.

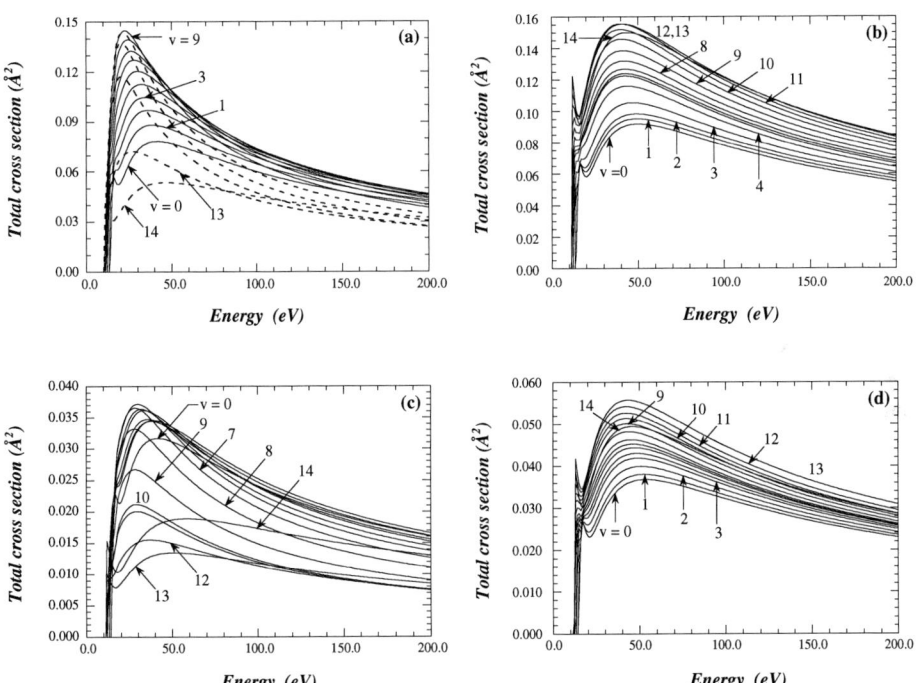

FIGURE 2. Cross section as a function of energy, for different initial vibrational levels, for the process: (a) $H_2(X^1\Sigma_g^+, v_i) + e \rightarrow H_2(B'^1\Sigma_u^+) + e$; (b) $H_2(X^1\Sigma_g^+, v_i) + e \rightarrow H_2(D^1\Pi_u) + e$; (c) $H_2(X^1\Sigma_g^+, v_i) + e \rightarrow H_2(B''^1\Sigma_u^+) + e$; (d) $H_2(X^1\Sigma_g^+, v_i) + e \rightarrow H_2(D'^1\Pi_u) + e$.

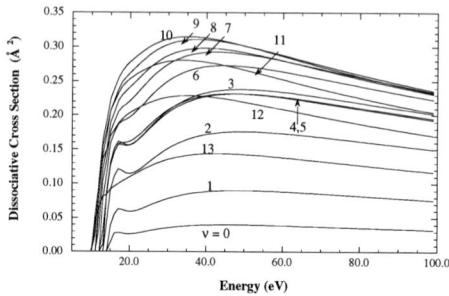

FIGURE 3. Cross section as a function of energy, for different initial vibrational levels, for the process: $H_2(X^1\Sigma_g^+, v_i) + e \rightarrow H_2(B, B', B''^1\Sigma_u^+; C, D, D'^1\Pi_u) + e \rightarrow H + H + e.$

Dissociation

The forbidden transition from the ground state to the pure repulsive $b^3\Sigma_u^+$ state represents the main dissociative channel and the related cross sections, for different initial vibrational quantum numbers, have been calculated by different authors [13, 14, 7]. As expected in forbidden transitions the $X \rightarrow b$ cross section is peaked in the threshold region, while in the high energy region the allowed transitions to the continuum of bound excited electronic states give a significant contribution.

A global dissociative cross section for direct dissociation through excited singlets, including Rydberg states, is reported in Fig. 3. The irregular dependence on the initial vibrational quantum number, determined by the behaviour of Franck Condon density [15, 11], shows that the dissociation is favoured by vibrational excitation.

Transitions populating the vibrational manifold of the lowest bound triplet states (processes (9),(10)) also lead to dissociation through radiative cascade to the b state and predissociation mechanisms, respectively.

$$H_2(X^1\Sigma_g^+, v_i) + e \rightarrow H_2(a^3\Sigma_g^+, v') + e \tag{9}$$

$$H_2(X^1\Sigma_g^+, v_i) + e \rightarrow H_2(c^3\Pi_u, v') + e \tag{10}$$

State-to-state excitation cross sections, obtained in the Born approximation, summed on the final vibrational manifold, are reported in Figs. 4a-b. The direct dissociation through these states is negligible, this being confirmed by the Franck Condon density evaluation.

Triplet-Triplet Transitions

Only few examples of cross section calculations involving two electronically excited states are available in literature [7], nevertheless these transition can be relevant in plasma modeling. Transitions initiated from the metastable $a^3\Sigma_g^+$ and $c^3\Pi_u$ states to

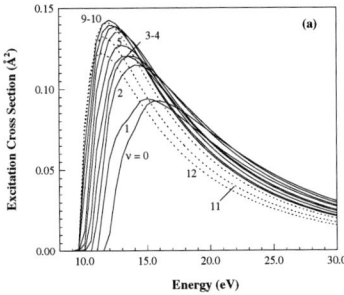 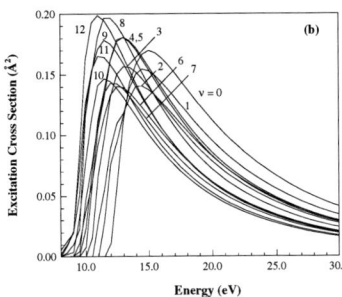

FIGURE 4. Cross sections as a function of energy for the process (a) $H_2(X^1\Sigma_g^+, v_i) + e \to H_2(a^3\Sigma_g^+) + e$; (b) $H_2(X^1\Sigma_g^+, v_i) + e \to H_2(c^3\Pi_u) + e$.

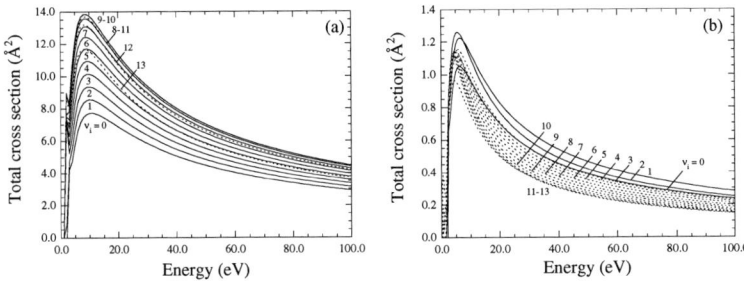

FIGURE 5. Cross sections as a function of energy for the process (a) $H_2(a^3\Sigma_g^+, v_i) + e \to H_2(d^3\Pi_u) + e$; (b) $H_2(c^3\Pi_u, v_i) + e \to H_2(g^3\Sigma_g^+) + e$.

other excited triplet states of H_2 molecules have been considered [16]. In particular process (11), known as Fulcher transition, is important in spectroscopic diagnostic methods, providing information about vibrational and rotational population of hydrogen plasmas [6].

$$H_2(a^3\Sigma_g^+, v_i) + e \to H_2(d^3\Pi_u) + e \qquad (11)$$

$$H_2(c^3\Pi_u, v_i) + e \to H_2(g^3\Sigma_g^+) + e \qquad (12)$$

$$H_2(c^3\Pi_u, v_i) + e \to H_2(h^3\Sigma_g^+) + e \qquad (13)$$

Total cross sections for transitions (11),(12) are shown, as a function of collision energy, respectively in Figs. 5a,b. The relative position of potential energy curves for electronic states does not favour the dissociative channel, which represents a negligible contribution to the total cross sections.

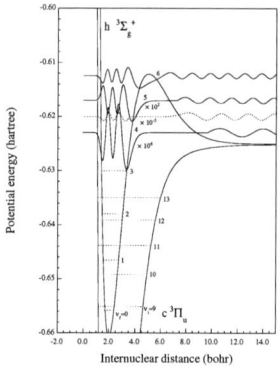

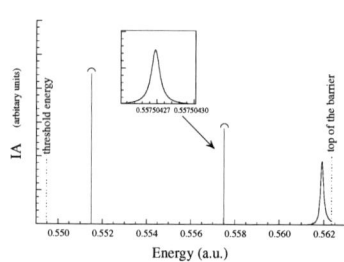

FIGURE 6. (a) Potential energy curves and vibrational levels for $c^3\Pi_u$ and $h^3\Sigma_g^+$ states of H_2. Some wavefunctions of quasi-bound states (full lines) and for a non quasi-bound level (dashed line) are also displayed. The left portion of the first three wavefunctions has been multiplied by 10^{-6}, 10^5 and 10^{-2} for a better representation. (b) Internal amplitude (14) as a function of continuum energy.

A special treatment is required for the transition to the $h^3\Sigma_g^+$ state, whose adiabatic potential energy curve exhibits a barrier sustaining three quasi-bound vibrational levels (Fig. 6a). These vibrational states, due to their quasistationary character, can lead to dissociation tunneling through the barrier. Energy positions of quasi-bound vibrational states is determined by using the method of *internal amplitude, (IA)*, defined as

$$IA(\varepsilon) = \int_{R_a}^{R_b} \frac{|\Psi_\varepsilon(R)|^2 \, dR}{[R_b - R_a]} \tag{14}$$

(R_a and R_b are the two classical turning points of the vibrational ε energy level). The *IA* is shown in Fig. 6b as a function of the potential energy ε. The peaks indicate three regions of resonances which, except for the last one very close to the top of the barrier, present a very sharp intensity. At a closer resolution becomes evident the energy enlargement, as shown in the inserted picture in Fig. 6b. In Fig. 6a are shown the vibrational wavefunctions of three quasi-bound states, at a selected energy corresponding to the maximum of the related *IA*. In the same plot the dotted line represent, as a comparison, the wavefunction of completely dissociative state. The cross sections for these states depend on the ratio between the tunneling and radiative times. The calculated values (obtained by evaluating the tunneling probabilities and radiative Einstein coefficients) are shown in Fig. 7. As expected, the last resonance, located at the maximum of the barrier, shows a strong dissociative character, being the contrary for the first state, placed inside the potential well, which behaves practically as a bound state. An ibrid character is displayed by the second resonance.

Total and dissociative cross sections for $c \to h$ transition, as a function of the energy and the initial vibrational quantum number, are presented in Fig. 8.

86

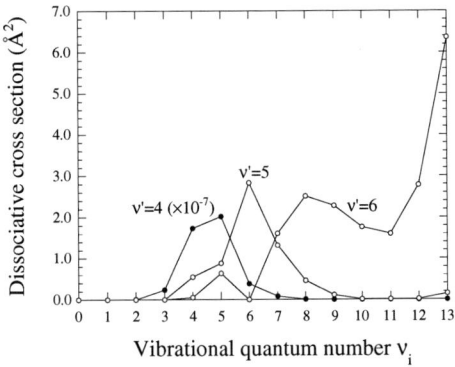

FIGURE 7. Quasi-bound state dissociative cross section for the process $H_2(c^3\Pi_u, v_i) + e \rightarrow$ $H_2(h^3\Sigma_g^+, v') + e$ as a function of initial vibrational quantum number, at the incident energy E=10 eV. The curve labeled as $v' =4$ has been multiplied by a factor of 10^7.

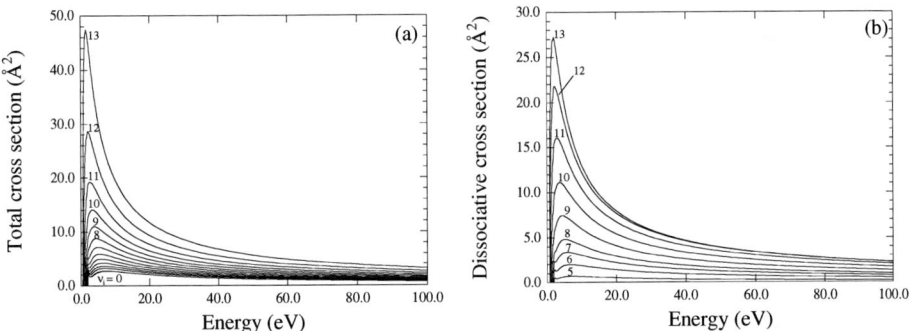

FIGURE 8. Cross sections as a function of energy for the process (a) $H_2(c^3\Pi_u, v_i) + e \rightarrow H_2(h^3\Sigma_g^+) + e$; (b) $H_2(c^3\Pi_u, v_i) + e \rightarrow H_2(h^3\Sigma_g^+, v) + e \rightarrow H + H + e$.

ATOM-MOLECULE COLLISION PROCESSES

Extensive dynamical calculations have been performed [17] on the LSTH (Liu-Siegbahn-Truhlar-Horowitz) PES [18] of state-to-state (*VT-processes*) and dissociation cross sections for collision of atomic hydrogen with rovibrationally excited molecular hydrogen.

All possible molecular rovibrational states (348) relative to the used PES have been considered and used as initial states, including quasibound ones (classically bound but quantally subject to tunnelling), which are of importance for recombination kinetics [19]. The code integrates the Hamilton equations of motion using Runge-Kutta fourth-order method with variable time step size controlled by a space interval of

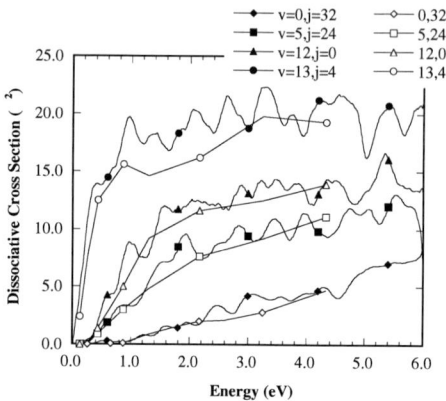

FIGURE 9. Dissociative cross section for selected initial states of H_2, as a function of collision energy, compared with results of ref. [21] (open symbols).

0.015 Å and a velocity interval of 0.01 Å/fs. Total energy error checks (6×10^{-4} eV) are also performed. The impact parameter is scanned using a stratified sampling method. The standard QCT method (quasiclassical trajectory method) has been modified by using continuous distributions of rovibrational actions in reactants instead of delta function distributions, weighting trajectories with a simple "square window" function centered in integer values with integer width [20].

Dissociative cross sections as a function of the relative collision energy for different (v,j) pairs are presented in Fig. 9. Cross section energy profile is characterized by a threshold energy, decreasing with the target internal energy, and by an increasing trend up to a maximum, whose value strongly depends on the initial roto-vibrational state. Comparing the cross sections with results calculated with a similar approach for few transitions by Dove and Mandy [21] (also reported in Fig. 9) a good agreement is found. State-selected dissociative rate coefficients, $K_d(v)$, (Fig. 10) have been obtained from cross section data, averaging on a Boltzmann distribution function of the rotational states.

Cross sections for vibrational de-excitation process

$$H + H_2(v, T_{rot}) \rightarrow H + H_2(v', \text{all } j') \tag{15}$$

summed over final j' values and averaged on the rotational temperature, have been obtained including both reactive and non-reactive collisions. The cross section sharply increases with translational energy up to a maximum and then decreases to an approximately flat trend. In Figs. 11a-b results for monoquantic de-excitation initiated from two different vibrational levels are shown for different rotational temperatures. Vibrational excitation emphasizes the peak value and shifts its position in the low energy region. The cross section dependence on the rotational temperature is more significant for low initial v, due to the wide rotational ladder and to the related large contribution at high

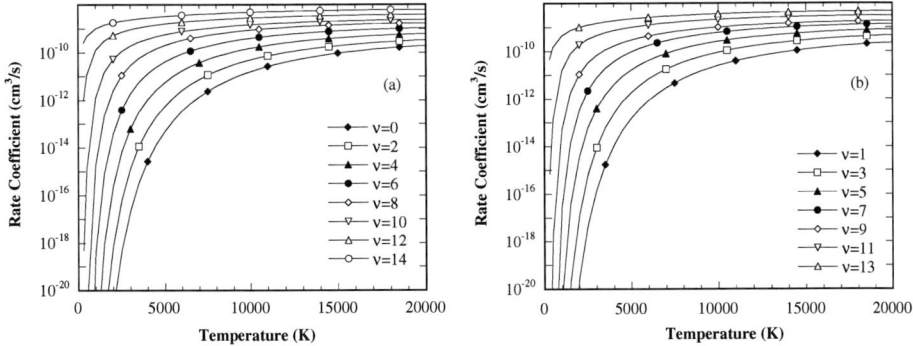

FIGURE 10. Dissociation rate coeffi cients averaged over Boltzmann distributions of rotational states plotted as a function of temperature for even (a) and odd (b) vibrational quantum numbers.

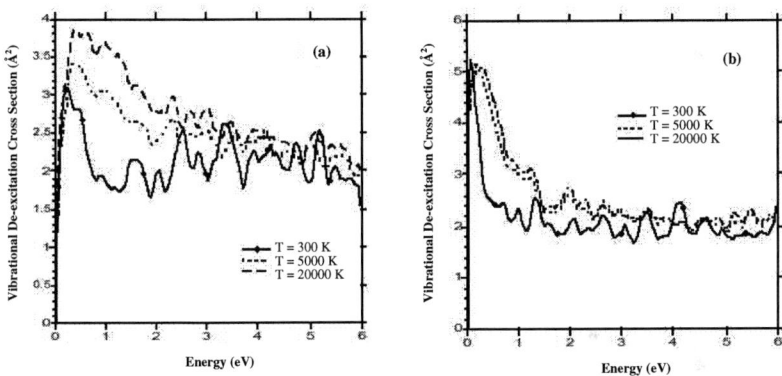

FIGURE 11. Monoquantic vibrational de-excitation cross sections as a function of translational energy, for different rotational temperatures. (a) $v=5$, (b) $v=9$.

rotational temperature. Derived rate coefficients are in good agreement with the rates calculated by Laganá [22].

In order to compare the calculated dynamical data with experimental results, the global dissociation rate coefficient has been derived, building up a kinetic model of the H_2 dissociation as occurring in a thermal bath of atoms at a temperature T. The atomic hydrogen concentration is taken to be 100 times higher than that one of the molecules. As a consequence only V-T processes involving atomic hydrogen can be assumed to significantly contribute to populate the vibrational ladder. The kinetic problem deals with the solution of the vibrational master equation including 15 vibrational levels of H_2 submitted to the action of processes

89

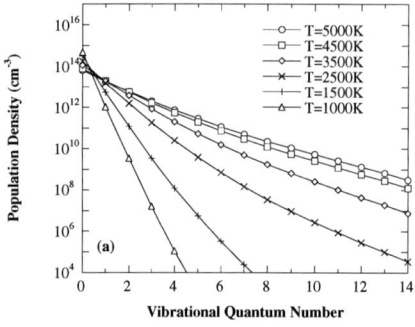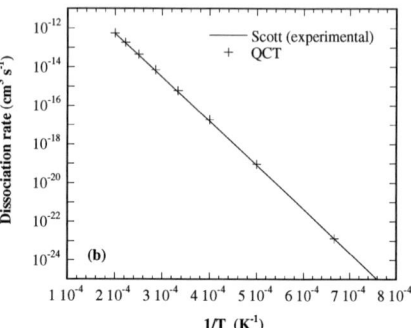

FIGURE 12. (a) Quasistationary vibrational distribution functions plotted versus vibrational quantum number for different gas temperature values. (b) A comparison of theoretical and experimental values of the global dissociation rate by Scott [23].

$$H + H_2(v) \rightarrow H + H_2(v') \tag{16}$$

$$H + H_2(v) \rightarrow 3H \tag{17}$$

In this scheme the global dissociation rate is the resultant of the dissociative processes initiated from vibrationally excited molecules, expressed as

$$v_d = N_H \sum_v N_v K_d(v) = N_H N_{H_2} K_d \tag{18}$$

K_d, the second order dissociation constant, can be obtained from the relevant $K_d(v)$ values and from the vibrational distribution function N_v. Vibrational populations have been determined by integration of the system of vibrational master equations, assuming the initial vibrational distribution as a delta function centered at $v=0$ level.

The temporal evolution of vibrational populations can be followed up to the quasistationary condition. Quasistationary vibrational distributions have a markedly Boltzmann character at all translational temperatures and distortions occur only for high vibrational levels (Fig. 12a). In Fig. 12b the global dissociation constant is displayed as a function of $1/T$, the experimental global dissociation rates quoted in [23] is also reported, showing the good agreement between experimental and theoretical results and indirectly confirming both the accuracy of the potential energy surface and the reliability of the QCT approach.

SURFACE PROCESSES

The interaction of atomic and molecular hydrogen on the reactor walls is relevant in affecting the H_2 vibrational distribution through recombination/dissociation reactions and diffusion processes on the surface.

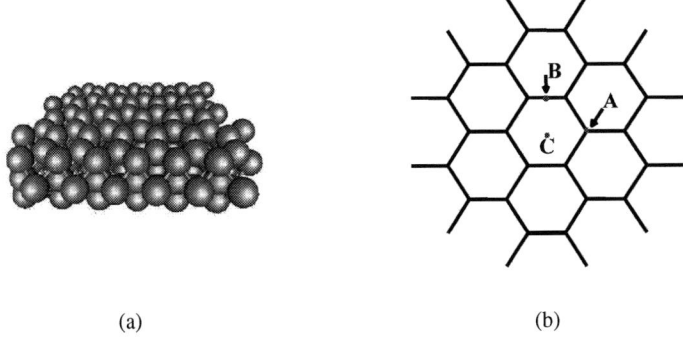

(a)   (b)

FIGURE 13. (a) Crystal model K62/62/62. (b) Graphite surface structure.

The recombination of atomic hydrogen on graphite has been studied in the framework of the semiclassical collisional method [24], according to which the dynamics of the chemical particles, H and H_2, interacting with the graphite surface is described classically by solving the relevant Hamilton's equations of motion while the phonons modes of the graphite surface are treated quantum-mechanically. The coupling between the classical degrees of freedom with the phonons dynamics is made via a time and surface temperature dependent effective potential V_{eff}, of the mean-field type.

The crystal model K62/62/62 consists of 186 carbon atoms displayed over three layers according to the appropriate lattice symmetry (Fig. 13a).

The adopted hydrogen-surface potential smoothly switches from H_2-graphite to H-graphite interaction potential according to the H-H interatomic separation. H_2 is physisorbed on graphite in the perpendicular geometry, as confirmed both by experimental and theoretical results [25, 26], the adsorption energy being negligible regardless the adsorption site. On the contrary the H-graphite interaction is characterized by chemiadsorption, the binding energy depending on the interacting site. Different semiempirical electronic structure calculations have been performed to explore the potential surface and it turns out that the strongest interaction occurs on top of carbon atom, i.e. at the lattice site A in Fig. 13b.

The interaction potential for atomic hydrogen approaching the surface perpendicularly at site A has been modelled fitting results by Fromhertz et al. [27] (Fig. 14a) and by Jeloaica et al. [28] (Fig. 14b). The more recent DFT calculations performed by Jeloaica et al. predict a well depth a factor two lower with respect to the value of ref. [27], affecting significantly the catalytic activity of the graphite surface. Reported results have been obtained employing the potential by Jeloaica et al.. Details on the parametric expressions for the interactions potentials can be found in refs. [29, 30]. Hydrogen recombination at graphite surface can occur through two alternative mechanisms:

- the direct Eley-Ridel two-step mechanism (*E-R mechanism*)

91

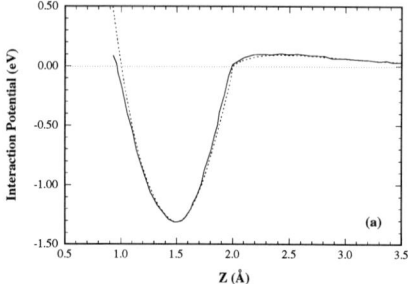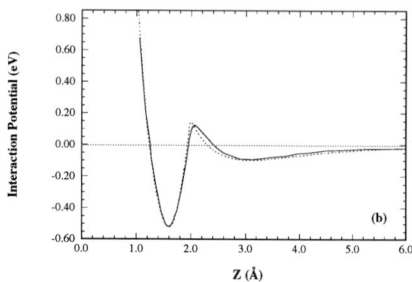

FIGURE 14. Hydrogen-surface interaction potential as a function of the atom surface distance, Z, at the site A. (a) ref. [27], (b) ref. [28]; dashed lines indicate fitting curves.

$$H_{gas} + \text{graphite} \rightarrow H_{ads} * \text{graphite} \qquad (19)$$

$$H_{ads} * \text{graphite} + H_{gas} \rightarrow H_2(v, j) + \text{graphite} \qquad (20)$$

- the indirect Langmuir-Hinshelwood mechanism (*L-H mechanism*)

$$H_{ads} * \text{graphite} + H_{ads} * \text{graphite} \rightarrow H_2(v', j') + \text{graphite} \qquad (21)$$

The L-H mechanism simulation of the diffusive processes of adsorbed hydrogen atoms across the surface is complex and requires the knowledge of the full topology of the interaction potential, therefore the E-R recombination reaction has been studied.

The adsorbed H atom is placed at the equilibrium distance of 1.5 Åand in thermal equilibrium with the surface at the temperature T_S, the molecular dynamic code simulates the H atom in the gas phase impinging on the surface with polar angles $(\theta; \phi)$ with a kinetic energy E_{kin}, giving the recombination probabilities, the nascent vibrational and rotational distribution of formed H_2 molecules and reaction probabilities for different reaction products.

Recombination probability critically depends on surface temperature: the increase of T_S corresponds to the decrease of the probability for the recombination. It can be appreciated by inspection of Fig. 15 where results for $T_S = 10K$ (temperature range for astrophysical applications) and for $T_S = 500K$ are presented. The energy distribution analysis shows that a fraction of the exothermic energy released in hydrogen recombination is shared among the internal states of the newly formed molecules, leading to non-Boltzmann nascent vibrational distribution, being the non-equilibrium character emphasized at high surface temperature. In Fig. 16 typical calculated vibrational populations of H_2 molecules for different surface temperatures are presented, at the impact energy of 0.03 eV.

The interaction with graphite actually can lead to activation of several reaction exit channels:

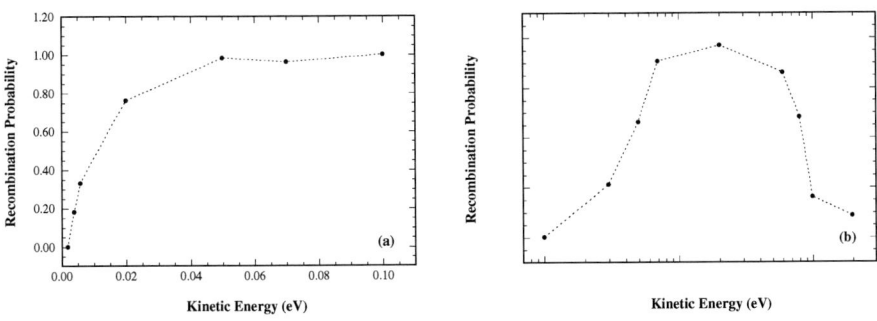

FIGURE 15. Recombination probability as a function of incident kinetic energy. (a) $T_S = 10K$; (b) $T_S = 500K$.

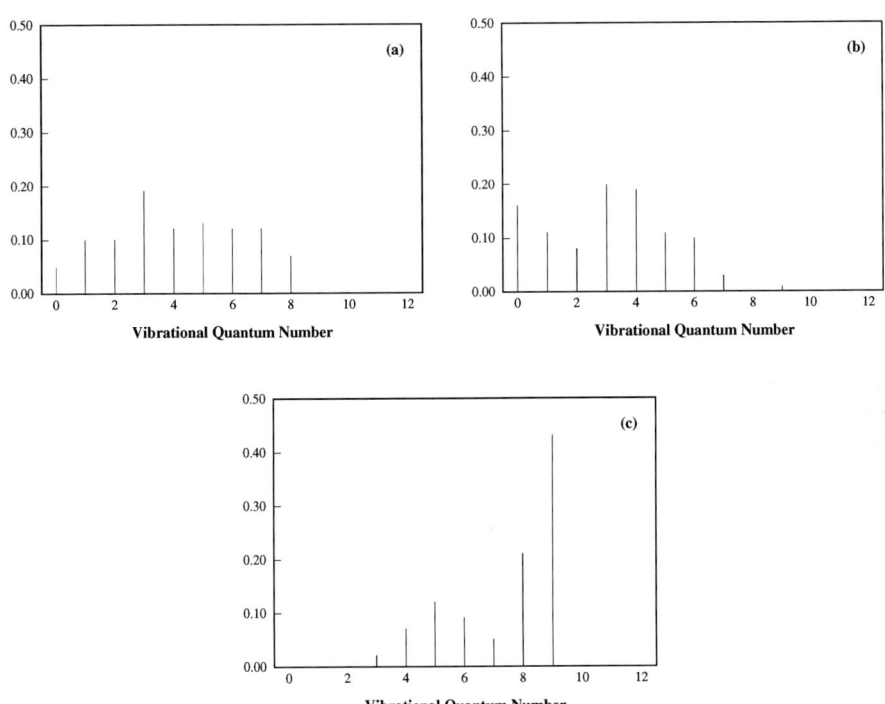

FIGURE 16. Population of vibrationally excited H_2 molecules in recombination process at the impact energy of 0.03 eV. (a) $T_S = 10K$; (b) $T_S = 100K$; (c) $T_S = 500K$.

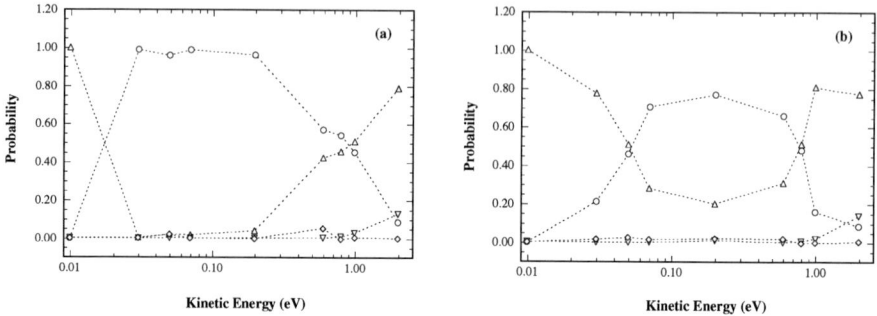

FIGURE 17. Probability for recombination (circles), adsorption of one atom (triangles), of both atoms (diamonds) and scattering of both atoms (down triangles) as a function of incident kinetic energy. (a) $T_S = 100K$; (b) $T_S = 500K$.

$$H_{ads} + H_{gas} \rightarrow \begin{array}{ll} H_2(v,j) & (a) \\ H_{gas} + H_{ads} & (b) \\ H_{gas} + H_{gas} & (c) \\ H_{ads} + H_{ads} & (d) \end{array} \qquad (22)$$

the non-reactive surface processes (22)b-d are relevant in the investigation of plasma walls performance in fusion devices, in fact the relative importance with respect to the recombination channel can be considered as an indication of the efficiency of surface damage processes experimentally observed. Probabilities for reactive and non-reactive channels in hydrogen-graphite interaction are presented in Fig. 17.

CONCLUSIONS

In the present report we have shown theoretical calculations of processes relevant to the modeling of negative ion sources. Emphasis has been given to the dependence of the process probability on the excitation of the internal degrees of freedom.

ACKNOWLEDGMENTS

This work has been partially supported by MIUR (Project No. 2003037912_010) and by ASI (contract I|R|055|02).

REFERENCES

1. M. Capitelli, R. Celiberto, and M. Cacciatore, in "Advanced in Atomic, Molecular and Optical Physics: Cross Section Data", edited by M. Inokuti, Academic Press, N.Y. and London, **33** (1994) 321.

2. R.K. Janev, in "Atomic and Molecular Processes in Fusion Edge Plasmas", edited by R.K. Janev, Plenum Press, N.Y. and London, (1995) 1.
3. J.N. Bardsley and J.M. Wadehra, Physical Review A **20** (1979) 1398.
4. J.M. Wadehra in "Nonequilibrium Vibrational Kinetics", edited by M. Capitelli, Springer-Verlag, New York and London, (1986).
5. C. Gorse, R. Celiberto, M. Cacciatore, A. Laganà and M. Capitelli, Chemical Physics **161** (1992) 211. C. Gorse, M. Bacal, R. Celiberto and M. Capitelli, Chemical Physics Letters **192** (1992) 161.
6. U. Fantz, B. Heger and D. Wünderlich, Plasma Physics Controlled Fusion **43** (2001) 1.
7. R. Celiberto, R.K. Janev, A. Laricchiuta, M. Capitelli, J.M. Wadehra and D.E. Atems, Atomic Data and Nuclear Data Tables **77** (2001) 161.
8. R. Celiberto and T.N. Rescigno, Physical Review A **47** (1993) 1939.
9. R. Celiberto, R.K. Janev and A. Laricchiuta, Physica Scripta **64** (2001) 26.
10. J.R. Hiskes, Journal of Applied Physics **51** (1980) 4592. J.R. Hiskes, Journal of Applied Physics **70** (1991) 3409.
11. R. Celiberto, A. Laricchiuta, U.T. Lamanna, R.K. Janev and M. Capitelli, Physical Review A, **60** (1999) 2091.
12. L.A. Pinnaduwage, W.X. Ding, D.L. McCorkle, S.H. Lin, A.M. Mebel and A. Garscadden, Journal of Applied Physics **85** (1999) 7064.
13. T.N. Rescigno and B.I. Schneider, Journal of Physics B **21** (1988) L691.
14. D.T. Stibbe and J. Tennyson, New Journal of Physics **1**, (1998) 21-29.
15. R. Celiberto, U.T. Lamanna, M. Capitelli, Physical Review A **50** (1994) 4778.
16. A. Laricchiuta, R. Celiberto, and R.K. Janev, Physical Review A **69** (2004) 022706.
17. F. Esposito, C. Gorse, M. Capitelli, Chemical Physics Letters **303** (1999) 636.
18. D.G. Truhlar, C.J. Horowitz, Journal of Chemical Physics **68** (1978) 2466; **71** (1979) 1514.
19. R.E. Roberts, R.B. Bernstein, C.F.Curtiss, Journal of Chemical Physics **50** (1969) 5163.
20. F. Esposito, PhD Thesis, Department of Chemistry, University of Bari (1999).
21. J.E. Dove, M.E. Mandy, Int. J. Chem. Kin. **18** (1986) 893.
22. A. Laganà, E. Garcia, "Quasiclassical rate coeffi cients for the H+H$_2$ reaction", University of Perugia, Italy (1996).
23. C.D. Scott, S. Fahrat, A. Gicquel, K. Hassouni, M. Lefebvre, J.Thermophysics and Heat Transfer **10** (1996) 426.
24. G.D. Billing, "Dynamics of molecule surface interactions", Wiley, New York (2000).
25. L. Mattera, F. Rosatelli, C. Salvo, F. Tommasini, U. Valbusa, G. Vidali, Surf. Sci. **93** (1980) 515.
26. D. Novaco, J. P. Wroblewski, Physical Review B **39** (1989) 11364.
27. T. Fromhertz, C. Mendoza, F. Ruette, Mon. Not. R. Astron. Soc. **263** (1993) 851.
28. L. Jeloaica, V. Sidis, Chemical Physics Letters **300** (1999) 157.
29. M. Rutigliano, M. Cacciatore and G.D. Billing, APID (Atomic and Plasma-Material Interaction Data for Fusion) **9** (2001) 267.
30. M. Rutigliano, M. Cacciatore and G.D. Billing, Chem.Phys.Lett. **340** (2001) 13.

Effects of Transverse Magnetic Field and Spatial Potential on Negative Ion Transport in Negative Ion Sources

T.Sakurabayashi, A.Hatayama and M.Bacal*

Faculty of Science and Technology, Keio Univ., Hiyoshi, Yokohama 223-8522, Japan
*Laboratoire de Physique et Technologie des Plasmas, Ecole Polytechnique, Laboratoire du CRNS, 91128 Palaisseau Cedex, France

Abstract. The effects of the magnetic field on the negative ion (H⁻) extraction in a negative ion source have been studied by means of a two-dimensional electrostatic particle simulation. A particle-in-cell (PIC) model is used which simulates the motion of the charged particles in their self-consistent electric field. Since most electrons are magnetized by the magnetic field, more H⁻ ions arrive instead of electrons in the region close to the plasma grid in order to ensure the plasma neutrality. In the present article, the effect of the electron diffusion across the magnetic field is also taken into account. Although the effects of the electron diffusion en the difference of the dynamics between electrons and H⁺ ions, the presence of the magnetic field still has a large contribution to the enhancement of the H⁻ extraction.

INTRODUCTION

Neutral beam injection based on the negative ions is one of the most promising candidates for heating and current drive in future fusion reactors. To generate intense beams of negative ions and optimize the source [1-3], understanding of the transport properties of negative ions (H⁻)[4-7] is indispensable. In the experiments, the extraction of H⁻ ions has been significantly improved by the magnetic field [8,9]. The extractor electrode has a pair of magnets, as it is shown in Fig.1. The magnets create a transverse magnetic field in the extractor, which is strong enough to suppress electrons from the negative ion beam. In addition, electrons inside the plasma source can be easily magnetized by the stray field of this electron suppression magnet (a few tens of Gauss) and be lost along the field line, while positive ions (H⁺) cannot be magnetized because of the larger Larmor radius. This difference in dynamics between electrons and H⁺ ions perturbs the plasma neutrality in this region and leads to the formation of a positive plasma potential. This modification in the plasma neutrality and resultant plasma potential due to the magnetic field are considered to be very important for the H⁻ ion transport and extraction.

In our previous paper [10] we confirmed by particle-in-cell (PIC) simulation the idea, that more H⁻ ions arrive to PG region in order to ensure the plasma neutrality. However, the effect of the electron diffusion across the magnetic field has not been taken into account. In ref. [11] the preliminary study was done with the effect of

CP763, *Production and Neutralization of Negative Ions and Beams*, edited by J. D. Sherman and Y. I. Belchenko
© 2005 American Institute of Physics 0-7354-0248-5/05/$22.50

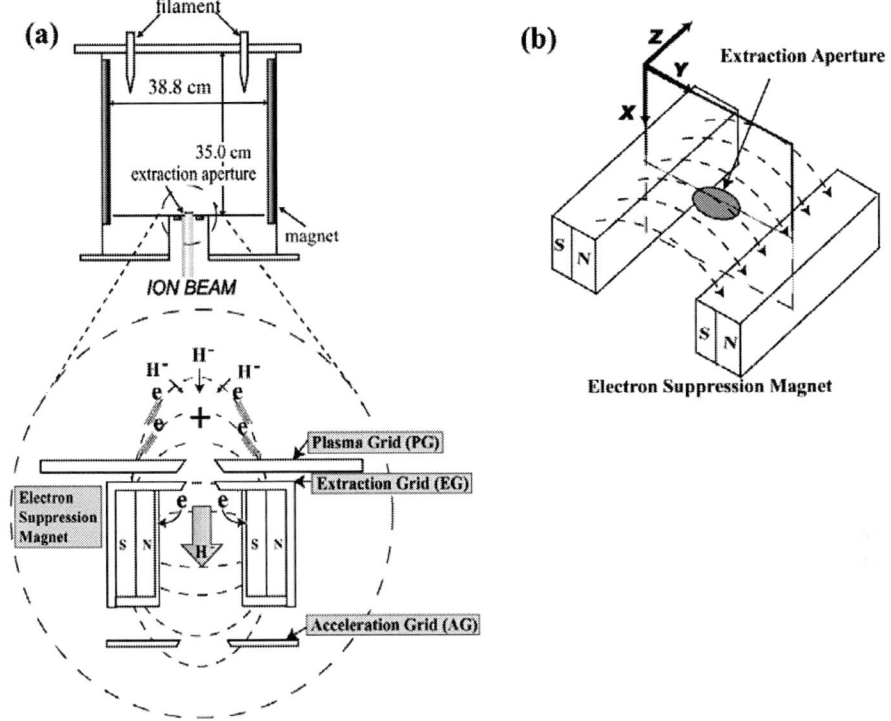

FIGURE 1. Source schematic diagram. (a) Cross-sectional view with the enlargement of the extractor. The extraction system has three grids. A pair of magnets creates the magnetic field. (b) Bird's–eye view of a pair of magnets. Magnetic lines of force are denoted by broken lines.

electron diffusion. The electron diffusion had small effects on H⁻ transport because most of the electrons were lost along the field line before they diffused across the magnetic field.

However, in ref. [10, 11] we neglected the variation of magnetic field along the field line for simplicity. In the real geometry the magnetic field is stronger near the electron suppression magnets (Fig.1). If we take into account this effect, some electrons may oscillate along the field line due to the effect of magnetic mirror. Due to this magnetic mirror effect, the effective time of electron loss along the field line could become large. Electrons trapped in the magnetic mirrors could have enough time to diffuse across the magnetic field before they will be lost along the field line. Thus, in order to perform a more realistic simulation, further analysis incorporating both, the effects of (i) the variation of magnetic field and (ii) the collision processes is required.

In this paper, we take into account these two effects for overall understanding of the effect of magnetic field on the spatial structure of electric potential and on the resultant H⁻ transport under more realistic model.

SIMULATION MODEL

The motion of the charged particles (H⁻ ions, H⁺ ions and electrons) in their self-consistent electric field is calculated by the PIC method [12-15]. The trajectories of the particles are calculated by numerically solving the equation of motion

$$m_s \, dv/dt = q_s (\mathbf{E} + \mathbf{v} \times \mathbf{B}) \tag{1}$$

where m_s, $\mathbf{v}$, q_s, $\mathbf{E}$ and $\mathbf{B}$ are the particle mass, velocity, charge, electric field and magnetic field for particles of species s, respectively. To obtain the charge density at the grid point, the particle charge is assigned to each grid point by linear interpolation. The particle densities n_s also were calculated from the charge densities on the grid points. Poisson's equation for the spatial electric potential ϕ

$$\nabla^2 \phi = -q(n_{H^+} - n_e - n_{H^-})/\varepsilon_0, \tag{2}$$

is solved by the finite difference method, where ε_0 is the permittivity of free space. The numerical algorithm employed to solve Poisson's equation is the SIP (Stone's Strongly Implicit Procedure) method [16].

The extractor used in the experiments consists of three electrodes, a plasma grid (PG), an extraction grid (EG) and an acceleration grid (AG), as shown in Fig.1 (a). The electrodes have an extraction aperture. A pair of magnets creating the magnetic field near the PG is located in the EG. The magnets are installed in parallel on both sides of the extraction aperture, as schematically shown in Fig.1 (b). The resultant magnetic lines are also shown in Fig.1 (b) with broken lines. We mainly focus on the limited region around the extraction aperture in this simulation. The coordinate system used in the analysis is also shown in Fig.1 (b). We assume a spatial uniformity in the z-direction near the extraction aperture, because the lengths of the magnets are relatively long in comparison with the extraction aperture, and the edge effect of the magnetic field in the z-direction can be neglected. Thus we confine our attention to the two-dimensional spatial profiles in the $x-y$ plane in Fig.1 (b).

The model geometry of the simulation is shown in Fig.2 which includes only the PG and EG. The magnetic field exists over the entire model region. In our previous study [10-11] a relatively simple model for the magnetic field was employed. The magnetic field component was assumed to have only y-component, namely B_y. A simple analytic distribution with the Gaussian profile for the x-direction was used. The variation of the magnetic field along the field line and the resultant magnetic mirror effect are considered to be important for electron diffusion. For this purpose, a pair of magnets with the infinite length in the z-direction is installed in the model geometry, as shown in Fig.2. The magnetic field has two components B_x and B_y in the

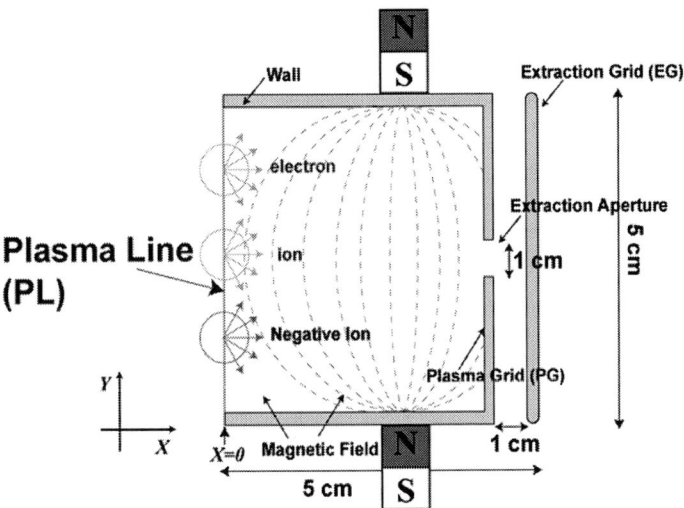

FIGURE 2. Schematic diagram of 2D particle simulation model used in this study. The left-hand boundary (PL: $x = 0$ cm) is the upstream boundary. The wall at $x = 4.0$ cm is used as PG, which has an extraction aperture of 1.0 cm . The wall at $x = 5.0$ cm is used as EG.

present model. Figure 3 shows the spatial profile of the strength of magnetic field $B = \sqrt{B_x^2 + B_y^2}$. The magnetic field is calculated at each point by using the analytical solution based on the magnetic charge model [17].

Electron diffusion across the magnetic field is taken into account by Monte-Carlo method in the following manner. The path length estimator algorithm [18] has been employed for each time step of integration to judge if the collision event takes place or not. The probability for electrons to collide with H^+ ions and neutrals is estimated as $P_e = 1 - \exp(-\sum v_e^{tot} \Delta t)$. Here, Δt is the time step of integration and v_e^{tot} is the total collision frequency of electrons including two kinds of collisions: electron-

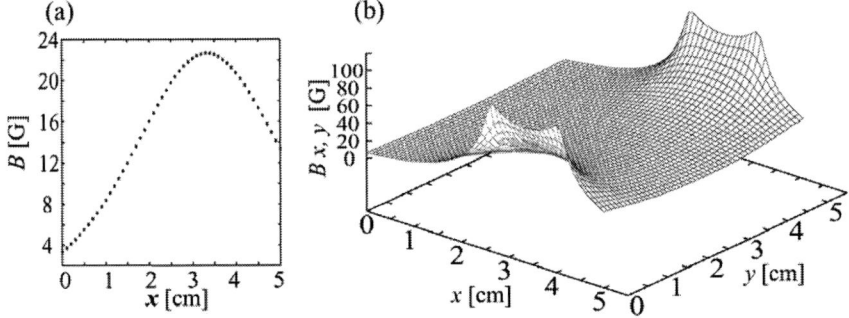

FIGURE 3. Spatial profile of the magnetic field: (a) in the x direction at $y = 2.5$ cm , (b) in the $x - y$ plane.

neutral and electron-ion collisions $v_e^{tot} = v_{en} + v_{ei}$. Electron-electron collisions were neglected, since the collisions of light particles are not important for the spatial diffusion across the magnetic field. The electron speed $\mathbf{v}_e$ is assumed to be almost unchanged before and after collision because of large mass difference in electron-neutral and electron-ion collisions. The scattering angles θ and ϕ of collision were chosen randomly with an isotopic distribution. The Monte Carlo collision model employed is relatively simple. For example, it does not take into account the small scattering feature of Coulomb scattering, i.e. electron-ion collision. However, at least, as far as the diffusion in the uniform B-field with $\left(\omega_c/v_e^{tot}\right)^2 \gg 1$ is concerned, this model gives the same result of a particle diffusion coefficient $D_\perp = r_L^2 v_e^{tot}$, as that obtained from the fluid model, where ω_c and r_L are the cyclotron frequency and Larmor radius, respectively.

The initial condition of the calculation region is an empty computation region. Let the left-hand boundary ($x = 0$) be the upstream boundary; we call this boundary "Plasma Line" (PL), shown in Fig.2. The plasma ($x < 0$) is assumed to contain sources, which maintain the plasma neutrality and stationary particle fluxes across the PL. Thus the electric field at the PL is zero and equal numbers of positive and negative particles per time step are injected into the calculation region from the PL. The boundary conditions for the particle flux Γ are $\Gamma_{H^+} = \Gamma^-$, $\Gamma^- \equiv \Gamma_e + \Gamma_{H^-}$ and $\Gamma_{H^-}/\Gamma^- = 0.1$, where Γ_{H^+}, Γ_e and Γ_{H^-} are the H^+ ion flux, the electron flux and the H^- ion flux, respectively. The charged particles are isotropically launched from the PL in the x-direction. The initial velocity distribution is a Maxwellian one with temperatures $T_e = 1.0$ eV, $T_{H^+} = 0.5$ eV, $T_{H^-} = 0.5$ eV. If particles cross the PL from right to left, they are reloaded with new velocities in addition to the particles successively injected per time step. As shown in Fig.2, the vertical wall at $x = 4$ cm is the PG with the extraction aperture of 1 cm. The PG and wall potentials are fixed $\phi = 0$. The particles reaching the PG or walls are removed from the calculation region. The wall at $x = 5$ cm is the EG, which has no extraction aperture and limits the model from the right. The EG conditions are the same, as those for the walls and the PG, except the EG potential ϕ_{EG}. As was discussed in Ref.[10,11], we have adjusted ϕ_{EG} to an appropriate value $\phi_{EG} = 10$ V in order to separate the pure effect on the H^- extraction of the transverse magnetic field and of the extraction voltage penetration to emission aperture .

SIMULATION RESULTS

In order to study the effects of a transverse magnetic field and electron diffusion across the field on H^- extraction, the following three cases are compared:
(A)- no effects of the magnetic field and electron diffusion,
(B)- with the effect of the magnetic field only
(C)- with effects of the magnetic field and electron diffusion.

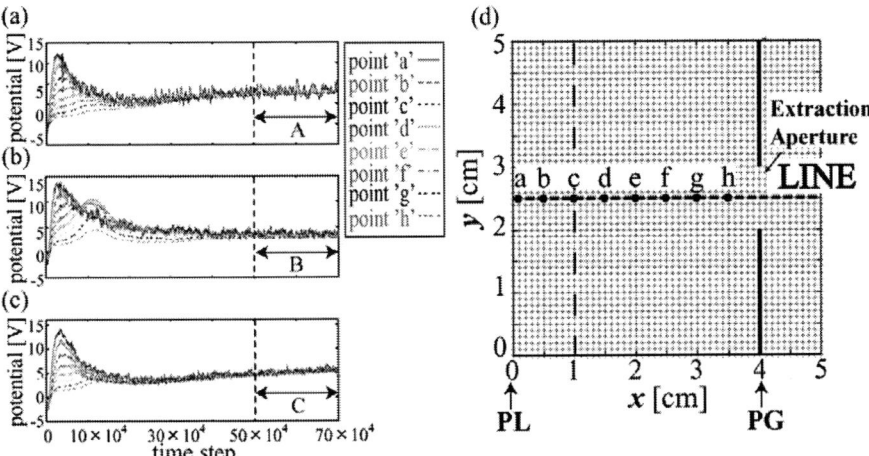

FIGURE 4. Time evolution of potentials (a) without the magnetic field and diffusion, (b) with only the magnetic field, c) with both effects of the magnetic field and diffusion at points a~h in (d)

Other conditions were the same, as those mentioned in the previous section for these three cases. Figure 4 (a, b, c), shows the time evolution of calculated electrostatic potentials for each point along the central LINE, as it is shown in Fig.4(d). Fig. 4 (a), (b) and (c) are the results for case (A), (B) and (C), respectively. After about 5×10^5 time steps, quasi-steady states have been reached for all the cases, as seen from Fig.4. Under these quasi-steady state conditions, the density profiles of H^+ ions, electrons and H^- ions are compared for cases (A), (B) and (C). In Fig.5, the density profiles are plotted versus the distance along the LINE for each case. The densities are normalized by the first cell H^+ density.

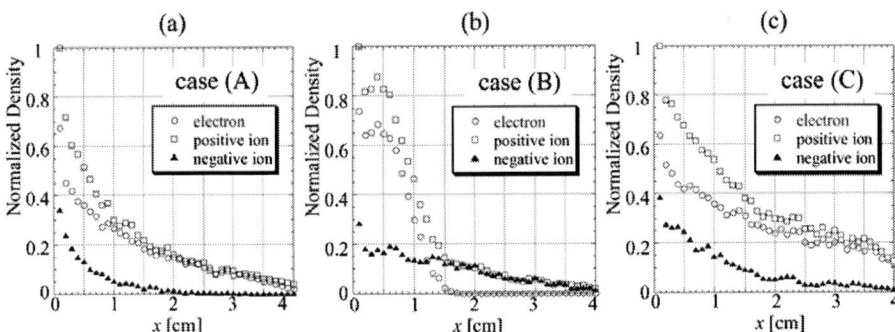

FIGURE 5. Variation of the charged particle densities versus the axial distance along the LINE shown in Fig.4 (d). (a), (b) and (c) are the results for case (A), (B) and (C), respectively. For case (A), the charge neutrality is maintained by mainly H^+ ions and electrons. For case (B), H^- ions are balanced with H^+ ions instead of electrons to maintain the plasma neutrality. For case (C), both negative ions and diffused electrons are balanced with H^+ ions to maintain the plasma neutrality.

101

As in ref. [10], the extraction of H⁻ ions for case (B) is significantly enhanced in comparison with case (A). Also, for case (C), significant increase of H⁻ extraction compared with case (A) has been observed in the simulation. Due to the diffusion across the magnetic field, electron diffusion has the effect to en the trapping effect by the magnetic field. However, the enhancement factor of H⁻ current for case (C) has been about the same as that for case (B). To understand these features and the physical mechanism of the enhancement of the H⁻ extraction for both cases, we will make more detailed comparisons of the numerical results in the following subsections.

Collisionless Cases Comparison

To understand the pure and intrinsic effect of the magnetic field on the H⁻ transport, we will make at first a comparison between two cases without the effect of electron diffusion, i.e., case (A) and case (B) in Fig.5. The comparison is useful as a basis for the understanding of the effect of electron diffusion discussed below.

The magnetic field drastically affects the density profiles of each species as seen from the comparison between Fig.5(a) and Fig.5(b). For case (B), electrons are magnetized by the magnetic field due to their small Larmor radius. Consequently, most of electrons are localized near the PL, as shown in Fig.5(b). Thus, the computational domain can be divided into the following two characteristic regions: the "Electron trapping region", the region from the PL to $x \simeq 1$ cm and the "Effective extraction region", the region from the right-hand side of the "Electron trapping region" to the PG (1 cm $\leq x \leq 4$ cm). In the effective extraction region, electron density n_e is almost zero, while H⁺ density n_{H^+} is comparable to H⁻ density n_{H^-} . Negative ions are balanced with H⁺ ions instead of electrons to maintain plasma neutrality.

The magnetic field also affects the spatial structure of the electric potential. Figure 6 shows the potential profiles for (a) without and (b) with the magnetic field, respectively. Here, in Fig.6 (a) and (b), we concentrate on the region 0 cm $\leq x \leq 3.5$ cm in order to see the difference between these two cases more clearly. As seen from Fig.6 (b), the potential has a characteristic positive peak at $x \approx 1$ cm for case (B) with the weak magnetic field, while there is no such a positive potential peak for case (A). Due to the effect of weak magnetic field, the spatial potential changes its spatial structure in such a way that it collects H⁻ ions. As a result, the H⁻ density in the effective extraction region increases, as shown in Fig.5(b).

The process of the potential peak formation has already been discussed in Ref.[10]. Here, we briefly summarize the main points. Electrons are magnetized by the magnetic field and localized in the electron trapping region. Then, a transverse electric field perpendicular to the magnetic field sets up to retard ion motion in the x -direction. In order to balance with electrons in the "electron trapping region", some ions are attracted toward the PL by the transverse electric field, and some ions with large velocity continue moving towards the PG. As a result, the magnetic field produces two H⁺ ion flows, i.e., back flow toward PL and the forward flow toward PG. The H⁻ ions collected in the effective extraction region move together with H⁺ ions to ensure the plasma neutrality and reach the PG. The position of the potential peak is almost

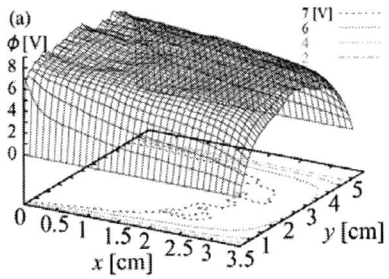 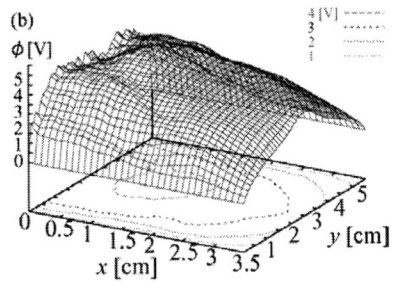

FIGURE 6. Structure of the potential in the region $0 \text{ cm} \leq x \leq 3.5 \text{ cm}$ (a) without the magnetic field and (b) with the magnetic field. The electron diffusion is not taken into account for both cases.

coincident with the point where the H^+ ion flows stagnate. Thus, the H^- ions collected in the effective extraction region move together with H^+ ions to ensure the plasma neutrality and reach the PG.

Effects of the Electron Diffusion Across the Magnetic Field

To understand the effects of electron diffusion, now we are coming back to Fig.5. As seen from the comparison between Fig.5(b) and Fig.5(c), the density profiles of each species (electrons, H^+ ion and H^- ion) are greatly changed by including the effect of electron diffusion. For case (B) without electron diffusion, electrons never enter the region close to the PG as shown in Fig.5(b). On the other hand, for case (C), electrons diffuse out of the "electron trapping region" and enter the "effective extraction region" due to the diffusion across the magnetic field as shown in Fig.5(c). As a result, the density profiles in Fig.5(c) for electrons and H^+ ions are more similar to those for case (A) without the magnetic field in Fig.5(a) than those for case (B) with the magnetic field in Fig.5(b).

However, even if the effect of the electron diffusion is taken into account, still relatively a large amount of H^- ions reaches the region close to PG for case (C) than case (A). Consequently, not only electrons, but also H^- ions play an important role to maintain the plasma neutrality near the PG for case (C).

Figure 7 shows a comparison of H^- density among case (A), (B) and (C). The negative ion densities are normalized by the H^- density in the first cell for case (C). In the region $1 \text{ cm} \leq x \leq 3 \text{ cm}$, the H^- density for case (B) becomes larger compared with those for other cases. However, if we concentrate on the region near the PG ($3 \text{ cm} \leq x \leq 4 \text{ cm}$), n_{H^-} for case (C) with the electron diffusion is comparable to that for case (B), or even slightly larger than that for case (B). The extraction current I_{H^-} (the H^- current that passes through the extraction aperture in Fig.2) for case (C) is comparable to that for case (B).

These features of H^- density and extraction current are closely related to the spatial structure of the electric potential. For case (B), electric potential has a peak at around $x \approx 1 \text{ cm}$ as shown in Fig. 6(b). As was discussed in the previous subsection and also

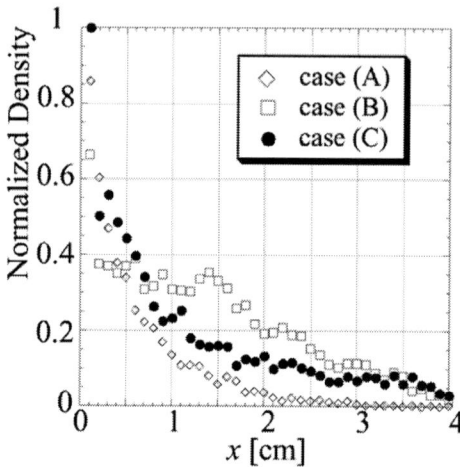

FIGURE 7. Comparison of spatial profiles of negative ion density among case (A), (B) and (C).

In ref.[10], the positive potential peak collects H$^-$ ions from the electron trapping region into the effective extraction region. Injected H$^-$ ions are accelerated into the region to the left of potential peak. After that they are decelerated at the right of the potential peak. As a result n_{H^-} becomes relatively large near the potential peak around $1 \text{ cm} \leq x \leq 2 \text{ cm}$ for case (B) as shown in Fig.7.

On the other hand, no such a clear potential peak at $x \approx 1$ cm was observed in the simulation for case (C) due to the effect of electron diffusion across the magnetic field. As a result, n_{H^-} for case (C) in the region $1 \text{ cm} \leq x \leq 2 \text{ cm}$ becomes small compared with that for case (B) as shown in Fig.7. Instead, for case (C), relatively small positive potential peak can be seen more close to the PG (at $x \approx 3$ cm) as shown in Fig.8. Due to this positive potential peak close to the PG, the negative ion density for case (C)

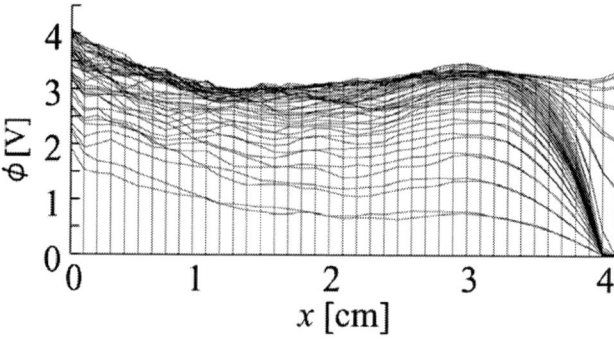

FIGURE 8. Side view of spatial potential structure in the region $0 \text{ cm} \leq x \leq 4 \text{ cm}$ for case (C). Relatively small positive potential peak locates near the PG $x \approx 3.0 \text{ cm}$.

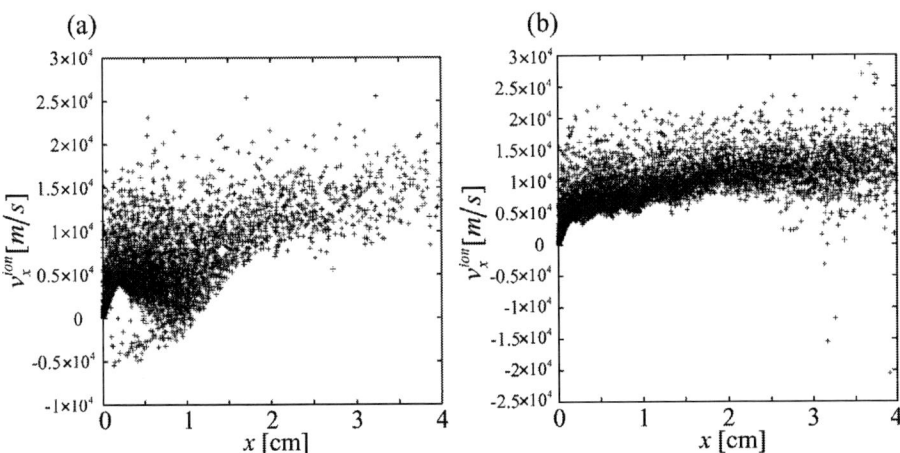

FIGURE 9. Distribution of x component of ion velocity along the LINE. (a) For case (B), positive ion flow is stagnated at $x \simeq 1$ cm, (b) For case (C), some ions are decelerated at $x \simeq 3.0$ cm.

becomes comparable to that of case (B) or even slightly larger in this region ($3\text{cm} < x \leq 4\text{cm}$) in front of PG as shown in Fig.7.

Figure 9 shows a typical snap shot of the phase space (x, v_x) plot of each H^+ ion, i.e., distribution of the x component of velocity v_x for H^+ ions along the LINE in Fig.4(d). Figure 9(a) and (b) correspond to the results for case (B) and case (C), respectively. As was pointed out in the previous subsection, for case (B) a clear back flow of H^+ ions arises with negative velocities ($v_x < 0$) toward the PL, where the positive potential peak exists as shown in Fig.6(b). On the other hand, for case (C), no such a clear back flow toward the PL is observed at $x \approx 1$cm. However, if we focus our attention to the region $2.5\text{cm} < x < 3.5\text{cm}$, a clear retardation of H^+ ions is observed around $x \approx 3$ cm as shown in Fig.9(b). The position is almost coincident with the location of the potential peak $x \approx 3$ cm in Fig.8. This modification of potential structure plays an important role to understand the increase in the H^- density in front of the PG in Fig.7 for case (C).

SUMMARY

The effect of the transverse magnetic field on the H^- extraction in the negative ion source is studied by a 2D PIC simulation. Electron diffusion across the magnetic field is taken into account. Without the effect of the electron diffusion, electrons are magnetized by the magnetic field and localized in the electron trapping region. On the other hand, positive ions H^+ cannot be magnetized due to their large mass. This difference in dynamics between electrons and H^+ ions has been verified to lead to the formation of the positive potential peak near the edge of the electron trapping region. The H^- ions are collected by the positive potential peak and move together with H^+ ions towards the PG to ensure the plasma neutrality.

With the effect of electron diffusion, both diffused electrons and negative ions are balanced with H^+ ions to maintain the plasma neutrality. As a result, the positive potential peak becomes smaller, but it moves closer to PG. Due to this positive potential peak close to the PG, the negative ion density becomes still comparable to that in the case without electron diffusion, or even slightly larger in front of PG. Consequently, even if we take into account the effect of electron diffusion across the magnetic field, the presence of magnetic field has a large contribution to the enhancement of the H^- extraction.

The simulations show that the intrinsic physical mechanism of the enhancement of H^- extraction observed in the simulation is possibly explained by the modification of plasma charge neutrality due to the transverse magnetic field and the formation of positive potential peak near the PG.

REFERENCES

1. M.Bacal, C.Michaut, L.I.Elizarov, and F.El Balghiti, *Rev.Sci.Instrum.* **67**, 1138(1996).
2. C.Courteile, J.Bruneteau, and M.Bacal, *Rev.Sci.Instrum.* **66**, 2533(1995).
3. F.El Balghiti-Sube, F.G.Baksht, and M.Bacal, *Rev.Sci.Instrum.* **67**, 2221(1996).
4. T.Sakurabayashi, A.Hatayama, K.Miyamoto, M.Ogasawara, and M.Bacal, *Rev.Sci.Instrum.* **73**, 1048(2002).
5. M.Uematsu, T.Morishita, A.Hatayama, and T. Sakurabayashi, *Rev.Sci.Instrum.* **71**, 883(2000).
6. A.Hatayama, T.Sakurabayashi, Y.Ishii, K.Makino, M.Ogasawara and M.Bacal, *Rev.Sci.Instrum.* **73**, 910(2002).
7. K.Makino, T.Sakurabayashi, A.Hatayama, K.Miyamoto, M.Ogasawara, *Rev.Sci.Instrum.* **73**, 1051(2002).
8. M.Bacal, J.Bruneteau, and P.Devynck, *Rev.Sci.Instrum.* **59**, 2152(1988).
9. P.Devynck, M.Bacal, J.Bruneteau, and F. Hillion, *Revue Phys.Appl.* (Paris) **22**, 753(1987).
10. T.Sakurabayashi, A.Hatayama and M.Bacal, *J.Appl.Phys.* **95**, 3937(2004).
11. T.Sakurabayashi, A.Hatayama and M.Bacal, *Rev.Sci.Instrum.* **75**, 1770 (2004).
12. C.K.Birdsall and A.B.Langdon, *Plasma Physics via Computer Simulation,* New York: McGraw-Hill, 1985.
13. J.P.Verboncoeur, M.V.Alves, V.Vahedi and C..K.Birdsall, *J.Comp.Phys.* **104**, 321(1993).
14. W.S.Lawson, *J.Comp.Phys.* **80**, 253(1989).
15. R.J.Procassini, C.K.Birdsall, and E.C.Morse, *Phys.Fluids* B **2**, 3191(1990).
16. H.L.Stone, SIAM *J.Numer.Anal.* **5**, 530(1968).
17. Y.Ohara, M.Akiba, H.Horiike, H.Imai, Y.Okumura and S.Tanaka, *J.Appl.Phys.* **61**, 1323(1987).
18. M.H.Hughes and D.E.Post, *J.Comp.Phys.* **28**, 43(1978).

2D PIC-MCC Code for Electron-Hydrogen Gas Interaction Study in H⁻ Ion Sources

Karim Benmeziane [(1)(2)], Robin Ferdinand[(1)], Raphael Gobin[(1)],
Gerard Gousset[(2)], Joseph D. Sherman[(3)]

*(1) Commissariat à l'Energie Atomique, CEA-Saclay, DSM/DAPNIA, 91191 Gif sur Yvette Cedex,
France*
(2) Laboratoire de Physique des Gaz et des Plasmas (Associé au CNRS), 91405 Orsay cedex, France
(3) LANSCE Division, Los Alamos National Laboratory, Los Alamos, NM87545, USA

Abstract. In order to make a reliable H⁻ ion source, a hybrid PIC 2D MCC 3D [1] fluid code has been developed. The aim of the code is to study the effect of electron injection into a cylindrical gas chamber. This new version takes into account the 2D space charge distribution. Thus, it is possible to calculate the H⁻ ion distribution everywhere in the plasma chamber. Many results have been brought as well as the best injected energy and the electron penetration length efficiency. Moreover, the calculations explain why it is more problematic to get an efficient volume production at high pressure (100 mTorr) than at low pressure (6 mTorr). The temporal H⁻ production evolution is finally discussed.

INTRODUCTION

A few years ago experimental investigation on an ECR H⁻ ion source were started at CEA-Saclay. Many results have been discussed [2,3,4]. Some conclusions have been made, but questions are still remaining. The ECR H⁻ ion source would not produce current more intense than a few µA if it kept as simple ECR source. The high frequency wave creates high-energy electrons not efficient for negative ion production. Moreover this source has the advantage to generate a high electron current density, thus a high proton current has been measured. They have been used to perform hard X ray, and large proton beam currents [5]. Among all H⁻ sources previously working around the world, two groups can be distinguished: those working via surface and volume processes. This source is being developed based on H⁻ volume production process. Surface process is efficient only if alkalai metal such as cesium is added. These elements bring the wall material work function down, thus allowing electron capture on the surface by atoms.

The ECR source studied produces a large amount of hydrogen atoms. It would be a very good H⁻ ions production surface source if the emphasis were not laid on a long lifetime. On contrary large source of atoms are not desired for H⁻ volume production study.

CP763, *Production and Neutralization of Negative Ions and Beams*, edited by J. D. Sherman and Y. I. Belchenko
© 2005 American Institute of Physics 0-7354-0248-5/05/$22.50

ECR ion sources have an important advantage, which is their long lifetime in continuous running mode. Nowadays, filament-like sources give significant H⁻ currents. Thus, the idea is to use the ECR source as an electron source. The beam extracted from the source is supposed to be injected into a second chamber which is the same gas chamber separated by a grid system. Those electrons would relax their energy, heat the plasma but no microwave is required in this second chamber. For many years, the process leading to H⁻ ion is well known. It is called the dissociative attachment (D-A) [6].

$$e_{slow} + H_2(v>9) \rightarrow H + H^- \tag{1}$$

Such reactions have a zero threshold for vibrational levels larger than nine. Then, it is essential to have hydrogen molecules in a vibrational state higher than 9 with large density. To make them (most of $H_2(v>9)$), electrons with energies around 20 eV are needed. To obtain an efficient source of H⁻ ions, injecting electron in a hydrogen gas chamber appears to be a good solution. For these reasons a stainless steel grid with 5 mm mesh gap has been introduced in the source and huge improvement have been observed. It is then possible to move that grid to different positions. This structure is supposed to stop the microwave energy. It is also possible to polarize the grid. A double grid system is expected to improve the electron injection into the second chamber. Many other applications of the grid might be realized. It is thought to build a grid in a magnetic material to decrease the effect of the transversal magnetic field in that second chamber. It would have the effect of an electron filter generally used in such a sources. A transversal magnetic field is used to avoid fast electrons in the production zone (only slow electrons are desired in this zone for the D-A). If a grid were made of a magnetic material an inductive magnetic field would appear. This field is in the so called 'driver zone' and is not wanted in the second chamber.

CALCULATION DESCRIPTION

To enlarge the understanding, a PIC 2D MCC3D code has been developed. The aim of the code is to understand phenomena under the interaction between an electron beam and hydrogen gas in the H⁻ production chamber, and especially if there is a specific place where the H⁻ ions are produced inside the chamber. Injecting an electron beam into a chamber is not as easy as it seems. The system becomes anisotropic which makes it quite different from other types of more symmetric sources (multicusps). To overcome the anisotropic configurations it is obligatory to favor a particle technique instead of a Boltzmann code which generally supposes a weak anisotropy. The code is articulated in two parts: the Monte Carlo code with collision (MCC) associated with a PIC code and a 'Fluid code'. The MCC consists in creating electrons with an initial energy on the front edge of the second chamber. The PIC code calculates the space charge field, which forces the movement of charged particles. The second kind of code named here 'fluid code' is used to determine the vibrational excitation of H_2 molecules and obtaining the H⁻ production rates.

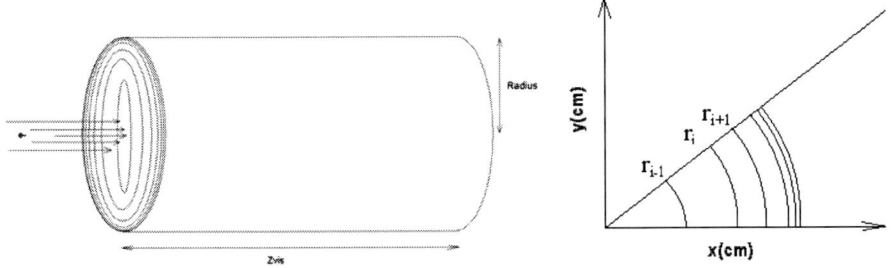

FIGURE 1. Diagram of the simulation status at the left.

The initial chamber, shown in Fig. 1, has 1cm radius and 5cm length. The electron beam is also 1cm radius. At the right is shown the radial mesh. The MCC code keeps track of electron and their interaction with the gas including 27 processes: two rotational excitations, three vibrational excitations, four triplet state excitations, eight singlet state excitations. These processes are all described in Tables 1 and 2.

TABLE 1. Chemistry of the PIC code, two rotational excitation, three vibrational excitation, four triplet state excitation, eight singlet state excitation, one ionization [7,8].

Process	Reaction	reaction (eV)
Elastic collision	$e + H_2 \rightarrow e + H_2$	0
rotational excitation	$e + H_2\ (X^1\Sigma_g^+, v=0, j=0) \rightarrow e + H_2\ (X^1\Sigma_g^+, v=0, j=2)$	0.044
rotational excitation	$e + H_2\ (X^1\Sigma_g^+, v=0, j=1) \rightarrow e + H_2\ (X^1\Sigma_g^+, v=0, j=3)$	0.073
vibrational excitation	$e + H_2\ (X^1\Sigma_g^+, v=0) \rightarrow e + H_2\ (X^1\Sigma_g^+, v=1)$	0.516
vibrational excitation	$e + H_2\ (X^1\Sigma_g^+, v=0) \rightarrow e + H_2\ (X^1\Sigma_g^+, v=2)$	1.0
vibrational excitation	$e + H_2\ (X^1\Sigma_g^+, v=0) \rightarrow e + H_2\ (X^1\Sigma_g^+, v=3)$	1.46
Electronic excitation $b^3\Sigma_u^+$	$e + H_2\ (X^1\Sigma_g^+) \rightarrow e + H_2\ (b^3\Sigma_u^+)$	10.0
Electronic excitation $c^3\Pi_u$	$e + H_2\ (X^1\Sigma_g^+) \rightarrow e + H_2\ (c^3\Pi_u)$	12.3
Electronic excitation $a^3\Sigma_g^+$	$e + H_2\ (X^1\Sigma_g^+) \rightarrow e + H_2\ (a^3\Sigma_g^+)$	12.0
Electronic excitation $e^3\Sigma_u^+$	$e + H_2\ (X^1\Sigma_g^+) \rightarrow e + H_2\ (e^3\Sigma_u^+)$	13.22
Electronic excitation $B^1\Sigma_u^+$	$e + H_2\ (X^1\Sigma_g^+) \rightarrow e + H_2\ (B^1\Sigma_u^+)$	12.7
Electronic excitation $C^1\Pi_u$	$e + H_2\ (X^1\Sigma_g^+) \rightarrow e + H_2\ (C^1\Pi_u)$	12.4
Electronic excitation $E^1\Sigma_g^+$-$F^1\Sigma_g^+$	$e + H_2\ (X^1\Sigma_g^+) \rightarrow e + H_2\ (E^1\Sigma_g^+ - F^1\Sigma_g^+)$	13
Electronic excitation $B'^1\Sigma_u^+$	$e + H_2\ (X^1\Sigma_g^+) \rightarrow e + H_2\ (B'^1\Sigma_u^+)$	14.8
Electronic excitation $D^1\Pi_u$	$e + H_2\ (X^1\Sigma_g^+) \rightarrow e + H_2\ (D^1\Pi_u)$	14.9
Electronic excitation $B''^1\Sigma_u^+$	$e + H_2\ (X^1\Sigma_g^+) \rightarrow e + H_2\ (B''^1\Sigma_u^+)$	15.5
Electronic excitation $D'^1\Pi_u$	$e + H_2\ (X^1\Sigma_g^+) \rightarrow e + H_2\ (D'^1\Pi_u)$	15.6
Ionization	$e + H_2 \rightarrow 2e + H_2^+$	15.4

Excitation, ionization and dissociation are thus considered. Two types of positive ions are included in the present model. H_3^+ and H_2^+ ion kinetics are especially very fundamental ion studies when low energy are involved:

$$H_2 + H_2^+ \rightarrow H_3^+ + H + 1.71eV \qquad (2)$$

To determine the relative importance of atomic positive ion concentration created by ionization of atoms, we take into account atom kinetics, their volume creation and their loss at the walls. Atomic hydrogen are thus very important in the global system

TABLE 2. Processes used to calculate hydrogen atoms by dissociation.

Process	Branching ratio
$H_2 (b^3\Sigma_u^+) \rightarrow H(1s) + H(1s)$	$\chi = 1$
$H_2 (c^3\Pi_u) \rightarrow H(1s) + H(1s) + h\nu$	$\chi = 1$
$H_2 (a^3\Sigma_g^+) \rightarrow H(1s) + H(1s) + h\nu$	$\chi = 1$
$H_2 (e^3\Sigma_u) \rightarrow H(1s) + H(1s) + h\nu$	$\chi = 1$
$H_2 (D^3\Pi_u) \rightarrow H(1s) + H(2l)$	$\chi = 0.298$
$H_2 (B''^1\Sigma_u^+) \rightarrow H(1s) + H(2l)$	$\chi = 0.967$
$H_2 (D'^1\Pi_u) \rightarrow H(1s) + H(2l)$	$\chi = 1$

kinetics, and are considered as shown in the Table 2 equations. Their loss takes place at the wall recombination giving hydrogen molecules [9]:

$$H(1s) + wall \rightarrow (1/2)H_2 (X^1\Sigma_g^+, v=0) \ (\gamma = 10^{-2}) \tag{3}$$

where γ depends on the wall material (for iron [10] $\gamma = 10^{-4}$, for stainless steel [11] $0.07 \leq \gamma \leq 0.15$). As result of the model prediction, the concentration of H atoms is too low to have a large concentration of H^+ ions by electron ionization. The $\frac{[H]}{[H_2]}$ order of magnitude can be under 10^{-2}. Moreover, due to the low energy of electrons the dissociation ionization by electron impact of H_2 molecules is also inefficient. Thus, H^+ ions production can be considered as negligible. Electron's cross section data of mechanisms considered in the present calculation are from two sources, J.Loureiro [12] for low kinetic energy and H.Tawara [13] for higher energy than 40 eV.

To determine the electron rate coefficient involved in the vibrational H_2 molecule kinetics, it is necessary to have some informations on electron energy distribution function 'EEDF'. The MCC code allows an estimation of EEDF in energy mesh as:

$$f_e(E) = \frac{n_e(r,t)}{\Delta V . \Delta E} . \tag{4}$$

Where n_e, ΔV, ΔE are the number of electrons and the volume, and the energy width of each cell, respectively. According the last definition, the function $f_e(E)$ is normalized to the local electron density $n_e(r,t)$. This function is thus the number of electrons per unit volume with energy between $E-\Delta E$ and $E+\Delta E$ (in eV). In the H^- ion source community it is well know that negative ion production is related with the incident power. Actually this power depends on the discharge current which itself is related with the electron density owing to the following expression: $I=e.n_e.v$, where v is the electron drift velocity.

As it is not possible to consider the movement of each electron, the use of computing techniques is necessary. To do the most of the calculation time, each computed particle so called 'macro-particle' or 'super-particle' is actually considered as many real particles (Bundle of particles) with a specific weight larger than 1. Thus, the number of calculated 'macroparticles' can be reduced and the number of real particles considered in the physics can be high. If this weight initially injected becomes too large, no space-charge compensation would appear and causes incorrect calculation. To avoid such difficulties the weight is then gradually introduced for stability convenience of the code. This number is calculated from the following exponential function: Weight(t) = $FW_*(1-e^{-t/\tau})$.

FW is the final weight to be achieved. The choice of this final weight is very important. If this factor is too large, time step will be too short and calculation time increases considerably and divergences could appear. An ideal one seems to be 10^5 real particles in a 'macro-particle'.

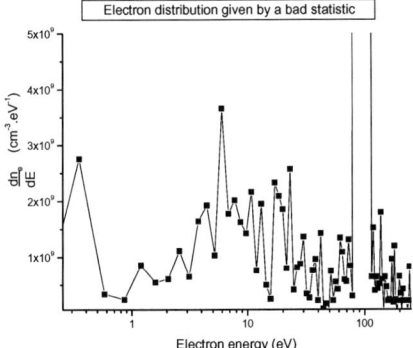 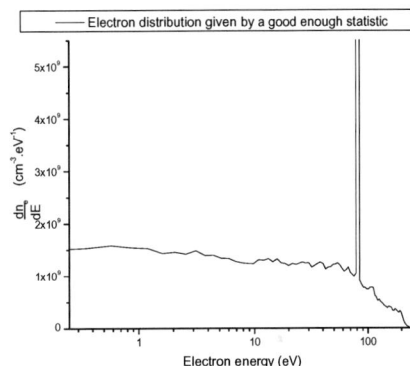

FIGURE 2. Bad eedf statistic given by a 10^7 weight.

FIGURE 3. Reasonable statistic obtained by 10^5 weight.

Figure 2 compared to Figure 3 shows how this is important. The EEDF have been calculated at the same plasma conditions. For the first distribution: 10^7 particles per 'macroparticle' are injected when they are only 10^5 in the second one. Indeed, if the weight were reduced the calculation precision would be better but the calculation time increases considerably. Calculation time is the only weakness of such a code, which is usually very reliable. This weight is a compromise between expected precision and short calculation time.

A PIC code has been used to determine the space charge field by solving the Poisson equation every 10^{-11}s which is the Maxwell time:

$$\frac{1}{r}\cdot\frac{\partial r.E_r}{\partial r}(r,z,t)+\frac{\partial E_z}{\partial z}(r,z,t)=\frac{e.(n_+-n_e)}{\varepsilon_0}\tag{5}$$

It is thus possible to calculate the electric field, which forces and modifies charge particle motion all over the chamber. The calculation is completed through a spatial discretization, which is constant along z-axis but varies along radius to maintain a constant ring area (Figure 1).

THE FLUID CODE

As result of the first part of the code, the electron rate coefficient of the reactions involving the $H_2(v)$ molecules kinetic are calculated in the second part of the model. This second part is a heavy particle kinetic code called fluid code in the rest of this paper. This code calculates the vibrational molecule distribution and their transport. The H^- ion production is then calculated. H^- ions are created from the dissociative attachment process previously described. H_2 molecules in the vibrational states are necessary to obtain a large source rate for H^- ions. This part includes four of the dominant processes such e-V, E-V, V-T V-V. All vibrational excitation source terms are described in Table 3.

TABLE 3. Vibrational molecule excitation processes with each coefficient

Reaction	Name
$e+H_2(w)\underset{\leftarrow}{\overset{\rightarrow}{\rightleftharpoons}}e+H_2(v)$	e-V
$e+H_2(v)\rightarrow e+H_2((B^1,C^1\Pi_u))$ $\rightarrow e+H_2(X^1\Sigma_g,v')+hv$	E-V
$H_2+H_2(v+1)\underset{\leftarrow}{\overset{\rightarrow}{\rightleftharpoons}}H_2+H_2(v)$	V-T Molecular contribution
$H+H_2(v)\underset{\leftarrow}{\overset{\rightarrow}{\rightleftharpoons}}H+H_2(v+k)$	V-T Atomic contribution
$H_2(w-1)+H_2(v+1)\underset{\leftarrow}{\overset{\rightarrow}{\rightleftharpoons}}H_2(w)+H_2(v)$	V-V

Once all source terms have been calculated, vibrational molecules can be considered by solving the master equation. The global source term is composed by a sum of all processes source term. It gives the following transport equation.

$$\left(\frac{\partial N_v}{\partial t}\right)_{e-V}+\left(\frac{\partial N_v}{\partial t}\right)_{E-V}+\left(\frac{\partial N_v}{\partial t}\right)_{V-T}^m+\left(\frac{\partial N_v}{\partial t}\right)_{V-T}^a+\left(\frac{\partial N_v}{\partial t}\right)_{V-V}=div\Gamma_{H2}\tag{6}$$

This equation can be solved by a Rung-Kutta method respecting the boundary conditions. Those conditions are mainly a flux equal to zero at the center, and the Miles condition at the edge.

112

$$\Gamma_H(r = R) = \frac{\gamma_H . N_{H2} . \alpha_H . v_{mH}}{4(1 - \gamma_H / 2)} \qquad (7)$$

This last condition is initially a condition for atoms, but it can be adapted for molecules knowing that Γ_H = -2.Γ_{H2}. The Miles condition becomes: $\Gamma_{H2(v)}(r=R) = -\frac{1}{2} . \frac{\gamma_H . N_{H2} . \alpha_H . v_{mH}}{4(1 - \gamma_H / 2)}$.

The H$^-$ ion calculation is possible by applying a balance equation [14]:

$$N_{H-} = \frac{n_e \sum_{v=0}^{v=14} N_v C_{att}(v)}{n_e . K_{rec} + N_H . K_{det} + \frac{1}{\tau_d}} \qquad (8)$$

$$C_{att}(v) = \int f_e(E) . v_e(E) \sigma_{ad}(v, E) dE \qquad (9)$$

The above equation has been used to obtain the results presented in this paper first part. The second part takes into account H- transport which is obviously not done by the formula above.

K_{rec} is the H$^-$ ion recombination coefficient. It is corresponding to the losses term for the reaction:

$$H^- + ions \rightarrow neutrals \quad K_{rec} = 2 \ 10^{-7} \ cm^3.s^{-1} \qquad (10)$$

K_{det} and K_e are H$^-$ ion detachment coefficient with hydrogen atoms and electrons respectively. The first one is corresponding to the losses term for the reaction:

$$H^- + H \rightarrow H_2 + e \quad K_{det} = 1.8 \ 10^{-9} \ cm^3.s^{-1} \qquad (11)$$

The second process corresponds to:

$$H^- + e \rightarrow H + 2e \quad K_e = 7.6 \ 10^{-9} \ cm^3.s^{-1} \qquad (12)$$

It is clear that atoms are very important to give a realistic calculation. Among all particles involved, excited molecules play an important role on the atom density. Thus, atoms are not only calculated in the PIC-MCC but some of them are determined thanks to the fluid code. Even if atoms from the H_3^+ conversion ($H_2 + H_2^+ \rightarrow H_3^+ H + 1.71eV$) are greatest, another type of atoms calculation are integrated by including an additional dissociation branch. It is the dissociation from the vibrational level number 15, all other molecules below this threshold give H$^-$ considering the conditions previously introduced. This vibrational dissociation is modeled as shown below:

$$H_2(15) + e \rightarrow H + H^* + e \qquad (13)$$

$$H_2(15) + H \rightarrow H + H^* + H \qquad (14)$$

$$H_2(15) + H_2 \rightarrow H + H^* + H_2 \qquad (15)$$

Each $H_2(15)$ molecules calculated automatically give two atoms.

RESULTS AND DISCUSSION

A 1D PIC has initially been developed, with an assumption based on negligible electric fields along the z-axis. A first calculation has been performed for 40 mA of 40 eV electron beam injected in 100 mTorr pressure hydrogen gas chamber at a temperature of 300 K. The EEDF obtained after two days calculation is represented in the Figure 4. The EEDF tail spreads over 40 eV and denotes the electron acceleration due to the space charge. This is typically the electron energy distribution function used to determine the vibrational distribution.

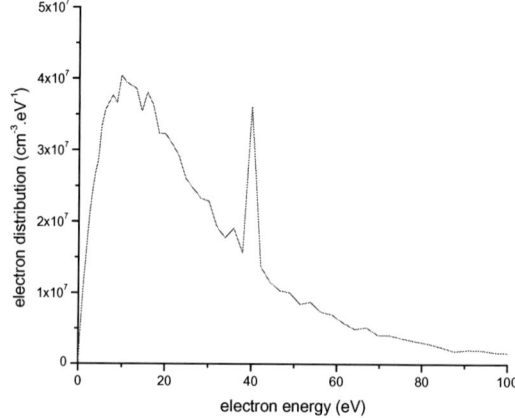

FIGURE 4. eedf obtained by injecting 40 mA electron beam at 40 eV in 100 mTorr of hydrogen gas.

Vibrational molecule calculation on the Figure 5 seems to be satisfying. The distribution converges at 0.146 ms corresponding to calculation times previously given. At 0.089 ms, the atomic population is not converged yet [11]. It is due to non-equilibrium between E-V processes and atomic V-T. This effect gets compensated after 0.125 ms.

114

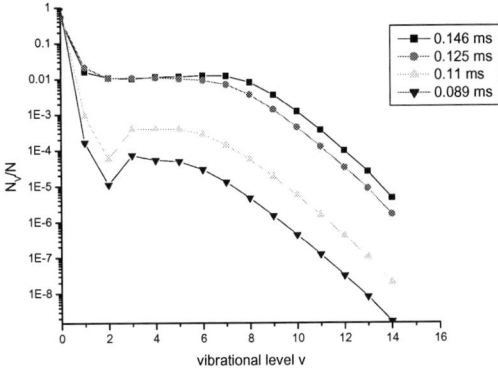

FIGURE 5. Evolution of the eedf with the time at 100 mTorr with a 40 mA electron beam at 40 eV.

The electron energy distribution function and the vibrational distribution are known at every point of the chamber. It is then possible to calculate the H⁻ ion source rate profile in 2D space even if the H⁻ ions transport is not done. For ion sources development, we can point out the source rate profile for negative ions. This profile is represented in Figure 6 by the ratio: H⁻ ion source term over the term of losses.

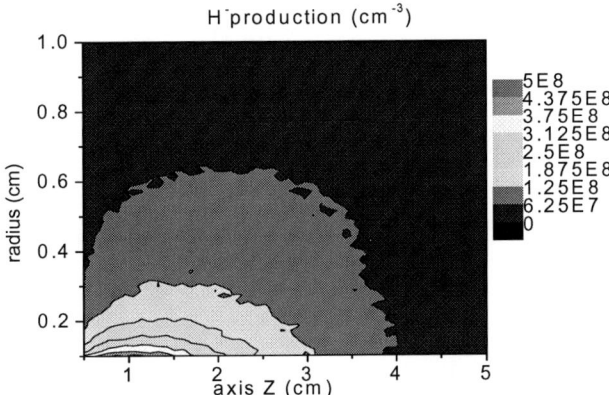

FIGURE 6. Distribution of H⁻ ions production in 2D view for 40 mA electron beam of 40 eV injected in a hydrogen gas chamber at a pressure of 100 mTorr. The H- transport has also been calculated and no large differences are noticed.

The maximum H⁻ ion production is spread up to 2 or 3 cm. It actually shows the electron efficiency through the z-axis. It also has been demonstrated by the Nietzche code [15] that H⁻ ions are destroyed 3 cm from the origin point they took place.

At the same time experimental measurements aimed to inject electrons into a second chamber confirm this result. This experiment introduces a polarized stainless grid inside an ECR ion source to control the energy of the injected electrons. The maximum H current is measured when the grid is placed between 2 and 3cm from the extraction electrode. A grid at 3 cm seems to be better. The experimental result is exposed at the end of this paper on the Figure 14. The Figure 7 shows how H ion production and electron are related, which is quite expected considering the equation used to calculate the H ion production. This result confirms the idea that 100 mTorr is a too high pressure for the H ion production. For such a source a ratio of $2.3 \ 10^{-2}$ is too small to be acceptable.

Results previously showed are obtained at 100 mTorr. At this pressure, many reasons can intervenes in the low H ion concentration. This pressure carries more atomic hydrogen and consequently brings a high destruction level. This plasma configuration increases the ionization and raises the mutual neutralization, which is a reason of H loss. Every efficient H ion source works at a very low pressure (between 1 and 10 mTorr) because of the pumping system, that is another reason why it is interesting to be able to get the pressure down.

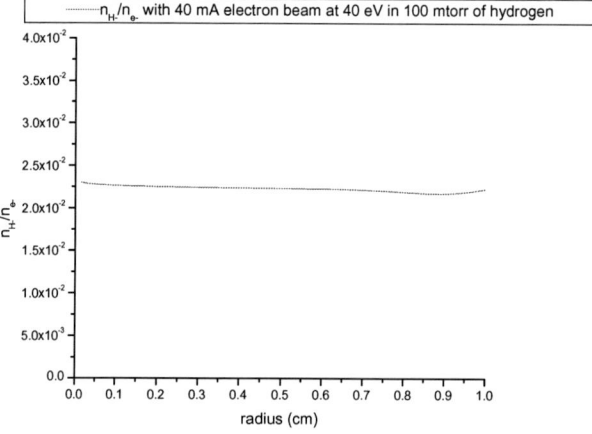

FIGURE 7. Average value of H ions over electron density ratio along the radius.

Many difficulties are coming with this purpose of reducing the pressure. Decreasing the pressure has the consequence to increase the free mean path of electron. If one wants a reasonable number of events taking place inside the chamber, it is important to enlarge it. For instance reducing the pressure by a factor of ten involves an enlargement of radius by a factor of $\sqrt{10}$. If so, cells areas become ten times bigger. Important consequences on the stability of the Poisson equation would be noticed. To keep cells size at the same dimension one has to increase cells number by a factor of

10. In that case the calculation time would be affected. To work this out, the solution is to include a solenoidal magnetic

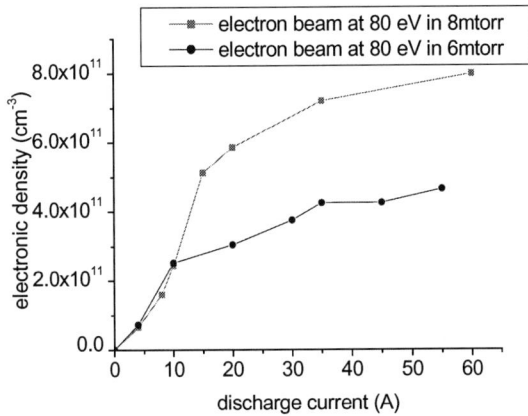

FIGURE 8. Electron density vs the current discharge for 6 and 8 mTorr by injecting 80 eV electron beam.

field in the calculation. For the moment a constant magnetic field equal to 1200 Gauss [3] is considered. It is in consequence, possible to achieve a pressure down to 6 mTorr. Figure 8 shows the results obtained for 6 and 8 mTorr since it has already been measured with filament source at the same conditions.

The analysis of the Figure 6 obviously reveals that H⁻ space distribution is not homogeneous. The non-existent electric field along the z-axis hypothesis is not correct. It has been then decided to solve the Poisson equation in two dimensions.

FIRST RESULTS FOR THE 2D CALCULATION

Very promising results are obtained after having adapted the 1D code to a 2D PIC. In addition, an H⁻ transport routine makes the calculation more realistic. One can observe a real improvement of the EEDF with the time (Figure 9) and can compare this curve with the one displayed on the Figure 3. Both are obtained in the same condition. The plasma thermalization goes a lot quicker in the 2D version (Figure 3). This improvement is characterized by a 2D EEDF magnitude higher at low energy range and lower at higher range than 80 eV (the EEDF spreading reduces at high energy).

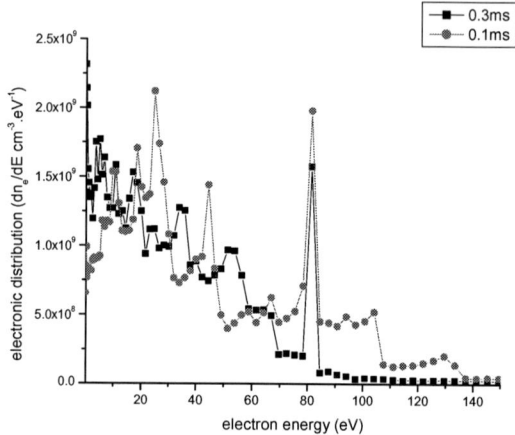

FIGURE 9. Evolution of the electron energy distribution function with the time at 6 mTorr with a 10 A. electron beam at 80 eV.

The vibrational distribution function is very satisfying as presented on the Figure 10:

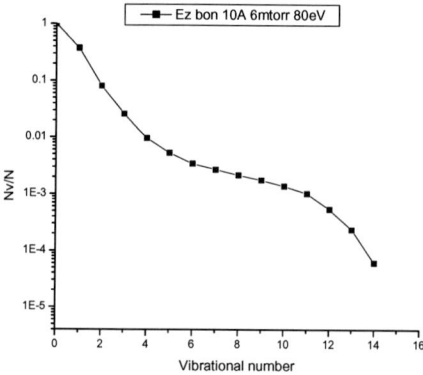

FIGURE 10. Vibrational distribution for 10 A of electron beam at 80 eV in 6 mTorr.

One can denote that is very important to get an idea about the most valuable energy. It is then possible to proceed to a comparison of the H^- ion production for different electron energy beams. Here the H^- transport has been calculated by using an exponential schema [16] to solve the equation: $div\Gamma_{H^-} = S_{H^-}$. The source term is:

$$S_{H^-} = n_e.K_{rec}.N_{H^-} + N_H.N_{H^-}.K_{det} + \frac{N_{H^-}}{\tau_d} - n_e.\sum_{v=0}^{v=14} N_v.C_{att}(v) \qquad (16)$$

The energy range around 40 eV appears to be the most efficient

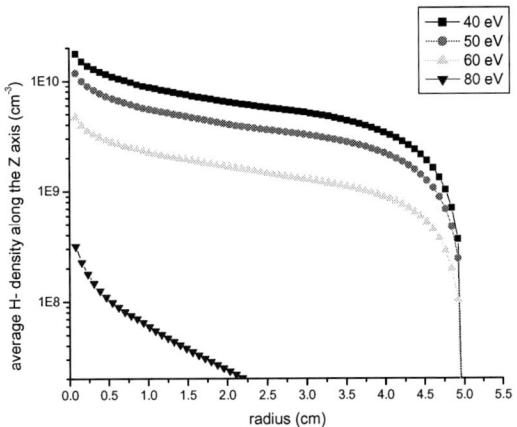

FIGURE 11. Average H- density at 0.3 ms for 10 A electron beam at 80 eV.

The H⁻ over electron ratio, probably gives the best idea of the efficiency of the electron. Injecting electrons at around 40 eV allows the best H⁻ ion density as possible.

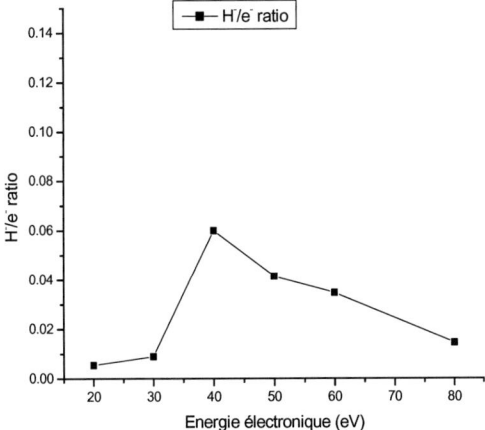

FIGURE 12. H⁻/e⁻ ratio for 10 A electron beam 6 mTorr for different energy.

ELECTRON PENETRATION EFFICIENCY

As shown in the Figure 13 the electron penetration is efficient up to the 3 cm vicinity. The calculation has been later confirmed by measurements presented in Figure 14.

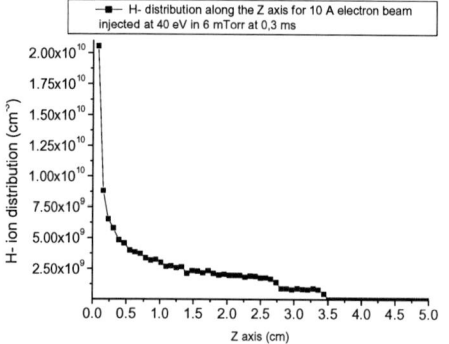

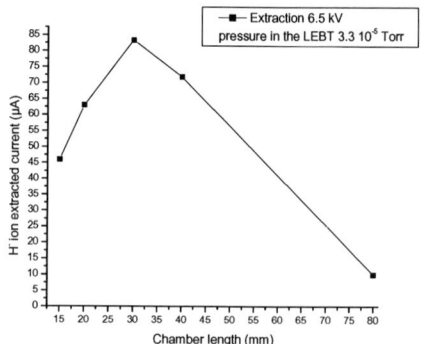

FIGURE 13. H- ion production vs the chamber length.

FIGURE 14. Experimental extracted current vs the chamber length.

As shown in the Figure 11 the H⁻ ion production is constant along 3 cm with a chamber radius of 5 cm. If the assumption based on the non-extraction of H⁻ after 3 cm is correct, the present radius chamber (5 cm) is large enough the keep a high negative ion density along 3 cm. Indeed, the distribution depends on the boundary condition and a larger chamber can provide H- for a larger distance, but shorter radius gives higher injected power density. As it is well known, the injected power density plays an important role in the H⁻ production.

CONCLUSION

The code gives some interesting results, which are conforming to experimental measurement done in such a configuration by filament sources. It would help to define where the H⁻ ions are produced and place the extraction system at the right place. As it has been calculated an ideal system would be an extraction system placed after 3 cm from the electron injection. If the secondary chamber length were larger than 3 cm, it would not be as much efficient because of two reasons: H⁻ losses risk and electron energy efficiency not optimized enough. An ideal chamber radius of 5 cm has been determined as well. Experimental studies have been done in parallel with the ECR H⁻ ion source at CEA-Saclay. After introducing a simple stainless steel grid inside the chamber, the H⁻ current has been increased by a factor higher than 200. This grid stays intact after many months of non-stop work. Very promising results are expected, and may be possible by using a double grid system to obtain an extraction and injecting system.

ACKNOWLEDGMENTS

The authors want to thanks Ziane Kechidi, and Bruno Pottin who kindly offered their computer for running the code. A. France for his advices on the paper. Indeed, the technical staff at CEA Saclay for doing an amazing works on the source. Without them, there is no point to do this code: Francis Harrault and Yannick Gauthier.

REFERENCES

1. K. Benmeziane, R. Ferdinand, R. Gobin, and G. Gousset, Rev. of Sci. Instrum. 75(5), 1729(2004).
2. R. Gobin, K. Benmeziane, O. Delferriere, R. Ferdinand, F. Harrault, and J. D. Sherman, AIP Conf. Proc. 639, 177(2002).
3. R. Gobin, P.-Y. Beauvais, K. Benmeziane, O. Delferriere, R. Ferdinand, F. Harrault, J. D. Sherman, 8th EPAC Conf. Proc., (Paris, France), 1715 (2002).
4. R. Gobin, P.-Y. Beauvais, K. Benmeziane, O. Delferriere, R. Ferdinand, F. Harrault, J.-M. Lagniel, and J. D. Sherman, Rev. of Sci. Instrum. 73(2), 983(2002).
5. R. Gobin,et. al., 8th EPAC Conf. Proc., (Paris, France), 1712 (2002).
6. M. B. Hopkins, M. Bacal, and W. G. Graham, J. Appl. Phys. 70(4), 2009(1001).
7. G. P. Arrighini, F. Biondi, and C. Guidotti, Molecular Phys. 41(6), 1501(1980).
8. M. A. P. Lima, T. L. Gibson, McKoy, and [PFTV92] W. M. Huo, Phys. Rev A 38, 4527(1988).
9. J. R. Hiskes, Notas Physica 5, 348(1982).
10. V. P. Zhdanov and K. I. Zamarev, Catal. Rev. Sci. Eng. 24, 373(1982).
11. C. Gorse, M. Capitelli, M. Bacal, J. Bretagne, A. Lagane, Chem. Phys. 117, 177(1987).
12. J. Loureiro and C. M. Ferreira, J. Phys. D: Appl. Phys, 1680(1989).
13. H. Tawara, Y. Itikawa, H. Nishimura, and M. Yoshino, J. Phyis. Chem. Ref. Data 19(3), 617(1990).
14. C. Gorse, M. Capitelli, M. Bacal, J. Bretagne, Chem Phys. 93, 1(1985).
15. D. Riz and J. Pamela, Rev. of Sci. Instrum. 69(2), 914(1998).
16. D. L. Scharfetter and H. K. Gummel, IEEE Trans. Electron Devices 16, 64(1969).

Relevance of Volume and Surface Plasma Generation of Negative Ions in Gas Discharges

Vadim Dudnikov

Brookhaven Technology Group, Inc., NY, USA

Abstract. The relative contribution of volume and surface-plasma generation in emission of H⁻ ions in gas discharge sources is analyzed. At the present time, it is generally accepted that surface-plasma generation of extracted H⁻ ions dominates over volume processes in discharges with admixture of cesium or other catalysts with low ionization potential. We will attract attention to the evidence, that surface-plasma generation can be significantly enhanced in high density discharges without cesium after electrode activation by high temperature conditioning in discharge. With this optimization of conditions for surface-plasma generation of emitted H⁻, an emission current density was increased up to ~1 A/cm^2 in discharges without cesium. Diffusion of impurities with low ionization potential can be the reason for the observed enhancement of H⁻ emission. Such optimization allows considerable improvement of H⁻/D⁻ source characteristics. Volume generation of extracted H⁻ in high density discharges is suppressed by the high level of gas dissociation.

INTRODUCTION

When describing negative ion generation in sources with direct beam extraction from the discharge, two mechanisms of H⁻ ion generation, Volume Generation and Surface-Plasma Generation are being discussed [1-10]. Volume Generation of negative ions (VG) is provided by reactions occurring at collisions of electrons with molecules and molecular ions in plasma volume, which explains the name of this mechanism. Surface-Plasma Generation of negative ions (SPG) is provided by reactions occurring at the interaction of plasma particles with the surface of electrodes, that is the basis for the choice of the name for this process.

The basic reaction of volume generation of negative ions in plasma is dissociative electron attachment [1,2,6-13]. In this reaction an electron is captured by a molecule in metastable auto-ionization state which can decay with dissociation into a negative ion and a neutral fragment, or can release an electron with the transition of a molecule into an excited state. The probability of negative ion formation strongly depends on the duration of fragments separation, i.e. on relative velocities of fragments, or excitation of vibrational/rotational states. The cross section of electron attachment to H$_2$ molecule in its ground state is very small (10^{-21} cm^2 with the energy of electrons above 3.7 eV). The cross section of dissociative attachment for deuterium molecules is $2 \cdot 10^3$ times less, i.e. a strong isotopic effect is observed. A polar dissociation, dividing of a molecule into a positive and a negative ions, has a greater cross section (10^{-20} cm^2), and also a greater threshold energy (17.2 eV), that makes this process inefficient in

CP763, *Production and Neutralization of Negative Ions and Beams*, edited by J. D. Sherman and Y. I. Belchenko
© 2005 American Institute of Physics 0-7354-0248-5/05/$22.50

cold plasma. The cross section of negative ion destruction by electrons at these energies is $\sim 3 \cdot 10^{-15}$ cm^2, and cross section of negative ion destruction in collisions with positive ions is $\sim 10^{-13}$ cm^2, which are both several orders higher than cross sections of negative ion formation. For these reasons, the calculated concentration of H$^-$ ions in gas discharge plasma should be very low. In real discharges, the concentration of H$^-$ ions appeared to be much greater than the calculated one, taking into account the dissociative electron attachment to molecules in the ground state and dissociative recombination. In publications [11,12], it has been noted that dissociative attachment to vibrational/rotational excited molecules has much greater cross sections and lower threshold energy. This circumstance could explain the increased concentration of H$^-$ ions in gas discharges.

The review of the situation with H$^-$ ion extraction from various discharges existing by 1973 is given in [13]. Alongside with H$^-$ ion beam current I$^-$, the intensity of accompanying electron beam I$_e$ is of a great significance. By 1973, from a duoplasmatron with axial extraction (standard ion source for positive ion production by ionization in plasma), up to 0.07 mA of H$^-$ current had been obtained at electron current 65 mA, the ratio being $\gamma = I_e/I^- \sim 1000$. From a duoplasmatron with a displaced emission aperture, up to 2.2 mA of H$^-$ ions had been obtained at electron current 650 mA, the ratio being $\gamma \sim 300$. From duoplasmatron with tubular discharge, up to 6 mA of H$^-$ ions had been obtained at electron current 90 mA, the ratio being $\gamma \sim 15$. In the Ehlers type ion source with a Penning discharge, up to 4.6 mA of H$^-$ had been extracted along with ~ 5 times higher electron current. From a planotron (magnetron) type source with magnetron discharge, up to 22 mA of H$^-$ ions had been obtained at electron current ~ 100 mA, the ratio being $\gamma \sim 4$-5. These values of H$^-$ currents could be explained by the generation of H$^-$ ions in the volume of plasma due to dissociative attachment of electrons to vibrational/rotational excited molecules in plasma volume. For this reason, these sources were named Volume Sources, and it was considered in a later review [7-9,12,13] that volume generation of H$^-$ ions dominates in them. The importance of Surface-Plasma Generation of negative ions in discharges without cesium was mentioned in publications [3-5,15-17,24,25]. Now, there are additional evidences allowing to conclude that the increase of H$^-$ ion current, and the improvement in the electron current to H$^-$ ion current ratio in these cases is connected to the enhancement of surface-plasma generation of H$^-$ ions.

SURFACE-PLASMA NEGATIVE ION SOURCES

The situation with negative ion production changed cardinally in 1971 after the observation of substantial enhancement in negative ion emission after cesium addition to the discharge (cesium catalysis) [14-17,24,25]. By the addition of cesium vapor into the ion source gas discharge chamber and the activation of electrode, the intensity of H$^-$ ion beam extracted from the discharge was increased considerably. With cesium addition, it was possible to essentially decrease gas density in the source, and the current of co-extracted electrons was decreased considerably. The intensity of

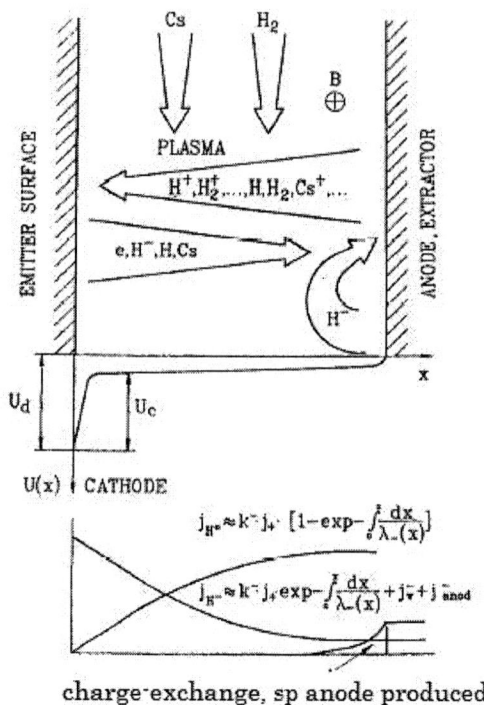

$$j_{H^-} \approx k^- j_+ \cdot \left[1 - \exp{-\int_0^x \frac{dx}{\lambda_-(x)}}\right]$$

$$j_{H^-} \approx k^- j_+ \exp{-\int_0^x \frac{dx}{\lambda_-(x)}} + j_{\overline{v}}^- + j_{anod}^-$$

charge-exchange, sp anode produced

FIGURE 1. Schematic illustration of surface-plasma negative ion generation.

co- extracted electrons became even less than H⁻ ion current. In order for this to occur, the density of electrons should become much less than the density of H⁻ ions around the emission surface. Such behavior of emission of H⁻ ions and electrons did not find an explanation within the framework of known elementary processes in plasma volume. When registering the energetic spectra of H⁻ ions from the discharge in planotron (flat magnetron), the ions accelerated by discharge voltage (originated on the cathode) and the ions formed in the anode area [15-17,24,25] have been detected. On the basis of these experiments, a conclusion has been made that the increased efficiency of H⁻ ion generation in the discharge with cesium addition is caused by a substantial increase in the secondary emission of negative ions from the electrode surface bombarded by plasma particles. Cesium adsorption lowers the surface work function, and this increases the probability of sputtered and reflected particles to escape in the form of negative ions. This probability increases with the work function decrease and the increase of velocity of particles moving away from the electrode surface. The theory of negative ionization of particles at collisions with an electrode surface has been developed in works [18,19]. The emission of negative ions increases with the increase of negative potential of the electrode contacting with plasma, because the energy and intensity of positive ions bombarding the electrode surface increase, the quantity of the sputtered and reflected particles increases, and so does the velocity of their movement away from the electrode surface. The potential drop near the electrode accelerates emitted negative ions without a space charge limitation. It also increases the free path of ions in plasma without destruction and allows their extracting through the emission aperture. Cesium is locked on the surface of the negative electrode as it comes back from plasma in the form of ions. Electrode surfaces with a potential close to the potential of plasma are also bombarded by fast plasma particles and emit negative ions. The efficiency of negative ion generation on these electrode surfaces is lower than on electrodes with negative potential, but can be high enough. The discovered mechanism of effective negative ion generation has been named the Surface-Plasma Generation of negative ion (SPG) because its functioning is based on the interaction of plasma particles with an electrode surface, and the optimization of this process defines characteristics of negative ion generation. The Surface-Plasma Generation (SPG) of negative ions is schematically represented in Figure1. Ion sources based on this process have been named Surface-Plasma Sources of negative ions (SPS). Rather quickly, the emission density of H⁻ beams from SPS

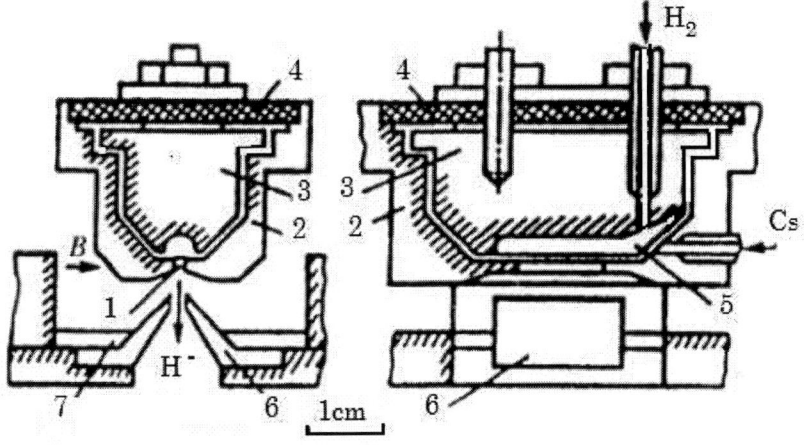

FIGURE 2. Schematic of Semiplanotron SPS [4,15]. 1- emission aperture; 2- anode; 3- cathode; 4- cathode insulator; 5- discharge channel; 6- extractor; 7- magnet with magnetic insertions.

was increased up to 3.7 A/cm^2, and after the development of geometrical focusing, it was increased up to 8 A/cm^2. Various versions of SPS with surface-plasma generation of negative ions on the cathode, on the special emitter or on the anode of the discharge (anode SPG) have been developed. Probably most clearly, the features of SPG NI are displayed in semiplanotron SPS. The schematic of semiplanotron SPS [1,4,15,25] is given in Figure 2. The discharge in crossed ExB fields is supported between the cathode (3) and the anode (2) in narrow (2-3 mm) semi-cylindrical discharge channel (5) opposite to the emission aperture (1). The discharge is ignited in a deeper grove in the right end of the cathode where the working gas and cesium are being injected. Plasma drifts in the crossed fields from the right to the left. Electrons oscillate along the magnetic field between the walls of a semi-cylindrical discharge channel, and positive ions bombard its surface, initiating emission of the sputtered and reflected particles. Part of these particles leaves the surface in the form of negative ions. The probability of particle escaping as negative ions increases essentially with the reduction of surface work function due to adsorption of cesium or other substances with low ionization potential. Negative ions are accelerated by the potential difference between electrode surface and plasma and are focused on the emission aperture due to the cylindrical form of the emitter. During the movement through plasma, part of the ions loses electrons in collisions with plasma particles and converts into fast neutrals, as shown in the schema in Fig. 1. Part of fast negative ions transfers electron to cold atoms, transforming into cold negative ions. This charge exchange cooling of negative ions increases beam brightness. The shielding of emission aperture by equi-potential grove (collar) and a strong transverse magnetic field form a magnetic filter, suppressing plasma diffusion to the emission aperture and extraction of accompanying electrons. The walls of equi-potential cavity are bombarded by fast ions and atoms from the cathode and by less energetic (with energy 5-10 eV) particles from the discharge.

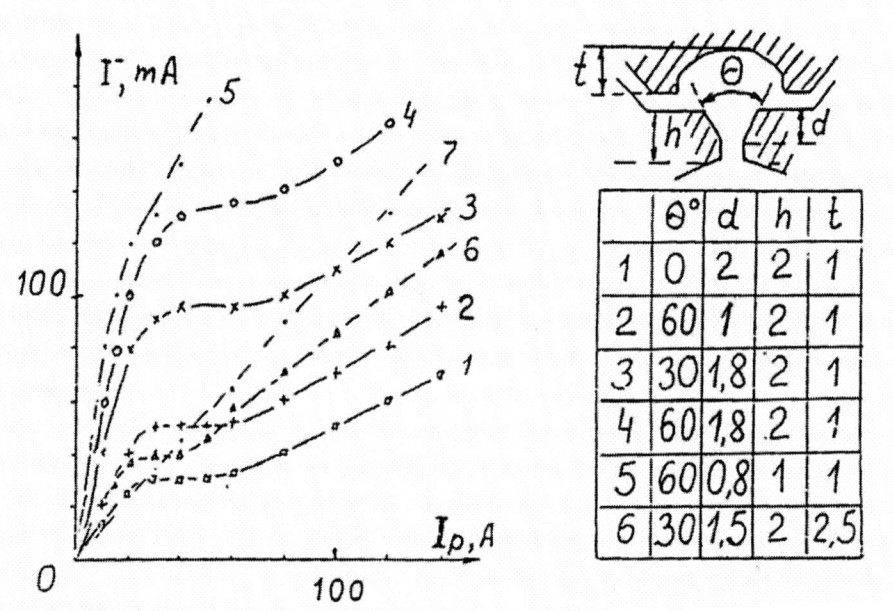

FIGURE 3. Dependence of H⁻ current on the discharge current for different geometries of discharge channel and magnetic filter in semiplanotrons. Dependence (7) is obtained in SPS with Penning discharge. Emission slit area is 0.5x10mm² for all sources.

This bombardment causes emission of negative ions from the anode surface around the emission aperture (anode SPG, shown on Fig. 1). Low density of electrons in this

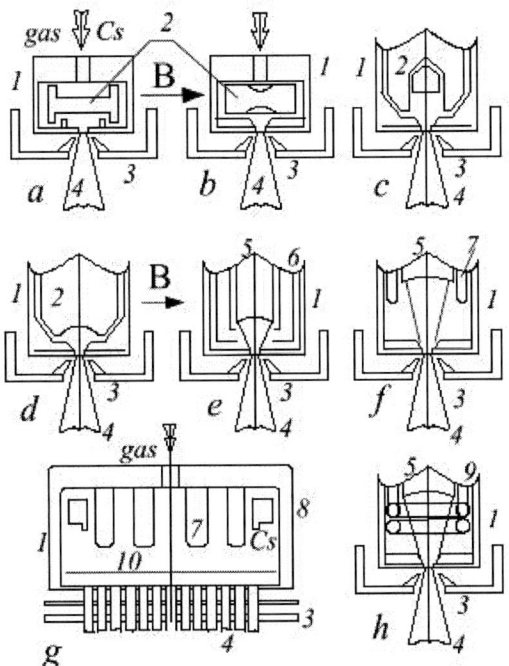

FIGURE 4. Electrode configuration for basic versions of SPS. (a) Planotron (flat magnetron) with the flat cathode-emitter; (b) Planotron with geometrical focusing (cylindrical and spherical); (c) SPS with Penning discharge (Dudnikov type source); (d) semiplanotron; (e) SPS with the hollow cathode and the independent emitter; (f) SPS with large volume and with the emitter; (g) large volume SPS with anode generation of NI; (h) large volume SPS with RF generation of plasma. 1-anode (the discharge chamber); 2-cold cathode-emitter; 3-extractor with magnetic system; 4-NI beam; 5-emitter with negative biased; 6-hollow cathode; 7-thermocathodes; 8-magnetic wall; 9-RF antenna (coil); 10-magnetic filter. Working gas and cesium are brought into the gas discharge of SPS.

area and low electron temperature promote survival and effective extraction of negative ions from this area. Emission characteristics of semiplanotron at various configurations of the discharge channel and the magnetic filter are shown in Figure 3. Graphs of the dependences of H⁻ ion beam intensity on the discharge current have a N-shaped form with three sections: linear growth at small discharge currents, saturation or a falling section at medium currents, and linear, but slow growth at the high currents. Linear dependence (7) is obtained in SPS with Penning discharge. At small discharge currents most of the extracted NI is provided by SPG on the cathode. In this section, the efficiency of SPG is the highest (up to 6 mA on 1 A of discharge current at discharge voltage of 100 V, up to 60 mA per kW). With the increase in discharge current, the flow of NI from the cathode is attenuated by destructions in plasma, and less effective anode SPG cannot compensate this attenuation that forms a falling section. At high discharge currents, linear growth of NI current is provided by anode SPG. Emission of positive ions is much less than the emission of NI; the contribution of volume generation of NI is relatively low. Anode SPG provides linear dependence of negative ion current on the discharge current in Penning SPS shown under number (7). The current of accompanying electrons in these measurements was less than the negative ion current in normal magnetic fields B~1 kG. In weak fields (B ~ 0.1 kG), the electron current could be 10-20 times greater then the negative ion current. In semiplanotron, SPG of negative ions is realized most effectively. Presently, many versions of SPS optimized for various applications are developed. Classification of various versions of SPS is presented in [5,15,16]. Configurations of some SPS are shown in Figure 4. Currents of accompanying electrons I_e in SPS are comparable to H⁻ ion currents I⁻, or is less, that is, ratio $I_e/I⁻ = \gamma_{sp} <\sim 1$, whereas at volume generation of negative ions, $\gamma_v \sim 500$-1000. Estimations of the relative contribution of volume and surface-plasma generation of negative ions in various discharges are discussed in many publications [27,28,30-32]. By present time, it has become a commonly supported opinion that surface-plasma generation of negative ions (SPG) absolutely dominates in discharges with addition of cesium, or barium [1-10,15-17,24,25,28]. However, until recently sources with discharges without cesium or other catalytic additives with low ionization potential were referred to as sources with volume generation of negative ions [1-10,21-23,28,30-32].

TANDEM LARGE VOLUME NEGATIVE ION SOURCES

Tandem ion sources with a magnetic filter and magnetic cusp wall used for the production of positive ion beams with high proton component have been adapted for optimization of volume generation of H⁻ ions [1,2,7-10,20]. In these sources, plasma is generated in a driver zone by DC filament discharge or by RF discharge in a chamber with a magnetic cusp wall. Plasma diffusion to the emission aperture (into the area of extraction) is suppressed by a transverse magnetic field of the filter dividing the source into two zones. The configuration of such source is shown in Figure 4, (g). It is believed that in the generation zone fast electrons excite molecules into high vibrational and rotational states, and these molecules diffuse to the extraction area with cold electrons. In the extraction area, H⁻ ions are formed by dissociative attachment of cold ~1eV electrons to the excited molecules. The generated H⁻ ions are extracted together with electrons through the emission aperture and are formed into a beam. At low electron temperature, the rate of negative ions destruction decreases, and they can be extracted from a larger generation area. Features of these sources were discussed in many publications, references to which are presented in reviews [1,2,7-10,31-33]. It is considered that the increase of proton components in tandem sources is caused by effective dissociation of molecular ions in the area of the magnetic filter by cold electrons. Probably, it is more essential that the magnetic filter suppresses the plasma transportation to the emission aperture, and to obtain previous emission current density it is necessary to have much greater plasma density and power density in the discharge contributing to a more complete dissociation of molecules. It is necessary to note that the same configuration of the discharge, with a dense plasma in generation zone separated from the extraction area with low density cold plasma of by a magnetic filter and a collar (anode ribs), was used in the first versions of Ehlers-type source and

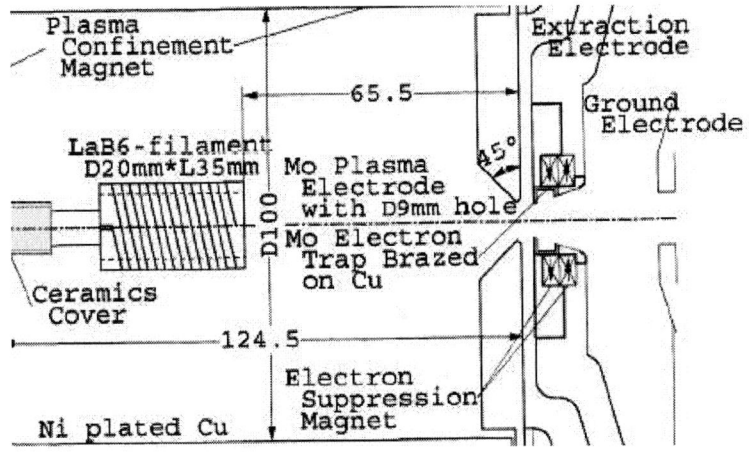

FIGURE 5. Design of the KEK/ JAERI Large Volume SPS from [29].

in the first versions of compact SPS, shown in Fig. 4 [1,3,4,8,15,16,22-24]. Due to empirical optimization of tandem sources, it was possible to obtain significant H⁻ ions currents at rather small currents of accompanying electrons. The parameters of tandem sources made and tested in different laboratories are shown in Table 1 from [8,9].In this table P_d is discharge power, P_{eff} is an energy efficiency Γ/P_d, df is duty factor. These sources typically have H⁻ ion current Γ up to ~40 mA with emission current density up to J~0.1 A/cm². The accompanying electron current to H⁻ ion current ratio I_e/Γ varies from ~26 down to ~4. The last value is close to the one obtained in SPS with cesium.

TABLE 1 .Tandem sources with the discharge without cesium.

Source	Γ,mA	P_d (kW)	P_{eff}(mA/kW	$\gamma=I_e/\Gamma$	df (%)
BNL (H₂)	35	25 (Ta cathode)	1.40	20	0.5
DESY (H₂)	38	27 (RF 2 MHz)	1.41	26	0.08
JAERI (H₂)	14	30 (W cathode)	0.47	10	5
SSC (H₂)	30	40 (RF 2 MHz)	0.75	20	0.35
ANLH₂+Ta	4-5	0.7 (ECR2.45GHz)	5.7-7.1	20	100
TRIUMF	20	4.8 (Ta cathode)	4.2	4	100

In the sources presented in Table 1, which authors classify as volume sources, by empirical optimization, relatively high emission current density and small ratios of electron current to H⁻ ion current (γ~ 4) were obtained. This means that there is a low density of electrons on the emission surface and low rate of volume generation of extracted negative ions. This implies the assertion, that in optimized H⁻ ion sources without cesium with high density discharge and low electron current, the surface-plasma generation of negative ions can dominates, as well as in discharges with cesium or others catalytic admixtures with low ionization potential.

In the tandem source described in [29], which the authors originally developed as a Volume Source, subsequent optimization of SPG on plasma electrode in discharges without cesium has allowed a considerable increase of H⁻ ion current (to 38 mA) and reduction of electron current. The design of this large volume source is shown in Figure 5. The increase of H⁻ current and suppression of electrons is reached by the optimization of plasma electrode (PE) surface around the emission aperture. The optimization of PE temperature and the activation by discharge was very important for the enhancement SPG of H⁻ ion. These experiments present a good confirmation of SPG domination in the optimized discharges without cesium in tandem large volume sources. The volume generation of H⁻ in this discharge should be suppressed by high degree of gas dissociation.

ION SOURCES WITH DISCHARGES IN CROSSED FIELDS

In this section, we will consider the features of negative ion generation in compact sources with a high discharge current density in crossed fields. In a planotron (flat magnetron) with cesium, an H⁻ beam with emission current density of up to 3.7 A/cm² was obtained [15,17] in the Institute of Nuclear Physics, Novosibirsk, in 1972. Later, a special "pure" planorton without traces of cesium or other alkaline admixtures, but with a "thick" cathode was manufactured. The schematic of this source is shown in Fig. 4 (*a*). The dependence of H⁻ emission current density on the discharge current in this source before the cesium admixture and with cesium is shown in Figure 6. From this source with the discharge in pure hydrogen, after the activation by high current discharges, H⁻ beam with emission current density of up to 0.75 A/cm² had been obtained. This current is much greater than the one expected from the volume generation. The electron current to H⁻ ion current ratio γ was up to 4 [17,24,25]. When the optimal supply of cesium had been reached, the emission current density increased up to 3.7 A/cm². On the basis of these results, a conclusion has been made that in this source without cesium high efficiency of negative ion generation is also caused by surface-plasma generation of negative ions, as well as in discharges with cesium. The discharge voltage in these experiments did not drop to 150-100 V as in discharges with cesium admixture, but was a little bit below the standard voltage of hydrogen discharges ~400-500 V. It was assumed, that surface-plasma generation of H⁻ ions had increased due to the decrease of electrode work function, related to the diffusion of potassium and other alkaline metals admixture from the volume of molybdenum on the electrode surface at intense heating. Usually, refractory metals such as tungsten, molybdenum, tantalum, etc. contain an admixture of alkaline metals (mainly potassium) with a high concentration. A little decrease in the work function which influences the discharge voltage only a little bit, can essentially increase H⁻ secondary emission and SPG, and make it the dominating mechanism of negative ion generation. Such decrease in the work function can be provided by uncontrollable alkaline pollution; for example, finger prints, drops of saliva, etc. Also, forinstance, the enamel used for the insulation of internal RF antenna coil in "volume" sources contains 15 %

of potassium by weight [7-9]. Sputtering/deposition of this coating is enough for catalysis of H⁻ ions SPG (increase of the secondary H⁻ emission up to 50 times [35,36]). It is necessary to note that after the addition of cesium or other catalysts, they are implanted into electrodes and are diffused again onto the surface at the activation by the discharge. It is not easy to clean out the source from the traces of these impurities.

An attempt to obtain an intense beam without cesium has been undertaken in works [26,27] using a discharge with a cathode from lanthanum hexaboride in Dudnikov-type source with Penning discharge. The schematic of this source is shown in Fig. 4 (c). With the discharge current of 50 A, and the emission aperture of 1 mm in diameter, there has been received an H⁻ ion beam with emission current density of 0.35 A/cm², and with co-extracted electron current to H⁻ ion current ratio γ of up to ~ 4. On the basis of indirect evidences in this work, a conclusion has been made that in these conditions H⁻ ions are formed due to volume processes [26,27].

Undoubtedly, in these conditions H⁻ ions were produced basically due to SPG on the surface of plasma electrode (the anode) near emission aperture (anode SPG). This is confirmed by the fact that the current of H⁻ ions increases linearly with the increase of the magnetic field up to 7 kG when the flow of electrons in the extraction area should have decreased many times, and the volume generation should have been suppressed.

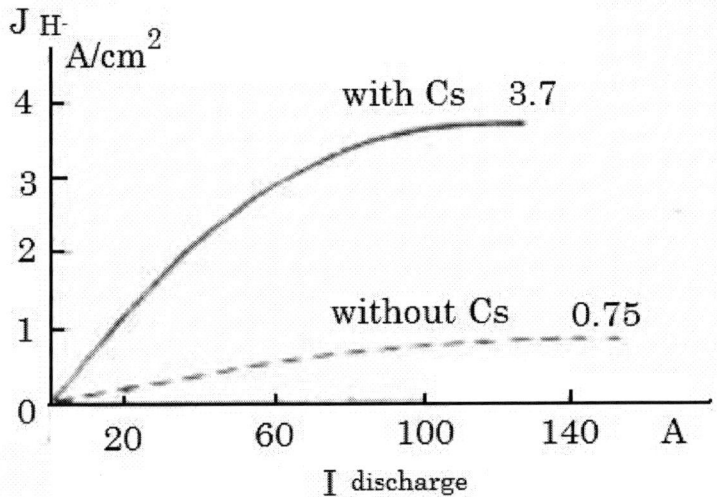

FIGURE 6. Dependence of emission H⁻ ion current density on the discharge current in planorton with pure hydrogen and with cesium addition.

For SPG, electrons are not needed in the extraction area; it is enough to have bombardment of the electrode surface by hyperthermal atoms with energies higher than threshold φ - A = 5.5-0.75 eV ~5-10 eV. The suppression of electron flow into the extraction area by the magnetic field reduces the destruction of negative ions by electrons and suppresses extraction of accompanying electrons. Almost complied dissociation of molecules into atoms caused by the high discharge current density suppresses volume generation of negative ions, but favorite for SPG.

A similar behavior has been observed in the research of H⁻ ion emission from Ehlers source with an insulated plasma electrode [23]. The schematic of this source is shown in Figure 7. The diameter of the discharge column with electron oscillation is 4.8 mm; the diameter of the plasma electrode is 9.6 mm. The output of H⁻ ions rose with the increase of magnetic field up to 5 kG. In usual conditions, with the potential of plasma electrode equal to anode potential, as well as in the previous source, up to 4.6 mA of H⁻ ions and 20 mA of electrons were extracted. With the decrease of the plasma electrode potential down to -6 V, the emission of H⁻ ions increased 2 times up to 9.7 mA (the emission current density increased up to 100 mA/см²), and the electron current increased up to 100 mA. At the increase of a plasma electrode potential of up to +5 V, emission of H⁻ ions decreased down to 3 mA (emission current density decreased down to 30 mA/см²), and the electron current decreased down to 15 mA. With a negative biasing of the plasma electrode relatively to the anode, the current of positive ions was almost 2 times less than the current of H⁻ ions. With positive plasma electrode the current of positive ions exceeded the current of H⁻ ions approximately 2 times. The proton consumption in positive ion beam was 90%, H_2^+ 4%, H_3^+ 6%. Dependency of H⁻ and D⁻ current on the gas flux had characteristic maxima with exponential attenuation at larger gas flows. This behavior is explained

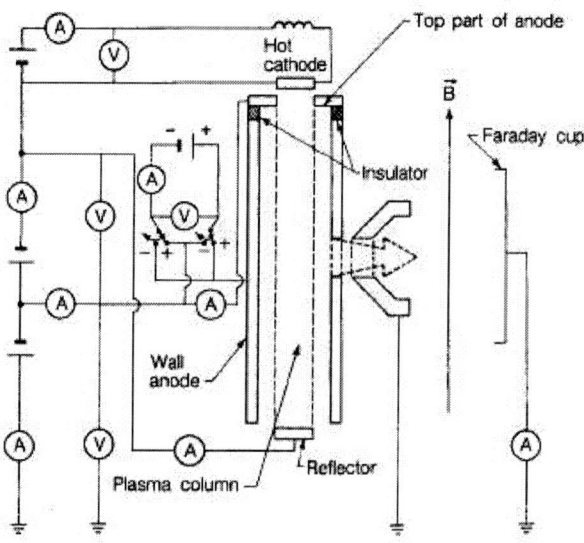

FIGURE 7. Schematic of modified Ehlers ion source with electron oscillation in magnetic field and insulated plasma electrode [23].

by the destruction of negative ions upon leaving gas after the extraction. The maximum current of D⁻ ions equal to 4.1 mA was 2 times less than the current of H⁻ ions. Authors concluded that H⁻ and D⁻ ions were generated by volume processes, but the presented dependences show that volume generation of negative ions should be strongly suppressed. Diffusion of plasma to plasma electrode is strongly suppressed, so the density of positive ions is less than the density of negative ions, so the density of electrons is considerably less than the density of negative ions that is impossible at the domination of volume generation. The electron current exceeded the current of H⁻ ions from 5 to 10 times, which also corresponds to significant domination of H⁻ ion density over the electron density. The gas in the researched conditions is strongly

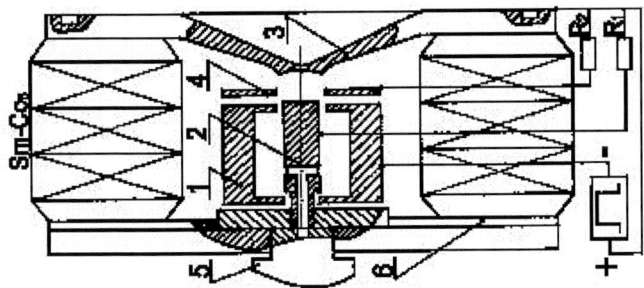

FIGURE 8. Schematic of ion source with a magnetron discharge in longitudinal magnetic field from [30].1-cathode; 2- anode; 3- plasma electrode; 4- intermediate electrode; 5-gas valve; 6- insulating plate.

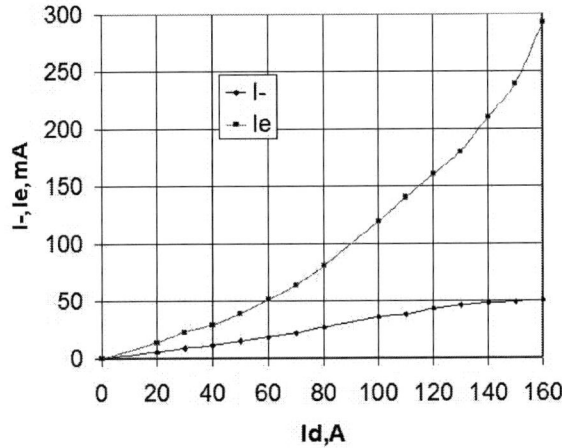

FIGURE 9. Dependence of H⁻ current I- in mA and electron current Ie in mA on the discharge current Id in A from [30].

dissociated (90 %), which also suppresses the volume generation of negative ions. The increase in negative ion emission at negative biasing on plasma electrode also proves domination of surface-plasma generation of negative ions and their extraction from the plasma electrode surface in the discussed discharge without cesium.

Similar research of H⁻ ion production in Ehlers source without cesium and with the cesium addition was carried out in work [28]. Calculations of H⁻ ion generation in discharges with cesium and without cesium were conducted at the same time. With similar experimental results, authors have come to the conclusion, that in discharges without cesium, volume generation of H⁻ ions dominates, and in discharges with cesium SPG dominates. The arguments mentioned above allow the conclusion that in discharges without cesium, SPG also dominates, but its efficiency is significantly lower than in discharges with cesium.

The authors of the source with magnetron discharge in longitudinal magnetic field [30] classify it as a volume source. The schematic of this source is shown in Figure 8. The dependence of extracted H⁻ current and electron current on discharge current are shown in Figure 9. The relatively small electron current to H⁻ ion current ratio γ~10-13, and its decrease down to 6 in the recent modification at the extraction along the magnetic field, are strong evidences of the SPG domination on the plasma electrode.

ESTIMATIONS OF VOLUME GENERATION CONTRIBUTION INTO H⁻ ION EMISSION

Developed in [31], the laser photo-detachment method for measurement of H⁻ ion density to electron density ratio allows reliable measurement of this ratio. The detection of the fact that H⁻ ion density to electron density ratio is not 10^{-5}, as it should be according to calculations with dissociative attachment to the molecules in the ground state, but is 10^{-1}, allowed assuming that volume processes can be used for intense H⁻ ion beam production in discharges without cesium [12,31]. In tandem volume source with low density discharge (emission current density of H⁻ is ~ 1 mA/cm^2) by electron filtration in "strong" magnetic field and electron collection by positive biased plasma electrode the ratio of electron current I_e to H⁻ current I^- was suppressed down to $I_e/I^- = γ~4$ (corresponded $n_e/n^- = α~0.5$ near the plasma electrode) [32]. In ref. [33] was detected $α ~ 0.1$. For tandem volume sources with higher emission current density is typically γ>20 (for DESY RF source γ~35 [10]). Introduction of cesium admixture as a rule increases H⁻ current several times (some time several tenth times) and decreases γ below 1 by enhancing SPG. Improvement of H⁻ generation and suppression of electron current in H⁻ sources was reached without admixture of cesium, but by activation of electrodes by high temperature heating. This activation can increase H⁻ current 2-3 times and decrease γ to 4-6. It was proposed, that this enhancement of H⁻ production was also caused by activation of SPG. Increase of secondary emission of H⁻ can be catalyzed by diffusion of small impurities with low ionization potential from bulk within the electrodes to the surface with lowering of the work function. In discharges with domination of SPG, H⁻ emission increases with negative biasing of plasma electrode.

The relative contribution of volume generation I_V^- into flux I^- of emitted H⁻ ions can be estimated on the basis of electron current I_e to H⁻ ion current I^- ratio. To estimate the contribution of volume and surface-plasma generation, we will divide the currents of emitted negative ions I^- and electrons I_e by the volume generated current I_v^-, I_v^e and surface-plasma generated current I_{sp}^-, I_{sp}^e,

$$I^- = I_v^- + I_{sp}^-, \quad I_e = I_v^e + I_{sp}^e. \tag{1}$$

Let us introduce the following denotations:

$$\gamma = I_e/I^-; \quad I^- = I_e/\gamma \quad \gamma_v = I_v^e/I_v^-; \quad \gamma_{sp} = I_{sp}^e/I_{sp}^-; \quad I_v^- = I_v^e/\gamma_v; \quad I_{sp}^- = I_{sp}^e/\gamma_{sp}; \quad (2)$$

In order to estimate the contribution of volume generation to H⁻ ion emission, we introduce the quantity β, defined by:

$$\beta = I_v^-/(I_v^- + I_{sp}^-) = I_v^-/I^- = (I_v^e/\gamma_v)/(I_e/\gamma) = (I_e/\gamma_v)/(I_e/\gamma) = \gamma/\gamma_v. \quad (3)$$

In volume generation of H⁻ ions, the ratio of electron density n_e to H⁻ ion density n_-, is $\alpha_v = n_e/n_- \sim 7\text{-}10$, and the ratio of electron current to H⁻ ion current, during extraction from dense plasma should be $\gamma_v = I_v^e/I_v^- \sim n_e v_e/n_- v^- \sim 500$ (v_e, v^- are electron and H⁻ ion velocities that differ ~50 times by value), as was in old sources [13]. In the case of surface-plasma generation, the emission of electrons is caused by secondary emission and thermo-emission. The coefficient of the secondary ion-electron emission is comparable with the coefficient of secondary emission of H⁻ ions, K~0.1 or less [34], and the corresponding electron current cannot exceed the H⁻ ion current significantly. In the bombardment of the emission surface by hyperthermal atoms, the secondary electron emission is much less than ion-electron emission. The thermo-emission electron current from the overheated plasma electrode with cesium can be relatively large, but in optimal conditions, the co-extracted electron current can be made much less than the H⁻ ion current. That is, the following ratio corresponds to the surface-plasma generation of H⁻ ions: $\gamma_{sp} = I_{sp}^e/I_{sp} < \sim 1$. At that, the relative contribution β of the volume generation of emitted H⁻ ions is equal to the ratio between actual γ in the researched discharge and $\gamma_v \sim 100\text{-}500$ without biasing of plasma electrode, $\beta = \gamma/\gamma_v$. For discharges with the ratio $\gamma < 10$, typical for good "volume" sources, the above estimation for the contribution of volume generation β is ~2- 10%. Edition research is necessary for more accurate estimation of β.

CONCLUSION

The analysis of H⁻ ion emission features from discharges in different ion sources is presented. The conditions for volume and surface-plasma generation of H⁻ in gas discharges coexist. However, the relative contribution of VG and SPG in the emission of H⁻ and in the support of volume H⁻ density can be very different. In discharges with high emission current density the free path of cold H⁻ is very short (~ mm). In these conditions, the volume density of H⁻ is determined by VG. But the production of emitted H⁻ is determined by SPG on the plasma electrode surface around the emission aperture. To enhance the efficiency of the emitted H⁻ production, it is necessary to optimize SPG on the plasma electrode. This can be achieved by the efficient transformation of gas molecules to flux of hyperthermal atoms with the energy of several eV. It is necessary to use a magnetic filter with a strong magnetic field for strong suppression of electron diffusion to the emission surface. It is useful to use a slit extraction system. It is necessary to optimize the shape and thermal property of the plasma electrode for the efficient extraction of generated H⁻ ions. It is essential to use the activation of the emission surface by discharge and by thermo processing. After such optimization, it is possible to produce a high density of H⁻ with much lower density of electrons on the emission surface, and to suppress the ratio of electron current to H⁻ current γ down to less than 5-10 in discharges without cesium. A high efficiency of such optimization was demonstrated in works [17,21-23,29]. The emission current density of H⁻ ion was increased up to ~1 A/cm^2 in discharges without cesium. The most probable reason for this secondary emission enhancement is the diffusion of an impurity with a low ionization potential from the bulk of the electrode to the surface. Refractory metals such as Mo, W, Ta have relatively high concentration of alkaline metals, mainly potassium. In experimental measurements, presented in [35,36,37], the secondary emission of D⁻ was increased up to 50 times by the deposition of sodium and potassium on the copper surface (work function φ was decreased from 4.8 eV down to 2.25 eV) and up to 100 times by deposition of cesium (φ was decreased from 4.8 eV down to 1.9 eV).

A further decrease of the work function of plasma electrode emission surface by an optimal deposition of cesium and activation (down to φ~1.6 eV) increases H⁻/D⁻ emission up to 4-5 times relative to the discharge with activated electrodes without cesium and decreases electron emission below H⁻ emission as in ref. [17].

ACKNOWLEDGMENTS

The author is grateful to his colleagues Yu. Belchenko, G. Dimov, G. Derevyankin for cooperation in production of basic results used in this work and to I. Doudnikova for her help in editing. I wish to thank Dr. Akira Ueno and Dr. Paul Farrell for useful discussions, J. Sherman and M. Bacal for useful remarks. This work was supported partly by DOE Grant No. DE-FG02-04ER83915

REFERENCES

1. H. Zhang, Ion Sources, Springer, 1999.
2. I. Brown, Editor, The Physics and Technology of Ion Sources, Second, revised Edition, Wiley VCH, 2004.
3. Yu.I.Belchenko, G.I.Dimov, V.G.Dudnikov, et.al, Journal of Applied Mechanics and Technical Physics, 28 (4), 568, (1987).
4. V.Dudnikov, Some effects of surface-plasma mechanism for production of negative ions, BNL 51304,UC-34a, p.137, (1980).
5. Yu.I. Belchenko, Surface negative ion production in ion sources, Rev. Sci. Instrum., 64 (6), 1385 (1993).
6. Charles W. Schmidt, "Review of Negative Hydrogen Ion Sources", Proc. of the 1990 Linear Accelerator Conference, Los Alamos Report (LA-12004-C), 259 (1990).
7. J.Peters, EPAC'2000. Peters, LINAC' 98, Chicago, 1998. Peters, Rev. Sci. Instrum., 71(2), 1069 (2000).

8. Joseph Sherman and Gary Rouleau, 17th International Conf. on the Application of Accelerators in Research and Industry, (Denton, Texan)AIP Conference Proc. 680, 1038, 2002.
9. R. Welton, Linac 2002 (2002).
10. J.Peters, Rev.Sci.Instrum., 75 (5), 1709 (2004).
11. V.Kuchinsky, V.Mishakov, A.Tibilov, A. Shukhtin, Optics and Spektroskopy, 39 (6), 1043, (1975).
12. M. Bacal, A. M. Bruneteau, W. G. Graham, G. W. Hamilton, and M. Nachman, J.Appl. Phys. 52(3), 1247 (1981).
13. K. Prelec, and Th. Sluyters, Formation of Negative Hydrogen Ions in Direct Extraction Sources, Rev. Sci. Instrum, 44(10),1451 (1973).
14. V. Dudnikov, The Method for Negative Ion Production, SU Author Certificate, C1.H013/04, No. 411542, Application, 10 March, 1972, granted 21 Sept.,1973, published 15 Jan.1974, Bul. No 2.
15. V. Dudnikov, Surface-Plasma Method of Negative Ion Production, Doctor Thesis, INP Novosibirsk, 1976 (unpublished).
16. V. Dudnikov, Rev. Sci. Instrum. 63(4),2660 (1992). V. Dudnikov, Rev. Sci. Instrum. 73(2), 992 (2002).
17. Yu. Belchenko, G. Dimov, V. Dudnikov, Izvestia Akademii nauk SSSR, Seriya Fizicheskaya, vol.37, no.12, p.2573-7, Dec.1973, ISSN0367-6765. Translated in: Bulletin of the Akademy of Sciences of the USSR Physical Series, v.37, no.12, p.91-5. 1974.
18. Michail Kishinevsky, Sov. Phys. Tech. Phys, 45, no.6, 1281 (1975), and 48, no.4, 773 (1978).
19. Leonid Kishinevsky, Isvestia An. SSSR, 38, 392(1974), translated Bul. Acad.Sci. USSR, Phys. Ser. V.38 (1974). Later these calculation were repeated by Cui [H.L Cui, J. Vac. Sci. Technol.,A, 9 (3), 1823 (1991)].
20. J. R. Hiskes, Comments At. Mol. Phys. 19(2), 59(1987).
21. K. W. Ehlers and K. N. Leung, Rev. Sci. Instrum. 53, 1423 (1982).
22. K, N. Leung, K. W. Ehlers, and R. V. Pyle, Rev. Sci. Instrum 56(3), 364(1985).
23. K.Jimbo, K.Ehlers, K.Leung, R.Pyle, Nucl.Inst.Methods, A 248, 282 (1986).
24. Yu. I. Belchenko, G. I. Dimov, and V. G. Dudnikov, Sov. Phys. Tech. Phys., 43, 1720 (1973).
25. Yu. I. Belchenko, G. I. Dimov, and V. G. Dudnikov, Proc. of the Symposium on the Production and Neutralization of Negative Hydrogen Ions and Beams, Brookhaven National Laboratory report BNL 50727, 79 (1977).
26. K. Leung, et al., AIP Conf. Proc.158, Edit J.Alessi, p.356, 1986.
27. K.Leung, at.al. Rev.Sci.Instrum., 58 (2), 235 (1987).
28. V.Goretsky, et.al, Rev.Sci.Instrum., 73 (3),1157 (2002).
29. Akira Ueno, Kiyoshi Ikegami, Yasuhiro Kondo, Rev.Sci.Instrum., 75 (5), 1714 (2004).
30. P.Litvinov, I.Savchenko, A New Plasma H⁻ Source. Proc. of the 7th International Symposium on the Production and Neutralization of Negative Ions and Beams, AIP Conf. Proc No.380, 272 (1995).
31. M. Bacal and G.W. Hamilton, Phys. Rev. Lett. **42**, 1538 (1979), M. Bacal et al, Rev. Sci. Instrum., **50**, 719 (1979), M.Bacal, Diagnostic Techniques for Negative Ion Source, Proc. of the 4th International Symposium on the Production and Neutralization of Negative Ions and Beams, AIP Conf. Proc No. 158, 120 (1987), Rev. Sci. Instrum., **71**, 3981 (2000).
32. A. Bruneteau, M. Bacal, P.Devinck, F.Hallion, Proc. of the 7th International Symposium on the Production and Neutralization of Negative Ions and Beams, AIP Conf. Proc No.380, 261 (1995). M. Bacal, J. Bruneteau, P. Devynck and F. Hillion, AIP Conf. Proceedings 158, p. 246 (also published in Rev. Sci. Instrum., **59**, 2152 (1986).
33. F. El Balghiti, F.G. Baksht and M. Bacal, AIP Conf. Proceedings 380, p. 421 (1995), Rev. Sci . Instrum. **67**, 2221 (1996).
34. M. Seidl, H. Cui, J. Isenberg, H. Kwon, B. Lee, and S. Melnichuk, Negative surface ionization of hydrogen atoms and molecules, J. Appl. Phys.,79(6), 2896 (1996).
35. P.J. Schneider, K.H.Berkner, W.G.Garaham, R.V.Pyle, and J.W.Sten, Proc. of the Symposium on the Production and Neutralization of Negative Hydrogen Ions and Beams, Brookhaven National Laboratory report BNL 50727, 63 (1977).
36. F.N.Hoffman, P.E.Oettinger, Proc. of the Internat. Symposium on the Production and Neutralization of Negative Hydrogen Ions and Beams, Brookhaven National Laboratory report BNL 51304,UC-34a, 119, (1980).
37. W. G. Graham, Proc. of the Internat. Symposium on the Production and Neutralization of Negative Hydrogen Ions and Beams, Brookhaven National Laboratory report BNL 51304,UC-34a, 53, (1980).

Volume Production of High Negative Hydrogen Ion Density in Low-Voltage Cesium-Hydrogen Discharge

F.G.Baksht, V.G.Ivanov, S.M.Shkol'nik

A.F.Ioffe Physical-Technical Institute, Russian Academy of Sciences, St. Petersburg, 194024 Russia

M.Bacal

Laboratoire de Physique et Technologie des Plasmas, Ecole Polytechnique, UMR 7648 du CNRS, 91128 Paleseau, France

Abstract. The work is dedicated to the theoretical and experimental investigation of the intense H^- volume-plasma generation in cesium-hydrogen discharge. A low-voltage (LV) mode of the discharge, a LV arc with a heated cathode, is studied. Theory of the discharge was created. It was shown theoretically that a very high H^- density ($\sim 10^{13}$ cm^{-3}) may be achieved in the discharge plasma. A comparison between the discharge theory and the experiments was fulfilled. The main plasma parameters were measured by probe method. The H^- density in plasma was determined from the experimentally measured absorption of laser radiation due to the photodetachment of electrons from H^- ions. The existence of a very high H^- density in plasma ($\sim 10^{13}$ cm^{-3}) was proved experimentally. Theoretical modeling shows that, with a high electron emission from the cathode (~ 10 A/cm^2), the maximum H^- density is located in the plasma near the anode. It allows extraction of narrow and bright H^- beam through a small opening in the anode.

INTRODUCTION

H^- ions are used in accelerators, fusion research, plasma technology etc. The volume-plasma sources are widely used for H^- generation. In these sources the H^- ions are produced in plasma due to dissociative attachment (DA) of electrons to rovibrationally excited H_2 molecules in the ground electronic state $X^1\Sigma_g^+(v)$ [1,2]. At present, in H^- volume-plasma sources, $X^1\Sigma_g^+(v)$ molecules are created, to a marked degree, because of radiative deexcitation of the singlet electronically excited states $B^1\Sigma_u^+$, $C^1\Pi_u^\pm$ etc. [3]. The excitation cross-sections of these states by electrons have noticeable values, if the electron energy $E > 40$ eV [4]. Therefore, comparatively high voltage discharges, where 50 eV $\leq E \leq 150$ eV, are used for H^- generation. It leads to diminution of H_2 density and to undesirable increase of H atom density due to H_2 dissociation by fast electrons. Because fast electrons destroy H^- ions [5], the volume-plasma H^- sources consist usually of two chambers [6]. In the first chamber, where excited molecules are generated, the beams of fast electrons exist. In the second chamber, where H^- ions are created by DA, the fast electrons are absent. Therefore the

CP763, *Production and Neutralization of Negative Ions and Beams*, edited by J. D. Sherman and Y. I. Belchenko
© 2005 American Institute of Physics 0-7354-0248-5/05/$22.50

H⁻ source represents a complicated device (e.g. see [7, 8]) with complex plasma composition because several kinds of positive hydrogen ions and electronically excited H_2 molecules penetrate the volume, where DA occurs.

The present communication is dedicated to H⁻ generation in low-voltage cesium-hydrogen discharge. Such volume-plasma H⁻ source was proposed in [9], where it was shown that, a high density of vibrationally excited molecules and heated thermal electrons, needed for DA, may be obtained by a simple method in a single discharge volume.

MAIN PECULIARITIES OF LV CESIUM-HYDROGEN DISCHARGE

The theory of LV cesium-hydrogen discharge was created at first [9-13]. A typical potential distribution in the discharge consists of a potential well for thermal electrons. The well is separated from the electrodes by very narrow near-electrode Langmuir layers, which dimensions $L_0 \sim 10^{-3} L$ [14], where L is the interelectrode gap. The cathode voltage drop φ_1 is limited by the condition $\varphi_1 < E_d/e$, where $E_d \cong 8.8$ eV is the threshold of H_2 direct dissociation by electron impact from the ground electronic and vibrational state $X^1\Sigma_g^+(v=0)$. At such φ_1 values, the processes of H_2 direct dissociation are almost eliminated in plasma as well as the processes of H_2 or H stepwise ionization via electronically excited states. It leads to a comparatively small contamination of the discharge by atomic hydrogen and therefore improves significantly the vibrational distribution function (VDF) of H_2 molecules, because the rates of v-t relaxation between H_2 and H [15] exceed the corresponding rates for pure H_2 [16] up to three orders of magnitude. It also prevents the formation of atomic or molecular hydrogen ions. Therefore the discharge plasma contains only Cs^+ ions, electrons and H⁻ ions, the thermal electron density n_e being very large ($n_e \sim 10^{13}$-10^{14} cm⁻³). n_e exceeds the density $n_e^{(1)}$ of cathode beam electrons up to two or three orders of magnitude. Therefore electron-vibration kinetics depends only on thermal plasma electrons.

Two modes of the discharge were considered theoretically: a discharge in dense plasma [9-11] and discharge in rare plasma [12-13]. In the dense plasma, free paths of charged and neutral particles are significantly less than the gap L and the energy relaxation length $L_\varepsilon = (D_0\tau_\varepsilon)^{1/2}$ of cathode electrons is also smaller than L. Here D_0 and τ_ε are the diffusion coefficient and the time of fast (beam) electron energy relaxation due to pair Coulomb collisions. In the discharge considered, where $n_e/N_{H2} \geq 10^{-3}$ and $\varphi_1 < E_d/e$, this relaxation occurs due to collisions between fast and thermal electrons [14, 17]. In this kind of discharge the electron energy distribution function (EEDF) is very close to a Maxwellian one, and the plasma is described by the set of hydrodynamic equations (continuity equations, equations of motion and energy for electrons and heavy particles) and corresponding boundary conditions [11,14]. The mechanism of thermal electron heating is Coulomb pair collisions between fast and thermal electrons in near-electrode plasma region. In rare plasma, free paths of the thermal particles in the majority of cases are larger than the gap L. Also, $L_\varepsilon > L$, i.e. the length of fast electron maxwellization is greater than the gap. In this case, the EEDF is non Maxwellian, and thermal electrons cannot be heated by pair Coulomb collisions between them and beam electrons. In this case, the main mechanism of thermal electron heating is the collisional damping of Langmuir waves excited in plasma by means of plasma-beam instability. Because of this mechanism, a significant part of initial beam energy is transferred to thermal electrons [14, 18].

In this report, comparatively high H_2 pressures ($P_{H2} \geq 1$ Torr) are considered, and the plasma is dense enough. Because of high n_e and P_{H2} values, VDF of H_2 molecules is mainly created due to the volume-plasma processes, the vibrational deexcitation on the electrode surface being almost inessential. Therefore the VDF does not depend on unknown probabilities of the vibrational deactivation of molecules on cesium-coated surfaces. It allows to compare experimental and theoretical data without fitting parameters.

Experimental investigation of LV cesium-hydrogen discharge was performed in [14, 19-24]. In the initial experimental works (e.g. see [14, 19-21]) the probe method of plasma diagnostics was developed and local parameters (n_e, T_e,φ) were measured. It was shown experimentally that all the plasma parameters, which are needed for intense H generation, were actually obtained in the discharge plasma. In particular, just in the first experimental works (e.g. see [14, 19-21]), it was shown that high electron density n_e and optimum electron temperature T_e for H generation by means of DA were actually achieved in the discharge plasma.

VOLUME-PLASMA DENSITY IN LV CESIUM-HYDROGEN DISCHARGE

In the recent works [22-24] the H density, N_{H-}, in plasma was measured experimentally. A diode with plane-parallel electrodes and interelectrode distance 3 mm was used in the experiment. The face ends of the cylinders 12 mm in diameter were used as the electrodes. A platinum foil, which was welded to the face end of the cathode cylinder, was used as an emitter in cesium vapor. A cylindrical probe was inserted into the gap through a hole in the center of the anode. The probe axis was parallel to electrode surfaces. The probe diameter and length were 0.1 mm and 2mm respectively. The probe was placed in the center of the gap. N_{H-} was determined from the experimentally measured absorption of laser radiation due to the photodetachment of electrons from H ions. A semiconductor laser of a continuous operation and a power $P \cong 0.1$ W was used. The spectral width of the radiation was approximately $\Delta\lambda \cong 2$ nm, and the maximum of the intensity was at $\lambda_0 = 816$ nm ($h\nu_0 = 1.52$ eV). Density of the excited Cs atoms were measured experimentally and calculated by theoretical methods. It was shown that the actual absorption of laser radiation, which was observed in the experiment, is larger than photoabsorption by excited Cs atoms by two or three orders of magnitude. The laser wavelength corresponds to the flat maximum of H photoionization cross-section ($\sigma_0 \cong 4 \cdot 10^{-17}$ cm^2). As a result, the values of electron density n_e, temperature T_e and potential φ were determined by probe method in the center of the gap, and H density was measured by the absorption of laser radiation in the near-cathode plasma. Simultaneously the distributions of all plasma parameters in the gap, including the H_2 VDF and densities of excited Cs atoms, were determined by the self consistent discharge theory. The following processes were taken into account in calculations of VDF: e-v exchange, which was calculated according to the method of [25] (the cross section σ_{01} of excitation of the $v = 0 \rightarrow v = 1$ transition by electron impact was borrowed from [26]); v-v and v-t exchange between H_2 molecules [27] (the corresponding rate constants were borrowed from [16, 28]); v-t exchange between H_2 molecules and H atoms, which was taken into account according [15, 29] and the processes of DA, which were considered according to [2, 25]. Several other processes were also taken into account as it was done in [30]. The experimental electron temperature in the center of the gap was taken as an initial value for theoretical calculations. Because in experimental conditions, the Cs density N_{Cs} in the gap is significantly less than the density N_{Cs} (T_{Cs}) just above the surface of liquid cesium, the theoretical parameter $N_{Cs}^{(0)}$, which is equal to total Cs density averaged over the discharge gap, was found from the additional condition, namely by equating the theoretical and experimental Cs^+ density in the gap center. The maximum deviation of experimental N_{H^-} values from corresponding points of the theoretical N_{H^-} curves, which was obtained in the present experiments, was less than or equal to 40 %. N_{H^-} densities up to $N_{H^-} \cong (0.5-0.6) \cdot 10^{13}$ cm^{-3} were obtained experimentally near cathode at cathode electron current density $j_s \cong (3-4)$ A/cm^2 and hydrogen pressure $P_{H2} \cong (1-2)$ Torr. According to the last theoretical calculations, the density up to $N_{H^-} \sim 10^{13}$ cm^{-3} may be obtained in plasma near anode. These results were obtained for comparatively high cathode emission current density $j_s = 10$ A/cm^2. Such j_s value cannot be obtained in stable mode of operation from the platinum cathode at optimum Cs pressure $P_{Cs} \cong 10^{-2}$ Torr, which was used in the present experiments, but may be easily obtained from LaB_6 flat cathode.

ACKNOWLEDGMENTS

This works was done with the support of the Scientific Program of the St. Petersburg Center of Russian Academy of Sciences. One of us (M. Bacal) gratefully acknowledges the support of the European Community (Contract HPRI-CT-2001-50021).

REFERENCES

1. Bacal M. and Hamilton G.W., *Phys. Rev. Lett.,* **42**, 1538 (1979).
2. Wadehra J.H. *Phys. Rev. A,* **29**, N°1, 106 (1984).
3. Hiskes J.R. *Appl.Phys. Lett.,* **57**, 231 (1989).
4. Hiskes J.R. J. *Appl.Phys.,* **51**, 4592 (1980).
5. Janev R.R., Langer W.D., Evans K.E. and Post D.E. *Elementary Processes in Hydrogen-Helium Plasmas.* Berlin. Springer (1987).
6. Bacal M., *Nucl. Instrum. Methods Phys. Res.*, **B37/38**, 28 (1989).
7. Kuroda T., et al, "Development of High-Current Hydrogen-Negative Ion Source for NBI in NIFS" in *Production and Neutralization of Negative Ions and Beams and Production and Application of Light Negative Ions, 1995,* edited by Krsto Prelec, AIP Conference Proceedings **380**, New York; American Institute of Physics, 201-213 (1996)
8. Tsumori K. et al, " Development of Large Scale Negative Ion Source for LHD-NBI " in *Production and Neutralization of Negative Ions and Beams and Production and Application of Light Negative*

Ions, 1997, edited by Claude Jacquot, AIP Conference Proceedings **439**, New York; American Institute of Physics, 93-104 (1998).

9. Baksht F.G., Ivanov V.G., *Pis'ma Zh. Tekh. Fiz.* **12**, N° 11, 672 (1986).
10. Baksht F.G., Elizarov L.I., Ivanov V.G., Yur'ev V.G., *Fiz. Plazmy*, **14**, N° 1, 91 (1988).
11. Baksht F.G., Elizarov L.I., Ivanov V.G., *Fiz. Plazmy*, **16**, N°7, 854 (1990).
12. Baksht F.G., Ivanov V.G., Kostin A.A., *Pis'ma Zh. Tekh. Fiz.*, **18**, N°12, 83 (1992).
13. Baksht F.G., Ivanov V.G., Kostin A.A. Zh. Tekh. Fiz., **63**, N° 9, 173 (1993).
14. Baksht F.G., Djuzhev G.A., Elizarov L.I., Ivanov V.G., Kostin A.A., Shkol'nik S.M., *Plasma Sources Sci. Technol*, **3**, 88 (1994).
15. Garcia E. and Lagana A., *Chem . Phys. Lett.*, **123**, 365 (1986).
16. Kiefer J.H., J. *Chem. Phys.*, **57**, 1938 (1972).
17. Baksht F.G., Ivanov V.G., *Fiz. Plasmy*, **12**, 286 (1986).
18. Baksht F.G., Ivanov V.G., Sov. *J. Plasma Phys.*, **12**, (1986).
19. Baksht F.G., Kolosov B.I., Kostin A.A., Tchuraev R.S., Yuriev V.G., *Mathematical Modelling of Processes in Low-voltage Plasma-Beam Discharge*, Moscow, Energoatomizdat (1990).
20. Baksht F.G., Dyuzhev G.A., Elizarov L.I., Ivanov V.G., Filatov A.G., Shkol'nik S.M. , *Zh. Tekh. Fiz.*, **62**, 148 (1992).
21. Baksht F.G., Dyuzhev G.A., Elizarov L.I., Ivanov V.G., Filatov A.G., Shkol'nik S.M., *Sov. Phys. Tech. Phys.*, **37**, 959 (1992).
22. Baksht F.G., Dyuzhev G.A., Elizarov L.I., Ivanov V.G., Kostin A.A., Shkol'nik S.M., *IEEE Trans. Plasma Science*, 21, 552 (1993)
21. Baksht F.G., Ivanov V.G., Kostin A., Nikitin A.G., Shkol'nik S.M., *Zh. Tekh. Fiz.*, **65**, 1927 (1995).
22. Baksht F.G., Ivanov V.G., Kon'kov S.I., Shkol'nik S.M., *Zh. Tekh. Fiz.*, **71**, 17 (2001).
23. Baksht F.G., Elizarov L.I., Ivanov V.G., Kon'kov S.I., Mitrofanov N.K., Shkol'nik S.M., *Fiz. Plazmy*, **29**, 1 (2003)
24. Baksht F.G., Ivanov V.G., Kon'kov S.I., Shkol'nik S.M. and Bacal M., *J. Phys. D: Appl. Phys.*, **35**, 1 (2002).
25. Skinner D.A., Bruneteau A.M., Berlemont P., Courteille C., Leroy R., Bacal M., *Phys. Rev. E.*, 48, N° 3, 2112 (1993).
26. Morrison M.A., Crompton R.V., Saha B.C., Petrovic, Z.L.,*Aust. J. Phys.*, **40**, N 3, 239 (1987).
27. Nonequilibrium Vibrational Kinetics (Ed. by M. Capitelli). Springer-Verlag (1986).
28. Kiefer J.H., Lutz R.M., *J. Chem. Phys.*, **44**, N° 2, 668 (1966).
29. Garcia E., Lagana A., *J. Phys. Chem.*, **90**, 987 (1986).
30. Baksht F.G., Elizarov, L.I., Ivanov V.G., Yuriev V.G., *Fiz. Plazmy*, **14**, N°1, 91 (1988).

EXTRACTION, TRANSPORT,
AND ACCELERATION

Beam Dumping Ghost Signals in Electric Sweep Scanners

M. P. Stockli,[1,2] M. Leitner,[3] D. P. Moehs,[4] R. Keller,[3] and R. F. Welton[1]

1) SNS, Oak Ridge National Laboratory, P.O. Box 2008, Oak Ridge, TN 37831, USA[*]
2) Department of Physics, University of Tennessee, Knoxville, TN 37996, USA
3) SNS, Lawrence Berkeley National Laboratory, 1 Cyclotron Rd., Berkeley, CA, 94720, USA[*]
4) Fermi National Accelerator Laboratory, P.O. Box 500, Batavia, IL 60510, USA

Abstract. Over the last 20 years many labs started to use Allison scanners to measure low-energy ion beam emittances. We show that large trajectory angles produce ghost signals due to the impact of the beamlet on the electric deflection plates. The strength of the ghost signal is proportional to the amount of beam entering the scanner. Depending on the ions and their velocity, ghost signals can have the opposite polarity as the main beam signals or the same polarity. These ghost signals are easily overlooked because they partly overlap the real signals, they are mostly below the 1% level, and they are often hidden in the noise. However, they cause significant errors in emittance estimates because they are associated with large trajectory angles. The strength of ghost signals, and the associated errors, can be drastically reduced with a simple modification of the deflection plates.

INTRODUCTION

The emittance of a particle beam describes the six-dimensional distribution of all position coordinates along the three configuration space directions and the associated velocity coordinates. The emittance is normally reduced into three subsets by projecting it into the two-dimensional planes {x-x′}, {y-y′}, and {z-z′}.

When measuring a transverse subset, either in x or y, the projection is accomplished with an entrance slit placed at a number of equidistant position coordinates that accepts a narrow band of the beam cross section. The corresponding trajectory angle distribution, x′ or y′, respectively, is determined for each main slit position from the downstream particle distribution, probed by a second slit or a wire harp.

Wire harps can measure each distribution in a single shot, but harps are subject to sagging, and can exhibit variations in the wire size, in surface conditions that affects the secondary electron emission rate, and in amplifier gain. A single secondary slit at the entrance of a suppressed Faraday cup, combined with some type of scanning mechanism, promises more reliable trajectory angle distributions if the beam remains stable during the time-consuming scan.

---

[*]SNS is a collaboration of six U.S. national laboratories: Argonne National Laboratory (ANL), Brookhaven National Laboratory (BNL), Thomas Jefferson National Accelerator Facility (TJNAF), Los Alamos National Laboratory (LANL), Lawrence Berkeley National Laboratory (LBNL), and Oak Ridge National Laboratory (ORNL).

CP763, *Production and Neutralization of Negative Ions and Beams*, edited by J. D. Sherman and Y. I. Belchenko
© 2005 American Institute of Physics 0-7354-0248-5/05/$22.50

Measuring a two-dimensional distribution with sufficient resolution implies that only a very small fraction of the entire beam current is measured at any given time. As a matter of fact, the current probe mostly measures the absence of a beam signal, the background. At all times almost all beam particles are intercepted by the entrance slit, where they generate a variety of secondary particles and cause a scattering of some of the primary particles back into the vacuum space. When some of those charged particles reach the current probe, they create a signal or contribute to a beam signal without truly representing the actual distribution of the two-dimensional beam emittance. Such signals are called ghost signals because they tend to be faint and are often observed under conditions where no real signal is expected. Slit scattering, for example, changes the trajectory angles and possibly the charge of the primary particles, besides generating secondary particles. Even small ghost signals can significantly alter measured emittances because they tend to appear at rather large coordinate values.

We have identified ghost signals that are generated in electric sweep scanners when the beamlet that passed the entrance slit impacts on a deflection plate rather than the second slit. A recent paper derived the impact conditions only for beamlets that enter the scanner on its axis [1]. In this paper we present a detailed analysis of the trajectory angles and location of impact for all ions that pass through the entrance slit. This analysis proves useful in designing a simple modification that reduces the strength of the ghost signals by about two orders of magnitude. Although the principles discussed apply to all electric sweep scanners [2], the analysis is restricted to the simple geometry favored by Allison scanners [3].

ALLISON SCANNERS

Over the last 20 years, Allison scanners [3] have been introduced in many laboratories to measure the emittance of low-energy ion beams [4]. Allison scanners feature entrance and exit slits that are rigidly mounted on the same support base, thus allowing for their relative alignment within tight tolerances. The space between the slits is occupied by a set of electric deflection plates as shown in Fig. 1. Charged particles that pass both slits are collected in the Faraday cup, which features secondary electron suppression. A grounded shield surrounds the assembly, intercepting any charged particles that could produce ghost signals [5].

A stepper motor moves the entire assembly through the beam to probe the different positions of the beam. At each stop, the beamlet that passed the entrance slit is scanned electrically across the exit slit to determine the distribution of the entry angles.

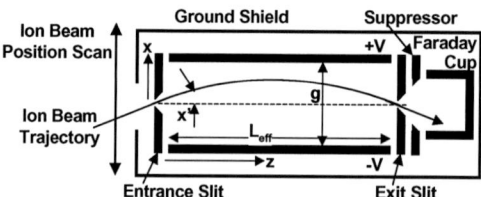

FIGURE 1. Schematic of an Allison emittance scanner.

After passing the entrance slits, ions with charge q and energy $q \cdot U$ enter the electric field between the deflection plates, which are charged to opposite voltages $-V$ and $+V$. The deflection voltage-to-entrance angle conversion depends primarily on the length of the deflection plates, L, and the gap, g, between them. Fringing field corrections [6] yield the more accurate effective length, L_{eff}, although the improvement is normally minor because the gap is small compared to the length of the deflection plates. The transverse position, x, is calculated from the transverse acceleration, a_x, as $x = \iint a_x \cdot dt^2 = -2 \cdot \iint dt^2 \cdot q \cdot V/g$. If $U \gg V$, then $2 \cdot q \cdot U \cdot dt^2 = m \cdot dz^2$ and $x = x'_0 \cdot z - V \cdot z^2/(2 \cdot g \cdot U)$, where $z (0 \le z \le L_{eff})$ is the axial distance from the entrance slit and x'_0 is the entry angle, the initial trajectory angle in radians, of the ion when it passes through the entrance slit $(x(z = 0) = 0)$. With this definition, positive voltages reduce the trajectory angle $x' = dx/dz = x'_0 - V \cdot z/(g \cdot U)$. Accordingly, positive ions are described by the voltage on the upper deflection plate, while negative ions are described by the voltage on the lower deflection plate. To pass through the exit slit $(x(z = L_{eff}) = 0)$, ions with an initial trajectory angle x'_0 require voltages of $V = 2 \cdot U \cdot x'_0 \cdot (g/L_{eff})$, or the voltage-to-angle conversion is $x'_0 = V \cdot L_{eff}/(2 \cdot g \cdot U)$.

The space between the deflection plates allows only for trajectories where x never exceeds $g/2$, which geometrically limits the angular acceptance to $x'_{Gmax} = 2 \cdot g/L_{eff}$. In addition, the deflection voltage can also limit the system to $x'_{Vmax} = V_0 \cdot L_{eff}/(2 \cdot g \cdot U)$, where V_0 is the maximum voltage generated by the bipolar deflection supplies, and U is the ion potential, defined as its energy per charge. The system's angular acceptance limit, x'_{max}, is given by the lower of the two: $x'_{max} = \min (V_0 \cdot L_{eff}/(2 \cdot g \cdot U), 2 \cdot g/L_{eff})$.

Several considerations guide the design of an emittance scanner: the most economical design matches x'_{Gmax} and x'_{Vmax} for U_0, the highest ion potential of interest, often limited by the maximum output of the ion source supply. This allows one to determine the minimum voltages, V_0, required from the two bipolar supplies: $V_0 \ge x'^2_{max} \cdot U_0$, where x'_{max} is the desired minimum angular acceptance of the system. The final V_0 is normally selected from a list of commercially available supplies.

The deflection length, L, should always be as long as convenient, assuring that the entire scanner fits through the scanner's mounting port and fits into the insertion gap.

Knowing U_0, V_0, and L allows for fine-tuning the design by selecting a gap, g, with $g \le (V_0/U_0)^{1/2} \cdot L/2$. The equal sign matches the geometrical and voltage limits and thus maximizes the angular acceptance of the scanner system for the highest ion potential.

If a larger gap is desired, one needs to increase the voltage V_0 of the bipolar supplies. Doubling the gap and the voltage maintains the system's angular acceptance, which will be voltage-limited. Quadrupling the voltage while doubling the gap doubles the system's angular acceptance with matched geometrical and voltage limits.

Knowing the design's final angular acceptance, x'_{max}, allows for designing slits free of slit-edge scattering. This is accomplished by tapering the downstream side of both slits with an angle that exceeds the angular acceptance, x'_{max}.

In this work we use an emittance scanner with $L = 115$ mm, $L_{eff} \approx 120$ mm, $g = 7$ mm, $x'_{Gmax} = 0.117$ rad. This practically matches $x'_{Vmax} = 0.120$ rad, the capability of our 1-kV bipolar supplies in analyzing our ion beams with up to 70 kV energy.

BEAM DUMPING INSIDE THE SWEEP SCANNER

So far the literature has described only the part of the beam that passes through both slits. Here, we consider a beamlet that passes through the entrance slit with entry angle x'_b, while the voltage is tuned for the Faraday cup to measure ions with entry angles x'_s, the sweep angle. The beamlet's equations of motion, written as $x = x'_b \cdot z - x'_s \cdot z^2 / L_{eff}$ and $x' = x'_b - 2 \cdot x'_s \cdot z / L_{eff}$, are used to determine the beamlet's impact location and angle. The analysis holds for positive as well as negative sweep angles and entry angles, with a positive value meaning that the angle increases the transverse position coordinate.

Figure 2 shows the beamlet's axial distance of impact from the entrance slit, z_i, as a function of the sweep angle x'_s and beamlet entry angle x'_b for our scanner. The results are symmetric in x'_b and therefore are shown only for $x'_b > 0$. The exit slit opening, where the beamlet enters the Faraday cup when $x'_s = x'_b$, can be seen as a small ridge in the center of the top plateau. The face of the exit slit acts as a beam stop as long as the absolute difference between the sweep angle, x'_s, and the entry angle, x'_b, does not exceed $|x'_s - x'_b| \le g/(2 \cdot L_{eff})$, appearing in Fig. 2 as the plateau at 120 mm. When the sweep angle is above this range ($x'_s > x'_b + g/\{2 \cdot L_{eff}\}$), the beamlet impacts on the lower deflection plate at a distance $z_{iL} = (x'_b + (x'^2_b + 2 \cdot x'_s \cdot g / L_{eff})^{1/2}) \cdot L_{eff} / (2 \cdot x'_s)$. Figure 2 shows the impact location gradually moving away from the exit slit when the sweep angle, x'_s, increases. When sweeping an angle below the range of the plateau ($x'_s < x'_b - g/\{2 \cdot L_{eff}\}$), the beamlet impacts on the upper plate at a distance of $z_{iu} = (x'_b - (x'^2_b - 2 \cdot x'_s \cdot g / L_{eff})^{1/2}) \cdot L_{eff} / (2 \cdot x'_s)$. Figure 2 shows that for the upper plate a decreasing sweep angle, x'_s, moves the impact location away from the exit slit at a pace that accelerates with increasing beam entry angle, x'_b. When $x'_b > g/L_{eff}$, a shadow starts to appear on the slit and the deflection plate, a discontinuity appearing in Fig. 2 as a nearly vertical cliff. This shadow is caused by trajectories with an apex within the

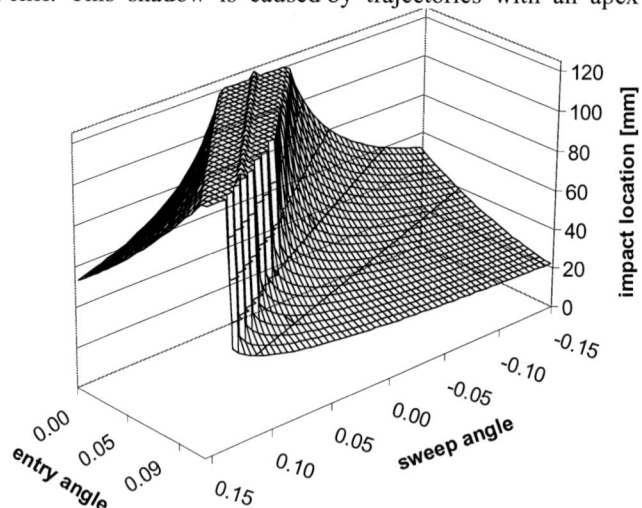

FIGURE 2: Distance of impact from entrance slit versus entry and sweep angle.

deflector length ($z_A < L_{eff}$), but outside the deflector gap ($x_A > g/2$). The length of the shadow increases rapidly with increasing entry angle, x'_b. When the entry angle reaches $2 \cdot g/L_{eff}$, the shadow covers half of the deflector plate and half of the slit. Larger entry angles are outside the scanner's useful range because the beamlet can no longer reach the opening in the exit slit.

Equally interesting are the impact angles that are determined from the trajectory angle at the location of impact, $x'_i(z = z_i)$. Impacting on the face of the exit slit, the trajectory angle is given by $x'_{ie}(z = L_{eff}) = x'_b - 2 \cdot x'_s$. The space restriction of $|x| < g/2$ limits the impact angle on the exit slit to $|x'_i| \leq (2 + \sqrt{2}) \cdot g/L_{eff}$, which is ± 0.20 or $\pm 11.3°$ in our case. Theses angles are formed by beamlets with entry angles $x'_b = \pm(1 + \sqrt{2}) \cdot g/L_{eff}$ when scanned with angles of $x'_s = \pm(3/2 + \sqrt{2}) \cdot g/L_{eff}$, which is beyond the useful range of the scanner. If scanning is limited to the geometrical acceptance $|x'_s| \leq x'_{Gmax} = 2 \cdot g/L_{eff}$, the trajectory angles on the exit slit are limited to $|x'_i| \leq 5 \cdot g/(2 \cdot L_{eff})$, which is ± 0.15 or $\pm 8.3°$ in our case. These angles are formed by beamlets entering with $x'_b = \pm 3 \cdot g/(2 \cdot L_{eff})$ while being scanned at $\pm x'_{Gmax}$. In either case, all impacts on the face of the exit slit are close to normal and therefore of little concern because all backscattered ions and secondary particles travel away from the Faraday cup.

Figure 3 shows the trajectory angle when the beamlet impacts on one of the deflection plates as a function of the sweep angle, x_s', and beamlet entry angle, x'_b, for our scanner. For clarification, all angles of trajectories that impact on the exit slit have been set to zero, and to -0.01 for trajectories passing through the center of the exit slit. When the beamlet impacts on the lower deflection plate, the trajectory angle is given by $x'_{iL} = -(x'^2_b + 2 \cdot x'_s \cdot g/L_{eff})^{1/2}$. Figure 3 shows the impact angle to start at $-g/L_{eff}$ for $x'_b = 0$ and $x'_s = g/(2 \cdot L_{eff})$, which is 0.06 or 3.3° in our case. The impact angle decreases gradually with increasing sweep angle and entry angle, making it less grazing.

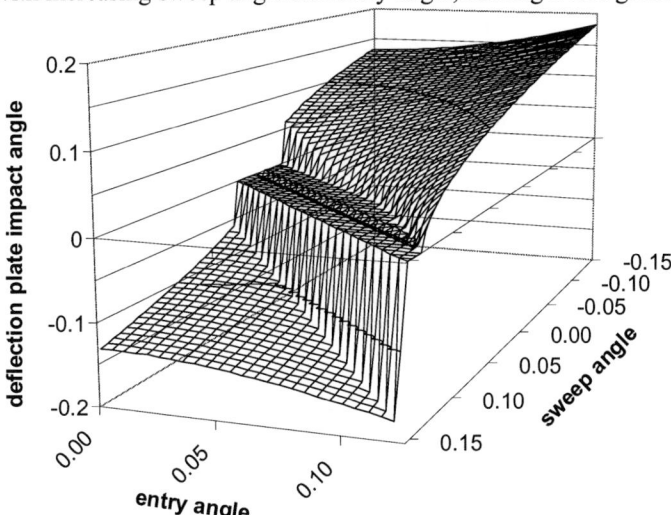

FIGURE 3: Impact angles [rad] on deflection plates versus beamlet entry angle and sweep angle [rad].

When the beamlet impacts on the upper plate the trajectory angle is given by $x'_{iU} = (x'^2_b - 2 \cdot x'_s \cdot g/L_{eff})^{1/2}$. Figure 3 shows the angle to start at g/L_{eff} for $x'_b = 0$ and $x'_s = -g/(2 \cdot L_{eff})$. It also shows that the impact angle on the upper plate always increases gradually with decreasing sweep angle, x'_s. But for impacts near the exit slit, the impact angle gradually decreases with increasing entry angle, x'_b, until it reaches zero at $x'_b = g/L_{eff}$. When the entry angle increases further the minimum impact angle remains at zero, but the location of impact moves away from the exit slit, as seen in Fig. 2.

If both angles are restricted to the scanner's geometrical acceptance limit, the beamlets with $x'_s = -x'_b = \pm 2 \cdot g/L_{eff}$ form the largest angles when impacting on a deflection plate, namely $|x'_{imax}| = (\sqrt{8}) \cdot g/L_{eff}$, which is 0.165 or 9.4° in our case.

Our concerns focus on beamlets impacting near the exit slit with rather small impact angles because grazing impact favors the emission of secondary particles as well as causing the primary particles to be scattered back into the vacuum space [7]. The rather low energy of secondary particles allows them to be absorbed quickly by the deflector plate with opposite polarity. Only beamlets with entry angles in excess of $x'_b > g/(2 \cdot L_{eff})$ hit the deflection plates in the absence of a deflection field ($x'_s = 0$); these impacts could allow some charged secondary particles to enter the Faraday cup and generate ghost signals. These ghost signals can be minimized or avoided by aligning the scanner axis with the center of the beam.

Primary particles that are scattered back into the vacuum space normally retain a large fraction of their momentum and can therefore reach the opening in the exit slit despite deflection fields. Scattering is often accompanied by charge exchange. Depending on the ion species and their energy, the ions may or may not change their charge and/or polarity. Negative ions often lose one or more electrons and so become neutral or positive, especially at higher energies.

GHOST-INFESTED EMITTANCE DATA

Our interest is the ion beam emittance injected into the RFQ of the SNS accelerator. Therefore, the emittance shown in Fig.4 was measured as the ion beam exits a duplicate low-energy-beam-transport section incorporated in our ion source hot spare stand [8].

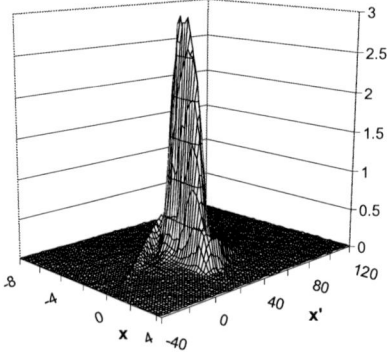

FIGURE 4. Measured beam current as a function of position x[mm] and trajectory angle x'[mrad].

150

The figure shows the x/x' distribution of a broad, slightly converging H⁻ beam with long, small tails probably caused by the electrostatic extraction and transport system. The background looks fairly uniform, with a rather low level of noise fluctuation and without obvious problem areas. Despite the normal appearance we will reveal ghost signals spread over wide areas.

Figure 5 shows the data from Fig. 4 as a density plot. The beam core and its two tails are seen in the center in gray tones that darken with intensity in steps of ~2% of the measured peak current. This structure is surrounded by an exclusively white zone, indicating that all signals exceed the noise variations. Farther away, the white is intermixed with black pixels, which represent signals with a polarity opposite to that of the real signals. The zero of the intensity scale was adjusted until the lower left corner appeared as a random pattern with an equal mix of black and white. This method highlights small deviations from random-noise background and thus can highlight ghost signals.

In the bottom third of Fig. 5, one finds a large black area that is cut in half by an extension of the lower tail. This area is obviously formed by reversed polarity signals that exceed the noise variations. The area is limited between –4.5 mm and 1.5 mm, the range in which relatively intense beamlets enter the scanner. The area starts at ~30 mrad below the entry angle of the beamlet center, which matches approximately the 29 mrad when the center of the beamlet starts to hit the upper deflection plate. As the beamlet entry angle decreases from ~40 to ~25 mrad, the impact angle near the exit slits increases from ~1.2° to ~2.4°, respectively. The area appears to extend over about 50 mrad. At the lower end, the center of the beamlet impacts near the center of the deflection plate at an angle of ~4.5°.

A similar but smaller black area appears above the main signal. It starts at ~40 mrad above the entry angle of the beamlet center, which slightly exceeds the 29 mrad where the beamlet center starts to hit the lower deflection plate. There, the beamlet center impact about 10 mm in front of the exit slits with impact angles of ~6°. The area appears to extend over about 40 mrad. At the upper end, the beamlet center impact ~40 mm in front of the exit slit with an angle of ~7°.

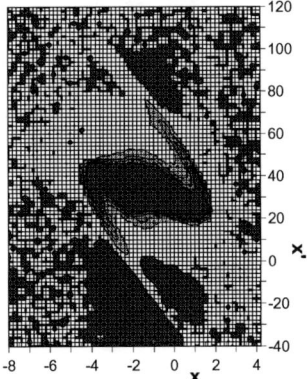

FIGURE 5. Density plot of the emittance data from Fig. 4.

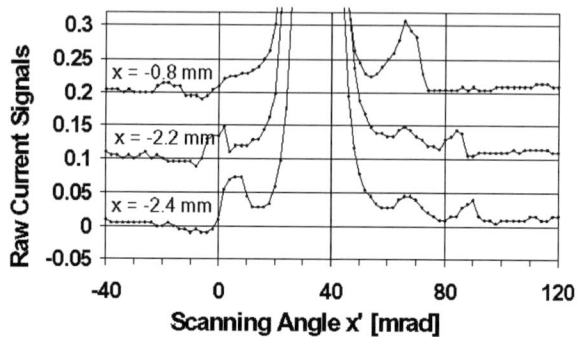

FIGURE 6. Raw current signals offset by 0.1 for 3 scans from Figs. 4 and 5.

Inverted small signals start to appear when the beamlet hits the deflection plate and fade away as the impact location moves away from the exit slit and the impact angle increases. The inverted signals are clearly caused by stripped H⁻ ions that are scattered back from the deflection plates. The process identification is based on the correlation of coordinates where the inverted signals are found.

The data shown in Figs. 4 and 5 consist of 5,022 current readings, 65 of which were inverted. Figure 6 shows the raw signals from electrical sweeps measured at 3 positions, containing 18 inverted signals. For clarity the current was offset by 0.1 and 0.2 for x = −2.2 mm and x = −0.8 mm, respectively. The data exhibit 5-mV steps from the 12-bit digitization of the ±10-V range. Of the 65 inverted raw signals, 52 read −5 mV and the other 13 read −10 mV. The extent of the ghost signals becomes clear only when a bias of ~11 mV is subtracted, which inverts many additional signals. After the bias correction the original −10 mV and −5 mV data correspond to −0.7% and −0.5% of the measured peak current, respectively. The low noise level of the Fig. 5 data enables direct observation of the ghost signals, which is impossible for the Fig. 7 data.

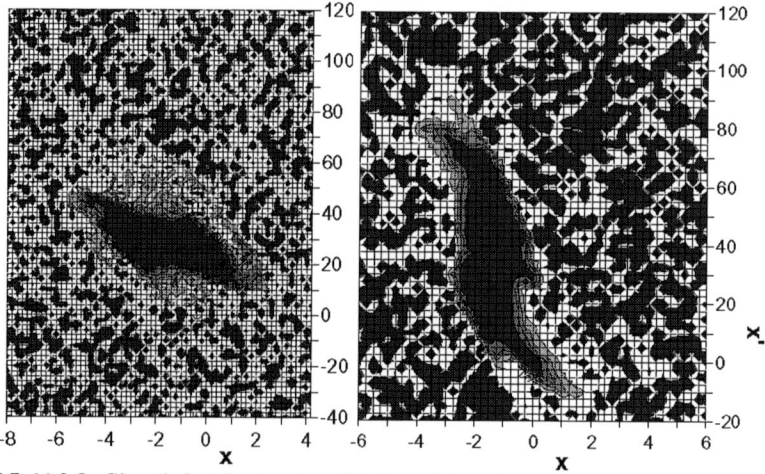

FIGURE 7. (a) left: Ghost-infested noisy data; (b) right: Infested data of a strongly focused beam.

Figure 7(a) shows emittance data with ~4% peak-to-peak noise. The noise fluctuations completely dominate the background and mask the inverted signals. Figure 7(b) shows a strongly focused beam with high angular divergence, where the long angular tails merge with the inverted signals. The only remaining sign of the inverted signals is found at the bottom of the figure, in a black cluster that is slightly larger than most others. The presence of inverted signals in both data sets will be proven later.

GHOST SIGNAL MITIGATION FOR ELECTRIC SCANNERS

Having identified and characterized the ghost signal-producing process, it is possible to develop a correction function that can be convoluted with the measured beam profile and applied to the measured data. Such a function would depend on the acceptance of the Faraday cup and therefore on the details of individual designs. In addition the probability of backscattering depends on the roughness and the condition of the surface, and therefore varies from scanner to scanner and may vary with time.

It is therefore generally preferred to reduce the production of ghost signals. This can be accomplished by increasing the ghost-free angular range of $g/(2 \cdot L_{eff})$ without shortening the length of the scanner. The increase in gap has to be accompanied by at least a linear increase of the deflection voltage. The data in Fig. 5 suggest that the gap and the deflector voltage would have to be increased by a factor of 4 to completely separate the angular ranges of the ghost signal from the range of the real signal. Such a large increase requires modifications that are difficult to incorporate into the present design but should be considered in future designs.

The ghost signals are generated by backscattered primary particles that retain their forward momentum after impacting on the deflection plates. Therefore, we have machined a staircase in the beam-facing surface of the deflector plates, as indicated in Fig. 8. The figure shows that particles impact almost normally on the faces of the stairs; this impact causes backscattered primary particles to move away from the Faraday cup. The staircase angles are critical parameters because primary particles impacting on the stair flats would have more grazing impacts with a significantly increased fraction of backscattered primary particles. As previously derived, a emittance scanner operated within its geometrical acceptance limit of $x'_{max} \leq 2 \cdot g/L_{eff}$ limits the trajectory angles at impact to $|x'_i| \leq (\sqrt{8}) \cdot g/L_{eff}$, or 9.4° in our case. To err on the safe side we selected 20° flats with 70° faces, with 1-mm steps 3 mm apart. After the steps were machined, a small final cut was made to obtain sharp edges that are ~25 μm wide.

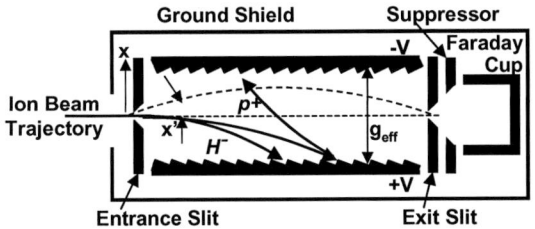

FIGURE 8. Staircase deflection plates to suppress ghost signals.

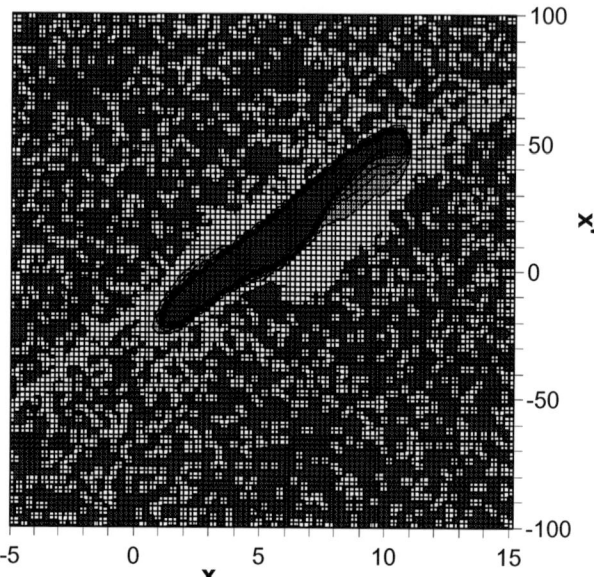

Figure 9. Emittance data obtained with stair-cased deflection plates.

The roughness of the edges and the ratio of edge width to separation suggest ghost signal suppression in excess of 99%. The effective gap of the modified deflection plates is approximated by the sum of the gap between the ridges and the height of one step.

Yu. Belchenko suggested deflection electrodes made from a screen of thin wires as an alternate mitigation [9]. Wires with diameters significantly smaller than 25 μm are commercially available. With increased complexity, larger suppression ratios may be possible if the stair-cased deflection plates provide insufficient suppression.

Figure 9 shows emittance data taken with the modified deflector plates in a density plot like that used in Fig. 5. The background appears as a uniformly random mixture of positive and negative data. However, this is not sufficient prove for the absence of ghost signals because the signals could be hiding in the noise.

STATISTICAL SIGNATURES OF INVERTED GHOST SIGNALS

Small signals buried in the noise can often be found with statistical averaging. The highest sensitivity can be achieved by averaging in groups that put the signals of interest together, which then can be compared with groups where no signal is expected. The peak of the discovered ghost signals is at least $g/(2 \cdot L_{eff})$ away from the peak of the real signals. This suggests grouping according to the distance from the beamlet center.

This is the same condition that guided our self-consistent bias estimation for emittance data. A bias of ~0.01 was subtracted from all data sets to produce the density plots that reveal the random background areas. However, this is not sufficient accuracy for the rms emittance, which is hypersensitive to bias when evaluated from all data. For example, for the ghost-free data of Fig. 9, the rms emittance calculated from all bias corrected data changes by 30% when the bias correction is changed by 0.0001, or 2% of the least significant bit of the digitizer.

Therefore, we developed a systematic method for self-consistently estimating the bias from the average background measured in the absence of real as well as ghost signals [10]. As farther away data are from the peak of the real signals as more likely they are pure background, consisting of bias and noise. However, good averaging requires the inclusion of many data, or most of the pure background. This requires the pure background to be separated from the other data through a boundary that encloses all other data tightly. Ellipses are a simple boundary that conforms relatively well with most emittance distributions. Being able to vary this boundary without significantly changing the resulting average of the pure background self consistently estimates the bias.

For this analysis a 10% threshold was applied to all data before calculating the Twiss parameters for each data set. Figure 10 shows the average current outside the ellipse as a function of the semi-axis product of the ellipse, with aspect ratios and orientation matching the calculated Twiss parameters. All data sets show excessively large average currents outside small ellipses because the average includes real signals.

The ghost-free data of Fig. 9 are represented in Fig. 10 by the solid line that indicates an ~800 mm·mrad ellipse to include all real signals and is consistent with a bias of zero. The low-noise, ghost-infested data of Fig. 5 are represented by the dotted line. This line undershoots to −0.001 before it recovers to 0 with an ellipse of ~2000 mm·mrad, which appears to include all real and all ghost signals. The noisy data of Fig. 7(a) are represented by the dot-dashed line, which also undershoots to −0.001. However its average becomes dominated by noise-induced fluctuation for ellipses above 800 mm·mrad. The strongly focused beam data of Fig. 7(b) are represented by the dashed line. It undershoots to −0.003 before noise induce fluctuations start to dominate.

The three undershoots are a clear signature of the peripheral area dominated by inverted ghost signals. While the self-consistent, elliptical exclusion method is a powerful tool to determine the bias of ghost-free data, it is unable to establish a self-consistent bias estimate for ghost-infested data.

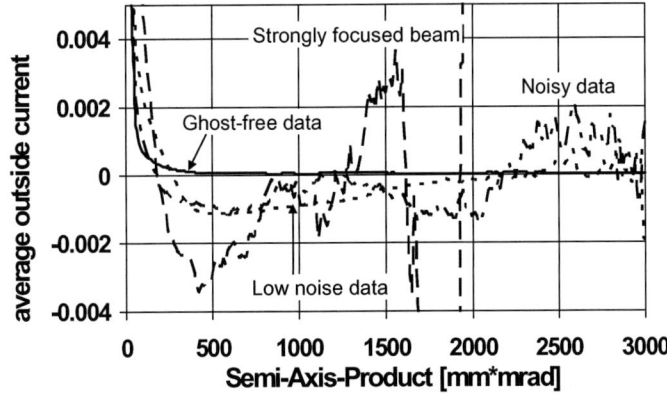

FIGURE 10: Elliptical exclusion estimates for the bias of all data sets

The Self-Consistent, UnBiased, Elliptical Exclusion method, SCUBEEx, was developed to obtain reliable and accurate rms emittance estimates [10]. This is accomplished by first calculating the average current outside the ellipse, and then subtracting this bias estimate from all data before calculating the rms emittance from the bias-corrected data within the ellipse.

The rms emittance estimates thus obtained, using the same ellipses as in Fig. 10, are shown in Fig. 11. Small ellipses exclude real data and therefore underestimate the rms emittance. When the solid line of the ghost-free data reaches ~800 mm·mrad, the result no longer changes significantly with increases in ellipse size. This self-consistently estimates the normalized rms emittance as 0.113 ±0.002 mm·mrad.

SCUBEEx, however, cannot self-consistently estimate the rms emittance of ghost-infested data. Small ellipses underestimate the rms emittance because the ghost signals reduce the signals in the tails. Large ellipses underestimate the rms emittance because the inverted contributions from the ghost signals with large x' start to dominate. The two effects discussed above may cause intermediate ellipses to underestimate the rms emittance, while at the same time the bias underestimation may cause an overestimation of the rms emittance.

The ghost-infested data show these trends in Fig. 11. The ghost infested lines reach apices around 500 mm·mrad before they fall off as a result of inverted ghost signals. These trends are very clear for the low-noise data of Fig. 5 represented by the dotted line. For the other two ghost-infested data sets, noise-generated fluctuations skew the estimate for ellipses above 800 mm·mrad.

The rms emittance of these four data sets is expected to vary greatly because the measurements were made under different conditions. The fact that the plateau of the ghost-free data has a value similar to that of the apices of the ghost-infested data is believed to be coincidental and should not be used to estimate the error introduced by the ghost signals.

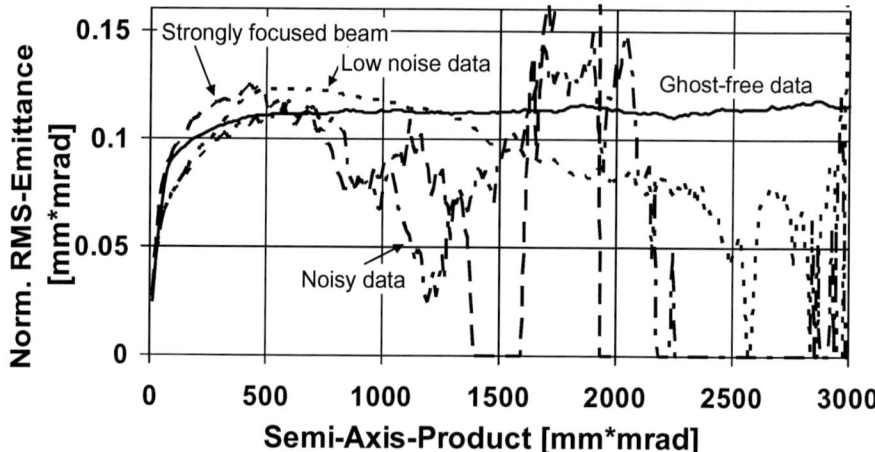

Figure 11. SCUBEEx rms emittance estimates

A self-consistent analysis of ghost-infested data is possible as long as the ghost signals do not overlap with the real signals. The bias needs to be self-consistently estimated from ghost-free background data (e.g. scanning positions where no beam enters the scanner). This estimate needs to be subtracted from all data. Next, a boundary needs to be found that includes all real signals, but excludes all ghost signals and most of the background [10]. Varying this boundary can again be used to establish self-consistency. We are planning to include such options in our emittance analysis code. The code features an automated SCUBEEx routine and supports all common data-treatment options. The code is available from our website [11].

Data infested with ghost signals overlapping the real signals require model-based corrections. The corrections are often of limited accuracy and thus are unable to completely eliminate the errors.

The opposite polarity of the ghost signals and the real signals may raise hopes that the signals could be separated with a threshold. This, however, is impossible because the two signal types overlap due to the noise convoluting all signals. Despite, thresholds are commonly applied because it is such a simple data-treatment option. It is therefore interesting to analyze the effect of thresholds on ghost-infested and ghost-free data.

Figure 12 shows the rms emittance as a function of the thresholds. Negative thresholds are useful in judging the noise level. However, rms emittances obtained with negative thresholds are very unreliable because they include all or almost all data, which causes the hypersensitivity to bias, as previously discussed [10].

The solid line in the graph on the left represents the ghost-free data of Fig. 9, while the dot-dot-dashed line was calculated for a Gaussian distribution with the same rms emittance. The comparison shows that even a small threshold of 1% causes an 11% error, much larger than the 1% error predicted for a Gaussian distribution.

The lack of self-consistent rms emittance estimates prohibits assessments of absolute errors for any of the ghost-infested data. However, the steep slope of the dotted line representing the ghost-infested low-noise data of Fig. 5 suggests errors that increase very rapidly with the threshold. This is probably caused by the ghost signals' reducing the small, real signals in the emittance tails. In addition, the line shows the noise-convoluted distribution of ghost signals stretching to the negative threshold of −0.7%.

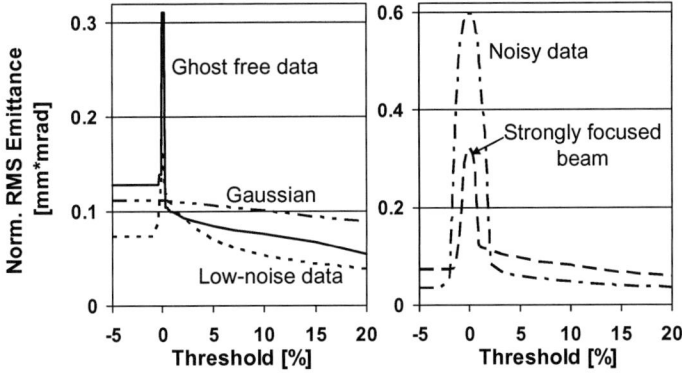

Figure 12. Rms emittance estimates using thresholds for low noise (left) and noisy data (right)

The right-hand graph in Fig. 12 plots the same information for the ghost-infested data from Fig. 7. Their elevated noise levels can be judged from the width of the noise peaks around 0. Again, the rapid change of the rms emittance values with increases in threshold show that the rms emittance is likely to be significantly underestimated or overestimated, depending on the selected threshold.

ACKNOWLEDGMENTS

We thank J. A. Mashburn for characterizing the sharpness of the modified staircased deflection plates. We value the skillful assembly and installation by S. Murray. We are grateful for the discussions and suggestions by many colleagues, especially S. Henderson and A. Aleksandrov. And last but not least we appreciate the support and proofreading by P. Kite, D. Shubert, and C. Moser. SNS is managed by UT-Battelle, LLC, under contract DE-AC05-00OR22725 for the U.S. Department of Energy.

REFERENCES

1. M. P. Stockli, M. Leitner, D. P. Moehs, R. Keller, and R. F. Welton, in *Proceedings of the 16th International Workshop on ECR Ion Sources*, M. Leitner, edt., AIP Conference Proceedings **749**, Melville, New York, 2005.
2. E.g., J. H. Billen, *Rev. Sci. Instrum.* **46**, 33-40, 1295 (1975).
3. P. W. Allison, J. D. Sherman, and D. B. Holtkamp, *IEEE Trans. Nucl. Sci.* **NS-30**, 2204-2206 (1983).
4. E.g., J. Rathke, M. Peacock, and J. Sredniawsi, in *LINAC 96*, 447-449; C. Michaut, J. Bucalossi, and D. Riz, in *Production and Neutralization of Negative Ions and Beams,* C. Jacquot, edt., AIP Conference Proceedings **439**, Melville, New York, 1998, 81-92; M. Dombsky et al., *Rev. Sci. Instrum.* **69**, 1170-1172 (1998); D. Wutte, M. A. Leitner, and C. M. Lyneis, *Physica Scripta* **T92**, 247-249 (2001); Y. J. Kim et al, *Rev. Sci. Instrum.* **75**, 1681-1683 (2004).
5. e.g. M. P. Stockli and S. Winecki, in *"Proceedings of the 12th. Int. Workshop on ECR Ion Sources"*, M. Sekiguchi and T. Nakagawa, edt., RIKEN, Japan, April 25-27, 1995, (Institute for Nuclear Study, Univ. of Tokyo, Tanashi, Tokyo 188, Japan, Report INS-J-182, September 1995), 90-94.
6. H. Wollnik and H. Ewald, *Nucl. Instr. Meth* **36**, 93-104 (1965) and its ref. 2.
7. J. Burgdörfer, in *Review of Fundamental Processes and Applications of Atoms and Ions*, C. D. Lin, edt., World Scientific, Singapore, 2002, 517-614.
8. R. Keller et al, *Rev. Sci. Instrum.* **73**, 914-916 (2002); R. F. Welton, M. P. Stockli, S. N. Murray, and R. Keller, *Rev. Sci. Instrum.* **75**, 1793-1795 (2004).
9. Yu. Belchenko, Budker Institute of Nuclear Physics, 63090 Novosibirsk, Russia, private communication, 2004.
10. M. P. Stockli, R. F. Welton, and R. Keller, *Rev. Sci. Instrum.* **75**, 1646-1649 (2004); M. P. Stockli, R. F. Welton, and R. Keller, in *PAC 03*, 527-529; M. P. Stockli et al. in *Production and Neutralization of Negative Ions and Beams*, M. P. Stockli, edt., AIP Conference Proceedings **639**, Melville, New York, 2002 pp. 135-159.
11. R. Welton et al. in *Production and Neutralization of Negative Ions and Beams,* M. P. Stockli, edt., AIP Conference Proceedings **639**, Melville, New York, 2002, pp. 160-174; https://www.sns.gov/APGroup/Codes/EAS/eas.htm

Extraction Probability of Negative Ions from Hydrogen Ion Sources - Effects of Filter Magnetic Field and Gas Pressure

Osamu Fukumasa and Ryo Nishida

Department of Electrical and Electronic Engineering, Faculty of Engineering,
Yamaguchi University, Tokiwadai 2-16-1, Ube 755-8611, Japan

Abstract. Trajectories of H⁻ ions are calculated numerically by solving the 3D motion equation, including effects of collisional destruction, elastic collisions and charge exchange collisions. According to these trajectories, extraction probability of H⁻ ions produced at any location inside the source and energy of extracted H⁻ ions are discussed as a function of gas pressure. Effects of production zone and filter magnetic field on extraction probability are also discussed. The probability for surface produced H⁻ ions keeps nearly the constant value, and that for volume produced H⁻ ions decreases with gas pressure. The kinetic energy of extracted H⁻ ions is reduced mainly by charge exchange collision.

INTRODUCTION

Negative ion based neutral beam injection is one of the most promising candidates for heating and current drive of fusion plasma. By seeding a small amount of cesium (Cs) vapor into the volume ion source, H⁻ production has been increased by a factor of 2-4 and optimum pressure decreases to 0.8-1.0 Pa [1]. Although Cs effects have been observed by many researchers, the mechanism remains to be discussed. We have studied source modeling [2-6] and Cs effects on enhancement of H⁻ production in a tandem two-chamber system, i.e. the source and the extraction regions. According to our numerical results, it is confirmed that the dominant process for enhancement of H⁻ production is surface production [5, 6].

For discussing the pressure dependence of extracted H⁻ current, we have also estimated the extracted H⁻ ions, by taking into account the stripping loss in the acceleration grid region only [4, 5]. But some H⁻ ions produced in the source aren't extracted because of collisional destructions. So, it is important to study the behavior of H⁻ ions in the second chamber, i.e. in the extraction region [7]. In addition, it has been reported that the beam divergence of surface produced H⁻ ions are nearly the same one as that of volume produced H⁻ ions [8]. However, the physical reason has not yet been clarified.

In this paper we will discuss the extraction probability of H⁻ ions by using both, model calculation [5] and H⁻ ion transport in the second chamber [7]. The preliminary results have been presented earlier [9], herewith H⁻ ion transport is further studied including effects of production zone and filter magnetic field. To clarify good beam

CP763, *Production and Neutralization of Negative Ions and Beams*, edited by J. D. Sherman and Y. I. Belchenko
© 2005 American Institute of Physics 0-7354-0248-5/05/$22.50

optics of surface produced H⁻ ions, we will also study both mean kinetic energy and the velocity distribution of extracted H⁻ ions.

SIMULATION MODEL AND PROCEDURE

To simulate H⁻ production in a tandem two chamber system, we have used the zero-dimensional code with source model, shown in Fig. 1 [3-5]. In the present study, with a coordinate system shown in Fig.1 and 2, negative ion trajectory in the second chamber is calculated numerically with width $L = 30$ cm. Magnetic filter is set at 2 cm

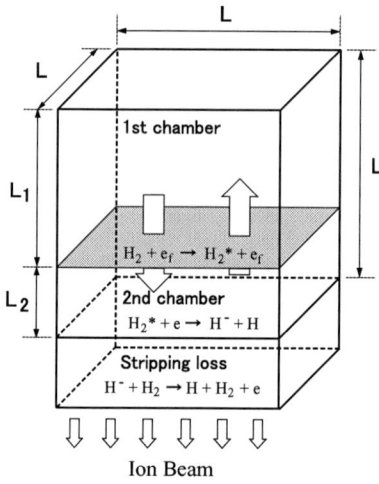

FIGURE 1. Simulation model for the tandem two-chamber system.

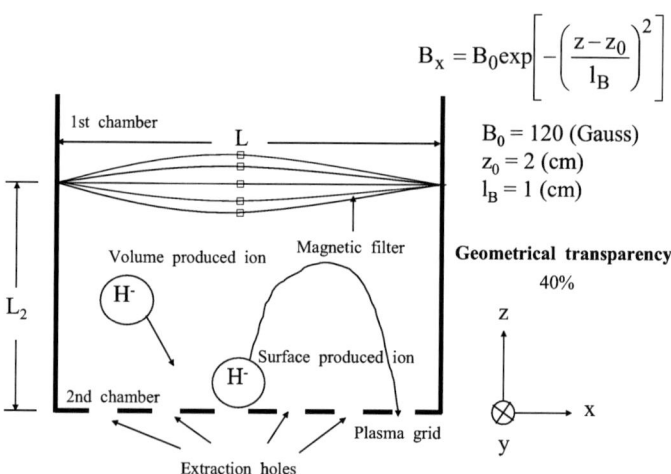

$$B_X = B_0 \exp\left[-\left(\frac{z - z_0}{l_B} \right)^2 \right]$$

$B_0 = 120$ (Gauss)
$z_0 = 2$ (cm)
$l_B = 1$ (cm)

Geometrical transparency
40%

FIGURE 2. Model geometry of second chamber used for tandem system shown in Fig. 1.

160

(= L_2) upstream from a plasma grid (PG). The spatial profile of magnetic filter is given by the Gaussian profile $B_x(y, z) = B_0 \exp[- (z-z_0)^2 / l_B^2]$, where $z_0 = 2$ cm, $l_B = 1$ cm and $B_0 = 120$ Gauss. Surface confinement magnet field is also present. Sixteen columns of permanent magnets are arranged to construct line cusp field.

When a negative ion is produced, it moves inside the source until destruction or extraction. Trajectories of H⁻ ions are calculated numerically by solving the 3D motion equation as follows:

$$M \, dV/dt = q(V \times B) + F_{col}, \qquad (1)$$

where M is mass of the H⁻ ion, q is charge, V is the velocity vector and B is the vector of magnetic field. The electric field is neglected in the above equation because it is negligibly small in the plasma region as compared with the electric field in the sheaths near the plasma grid and chamber walls. The second term on the right-hand side F_{col} is the collisional term, which is explained below. When x is the vector of the position, the definition of velocity vector can be described as

$$dx/dt = V \qquad (2)$$

We solved equations (A) and (B) in three dimensions using the Runge-Kutta-Gill method as the initial value problem. The collisions between H⁻ ions and other particles are calculated by the Monte Carlo method [7, 10]. The following destruction, charge exchange and elastic collisions are taken into account:

$H^- + e \rightarrow H + 2e$	electronic detachment (ED)	(3)
$H^- + H^+ \rightarrow 2H$	mutual neutralization (MN)	(4)
$H^- + H_2^+ \rightarrow H + H_2$		(5)
$H^- + H_3^+ \rightarrow 2H + H_2$		(6)
$H^- + H \rightarrow H_2 + e$	associative detachment (AD)	(7)
$H^- + H_2 \rightarrow H + H_2 + e$		(8)
$H^- + Cs^+ \rightarrow H + Cs$		(9)
$H^- + Cs \rightarrow H + Cs + e$		(10)
$H^- + H \rightarrow H + H^-$	charge exchange (CX) [11]	(11)
$H^- + H^+ \rightarrow H^- + H^+$	elastic collision (EC) with H⁺ ions.	(12)

Volume produced H⁻ ions were launched isotropically in all directions with an initial energy of 0.5 eV at any x,y location, except the axial points (z direction), where four launching points (i.e. z = 0.25, 0.75, 1.25 and 1.75 cm) were used. The surface produced H⁻ ions are launched from the PG an with initial energy of 0.5, 1 and 2 eV due to potential difference between plasma and plasma grid. When H⁻ ions have reached the PG or destroyed by collisional processes, the calculation is finished.

The background plasma profiles are assumed to be uniform, and these values are obtained by the previous model calculation [4, 5] and are used to estimate mean free paths for collisions mentioned above. To determine the electron density dependence of H⁻ production and particle densities, calculation is performed as a function of electron density $n_e(1)$ in the first chamber on the assumption, that other plasma parameters are kept constant [3-5]. A typical numerical result is summarized in Table 1. Plasma

161

n_e	Electron density	1.00×10^{12} cm^{-3}
n_H	H atom density	5.22×10^{13} cm^{-3}
n_{H2}	H$_2$ atom density	8.31×10^{13} cm^{-3}
n_{H^+}	H$^\square$ ion density	3.73×10^{11} cm^{-3}
n_{H2^+}	H$_2^+$ ion density	2.71×10^{11} cm^{-3}
n_{H3^+}	H$_3^+$ ion density	1.52×10^{11} cm^{-3}
n_{Cs^+}	Cs$^+$ ion density	5.85×10^{11} cm^{-3}
n_{Cs}	Cs atom density	4.41×10^{12} cm^{-3}
T_e	Electron temperature	1.0 eV
T_H	H atom temperature	0.5 eV
T_H^+	H$^+$ ion temperature	0.5 eV

TABLE 1. Plasma parameters used in this simulation when gas pressure $p = 5$ mTorr.

conditions for model calculation is as follows: the gas pressure $p = 5$ mTorr, the electron density ratio between two chambers $n_e(2)/n_e(1) = 0.2$, density of e_f in the first chamber $n_{fe}(1)/n_e(1) = 0.05$, electron temperature in the first and second chambers is, respectively, $\kappa T_e(1) = 5$ eV, $\kappa T_e(2) = 1$ eV, and magnetic filter position $L_1 : L_2 = 28 : 2$ cm (i.e. $z_0 = L_2 = 2$ cm).

NUMERICAL RESULTS AND DISCUSSION

The trajectories of H$^-$ ions are obtained by solving the 3D motion equation until ions are destroyed or extracted (i.e., reached to the PG). Typical orbits of H$^-$ ions in the second chamber of the negative ion source are shown in Fig. 3.

At first, characteristic features of H$^-$ ion trajectories (i.e. properties on H$^-$ ion extraction) are discussed. To this end, for a certain plasma conditions, a set of five calculations (one calculation for surface produced H$^-$ ions and four calculations for volume produced H$^-$ ions with different four z positions) are done. We used 10^3 test H$^-$ ions for one calculation. Table 2 shows the simulation result, where gas pressure is 5 mTorr. In the present case, 740 surface produced H$^-$ ions reached the PG and extraction probability is about 25.6 % (geometrical transparency of the PG is assumed to be 40 %). For volume produced H$^-$ ions, the probability to reach the PG depends strongly on upstream distance z from the PG. Then, mean value of the extraction probability is 4.2 %.

This probability depends on gas pressure. Extraction probability of volume produced H$^-$ ions decreases with gas pressure. These characteristic features are clearly shown in Fig. 4. Effect of magnetic filter field on H$^-$ trajectories is also discussed.

But, there is scarcely difference in extraction probability due to difference of filter field. Numerical result is shown in Fig. 5.

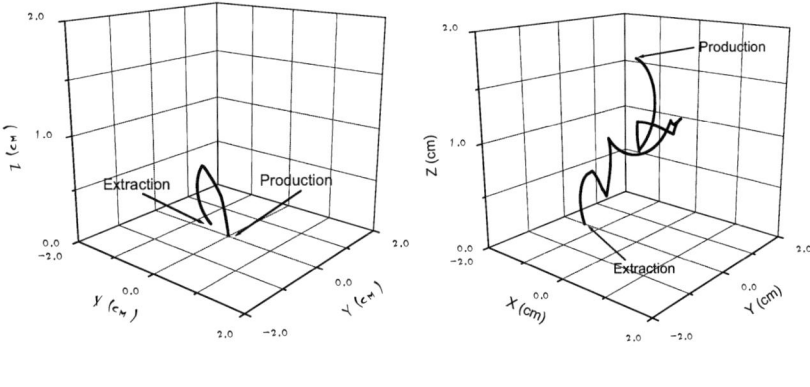

(a) surface produced H⁻ ions   (b) volume produced H⁻ ions

FIGURE 3. Examples of H- ion trajectories in the second chamber: (a) a surface produced H- ion (initial energy : 1 eV, birth point $(x, y, z) = (0, 0, 0)$), (b) a volume produced H- ion (initial energy : 0.5 eV, birth point $(x, y, z) = (0, 0, 1.75$ cm$))$.

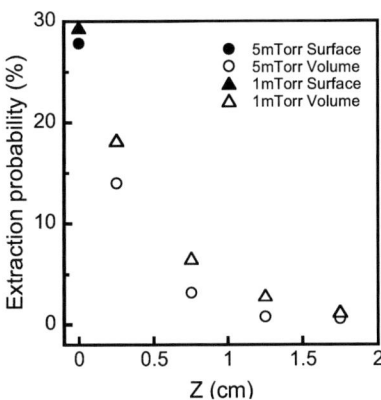

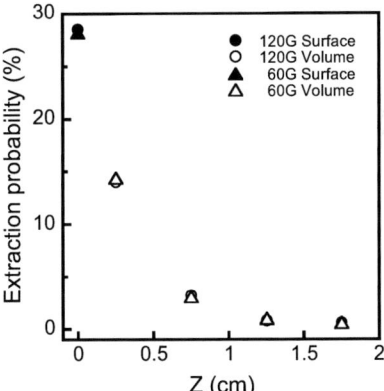

FIGURE 4. Extraction probability as a function of z. Parameter is hydrogen gas pressure, where $B_0 = 120$ G.

FIGURE 5. Extraction probability. Parameter is magnetic filter field, where gas pressure $p = 5$ mTorr.

163

Kinds of H⁻ ion loss		Surface produced H⁻ ions	Volume produced H⁻ ions			
			Birth point of volume production [cm]			
			0.25	0.75	1.25	1.75
Wall loss		27	40	60	59	59
Collsional destruction	e	13	34	54	129	273
	H^+	26	91	130	144	98
	H_2^+	34	48	91	78	72
	H_3^+	13	30	40	41	35
	H	52	133	167	159	132
	H_2	73	184	249	248	199
	Cs^+	53	65	95	88	92
	Cs	13	25	35	33	26
Total		277	610	861	920	927
Elastic collision	H^+	660	1384	2062	2041	1461
Charge exchange	H	1562	2852	4351	4527	3766
Extracted H⁻ ions		696	350	79	21	14
Average energy of extracted H⁻ ions [eV]		0.66	0.46	0.44	0.43	0.47
Extraction probability [%]		27.8	14.0	3.2	0.8	0.6

TABLE 2. Numerical results of H⁻ transport when $p = 5$ mTorr.

For simplicity, the modeling is made for a constant and equal plasma potential in the first and second chambers. With this choice of plasma potential, H⁻ ions are injected from the first chamber into the second chamber, but this effect is not considered in the present simulation and may modify a little the number of H⁻ ions reaching the PG. Namely, a little enhancement of extraction probability may be expected.

H⁻ ion transport (i.e. the extraction probability) depends on gas pressure. Discussing this point, the same calculations described above are done by changing gas pressure. In the present calculation, initial positions (i.e. birth points) of surface produced H⁻ ions are distributed at any location on the PG and those of volume produced H⁻ ions are also distributed at any location in the second chamber, i.e. three dimensional. Now, 10^3 test particles for surface produced H⁻ ions and 2×10^3 test particles for volume produced H⁻ ions are used, respectively. Numerical results are shown in Fig. 6. Extraction probability of volume produced H⁻ ions decrease with gas pressure nearly the same manner as that of surface produced H⁻ ions. It is remarkable, however, that extraction probability of surface produced H⁻ ions is much higher than that of volume produced H⁻ ions. Physical meaning is as follows: With increasing gas pressure, particle densities increase and mean free path of H⁻ ions decreases in its value. Therefore, transport of H⁻ ions in the extraction region decreases due to collisional effects. In particular, surface produced H⁻ ions injected into plasmas are reflected easily by elastic and charge exchange collisions and reach the PG. On the other hand, volume produced H⁻ ions are impended to reach the PG by collisional processes.

Kinetic energy (KE) of H⁻ ions are reduced by elastic (12) and charge exchange (11) collisions. According to the Table 2, for surface produced H⁻ ions with initial energy 1eV, KE of extracted H⁻ ions is reduced to 0.67 eV. On the other hand, for volume produced H⁻ ions with 0.5 eV, KE of extracted H⁻ ions is reduced to 0.445 eV, and lower than that of surface produced H⁻ ions due to difference in initial energy of H⁻ ions. Fig. 7 shows velocity distribution of extracted H⁻ ions. Although there is some difference between the velocity distribution of extracted H⁻ ions for surface produced H⁻ ions and that for volume produced H⁻ ions, this energy relaxation and velocity distribution are the cause for good beam optics of negative ion current with Cs seeding [8]. As is shown in Table 2, charge exchange collision is the most dominant collision process. With decreasing p, effects of elastic collisions become remarkable. Therefore, both elastic collision and charge exchange collision play important roles in energy relaxation of the extracted H⁻ ions.

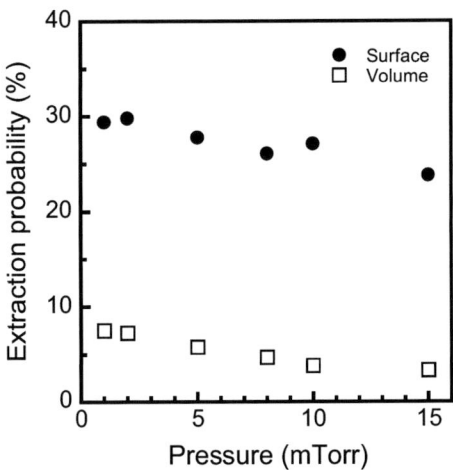

FIGURE 6. Pressure dependence of extraction probability: ● for surface produced H⁻ ions, □ for volume produced H⁻ ions with Cs.

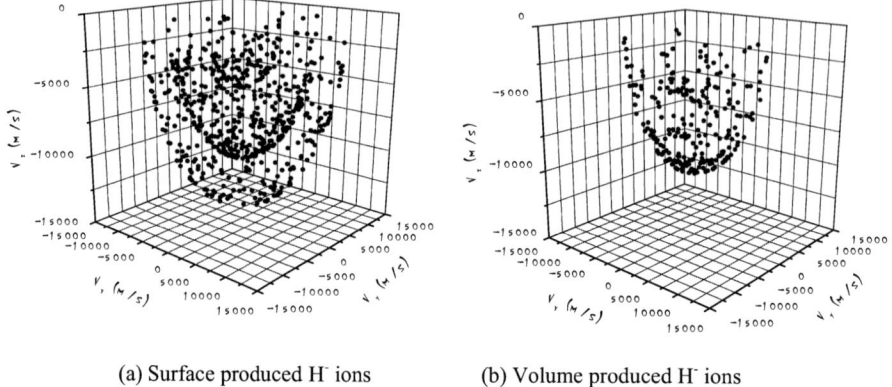

(a) Surface produced H⁻ ions   (b) Volume produced H⁻ ions

FIGURE 7. Velocity distribution of extracted H⁻ ions; (a) surface produced H⁻ ions, (b) volume produced H⁻ ions.

CONCLUSIONS

The probability for H⁻ ions to reach the plasma grid (i.e. extraction probability) is estimated. As a whole, extraction probability is relatively low. It is confirmed that extraction probability for surface produced H⁻ ions is much higher than that for volume produced H⁻ ions. Within the present numerical conditions, the extraction probability for surface produced H⁻ ions keeps nearly the constant value (i.e. 24-30%), and that for volume produced H⁻ ions decreases in its value from 8% to 3% with increasing gas pressure. The kinetic energy of the extracted H⁻ ions is reduced by both charge exchange collisions with H and elastic collisions with H⁺. There is a certain energy difference in extracted H⁻ ions between volume produced H⁻ ions and surface produced H⁻ ions. In the future, we will discuss the characteristics of extracted negative ion current with the use of the present numerical results and the results of our previous model calculation.

ACKNOWLEDGEMENTS

The authors would like to thank Prof. H. Naitou and S. Mori for their discussion and support in preparation of the present paper. A part of this work was supported by the Grant-in-Aid for Scientific Research from the Ministry of Education, Culture, Sports, Science and Technology, Japan. This work was also carried out as the collaboration research program (the LHD project) of National Institute for Fusion Science.

REFERENCES

1. Okumura, Y. et al., *Rev. Sci. Instrum.* **63**, 2708-2710 (1992).
2. Fukumasa, O., *J. Phys.* **D22**, 1668-1679 (1989).

3. Fukumasa, O., *J. Appl. Phys.* **71**, 3193-3196 (1992).
4. Fukumasa, O. and Monji, H., *Rev. Sci. Instrum.* **71**, 1234-1236 (2000).
5. Fukumasa, O., *IEEE T. Plasma Sci.* **28**, 1009-1015 (2000).
6. Fujioka, T., Fukuchi, T. and Fukumasa, O., *Proceedings of the 25th International Conference on Phenomena in Ionized Gases* **2** (2001), p.245.
7. Riz, D. and Pamela, J., *Rev. Sci. Instrum.* **69**, 914-919 (1998).
8. Miyamoto, K. et al., *Proceedings of the 18th Symposium on Fusion Technology* **1** (1995), p.625.
9. Fukumasa, O., Fukuchi, T. and Fujioka, T., *Proceedings of the 9th International Symposium on the Production and Neutralaization of Negative Ions and Beams* (2002), pp.75-81.
10. Ido, S., Hasebe, H. and Fujita, Y., *Jpn. J. Appl. Phys.* **32**, 4761-4767 (1993).
11. Hummer, D. G., Stebbings, R. F., Fite, W. L. and Branscomb, L. M., *Phys. Rev.* **119**, 668-670 (1960).
12. Makino, K., Sakurabayashi, T., Hatayama, A., Miyamoto, K. and Ogasawara, M., *Rev. Sci. Instrum.* **73**, 1051-1053 (2002).

Acceleration of 100A/m² Negative Hydrogen Ion Beams in a 1 MeV Vacuum Insulated Beam Source

M. Taniguchi, T. Inoue, M. Kashiwagi , M. Hanada , K. Watanabe, K. Seki, and K. Sakamoto

Japan Atomic Energy Research Institute, 801-1 Mukoyama, Naka-machi, Naka-gun, Ibaraki-ken, 311-0193 Japan

Abstract. In the ITER NB, conventional gas insulated beam source (GIBS) cannot be utilized because of the radiation-induced conductivity of the insulation gas. Thus a vacuum insulated beam source (VIBS), where the whole beam source is immersed in vacuum, has been developed at JAERI. Recently, voltage holding capability of the VIBS was drastically improved by the large stress ring, which reduces the electric field concentration at the triple junction. Up to now, a high current density H⁻ beam of 102 A/m² (140 mA) at 800 keV has been accelerated. The beam acceleration was quite stable and accomplished for several hundreds shots. The degradation of the voltage holding due to the beam acceleration and/or Cs seeding was not observed. Thus the development of vacuum insulated beam source has solved technical issues of high voltage insulation of 1 MV level under the presence of the H⁻ ion beams.

INTRODUCTION

The ITER Neutral Beam (NB) system requires a high energy and high current beam source, which can provide a 16.5 MW neutral beam injection per module. For this purpose, a negative ion beam source for the ITER NB was designed to produce 40 A D⁻ ion beams at 1 MeV. However, charged particles of ampere order have so far never been accelerated up to MeV range. For the demonstration of the ampere class negative ion beam acceleration up to 1 MeV, a five-stage electrostatic accelerator has been developed in the MeV test facility whose power supply capability is 1 MV, 1 A, 60 sec[1].

One of the key technologies to realize the high current beam acceleration to the MeV class energy is DC ultra high voltage insulation. In the original design of the ITER NB system, the beam source is surrounded by a pressurized SF₆ insulation gas. However, the gas molecule of SF₆ is easily ionized by the radiation from the tokamak plasma, which causes the radiation-induced conductivity (RIC)[2,3]. It was clarified that the RIC causes a power loss of MW order [4], and hence, the gas insulated beam source (GIBS) cannot be utilized in the ITER NB. To avoid such problems related to RIC, a vacuum insulated beam source (VIBS), where all the components of the beam source are immersed in vacuum, is being developed at JAERI.

CP763, *Production and Neutralization of Negative Ions and Beams*, edited by J. D. Sherman and Y. I. Belchenko
© 2005 American Institute of Physics 0-7354-0248-5/05/$22.50

At the beginning of the R&D of VIBS, the H⁻ current was limited due to the breakdown at the fiber reinforced plastic (FRP) insulator stack. To improve the voltage holding capability, a large stress ring which protects the triple junction was designed and tested in the MeV test facility. During beam acceleration, X-rays are generated in the accelerator by bremsstrahlung. In addition, Cs is introduced into the beam source to enhance the H⁻ density. These X-ray and/or introduced Cs might cause the degradation of the voltage holding capability of the accelerator. To clarify these problems, high current density beam acceleration up to 100 A/m^2 was performed. This paper reports the results of these tests.

VACUUM INSULATED BEAM SOURCE (VIBS)

Figure 1 shows a photograph and a cross-sectional view of the VIBS developed at JAERI. The VIBS consists of a negative ion source, 5-stage electrostatic accelerator and FRP insulator columns. Overall dimensions of the VIBS are 1.8 m in diameter and 1.9 m in height. The negative hydrogen ions are produced in a cesium seeded plasma generator, called "KAMABOKO" source, which is mounted on the top of the accelerator. The negative ions are extracted from the plasma generator by applying an extraction voltage of 4 – 9 kV between a plasma grid and an extraction grid.

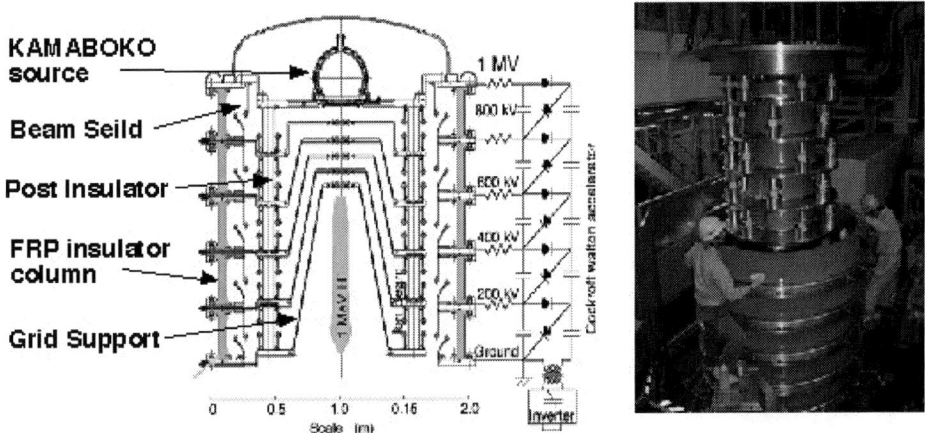

FIGURE 1. The schematics and the photograph of VIBS developed in JAERI.

The electrostatic accelerator consists of five acceleration stages, which are supported and insulated by post insulators made of Al_2O_3 ceramic. For the acceleration of the H⁻, 200 kV is applied between each grid and the H⁻ is accelerated up to 1MeV. High voltage of 1 MV is supplied by a Cockcroft – Walton type power supply.

As shown in Fig.1, the electrostatic accelerator is installed inside the cylindrical FRP stacks which act as an insulator column and a vacuum vessel. The accelerator is completely separated from the FRP insulator columns, and hence, there is a vacuum gap of 50 mm around the acceleration grids. This allows direct line of sight from the -

1MV potential to the ground through the long vacuum gaps of 0.5m - 1.8m. The pressure in the gap ranges in 0.02 Pa - 0.1 Pa during the operation of the ITER NB. Previous work has investigated the discharge characteristics under the above conditions. It was confirmed that glow discharge could not be generated in the operating pressure region [5], and the voltage holding test with a vacuum gap of 1.8m was experimentally studied [6]. The design of the VIBS was performed based on the results of this work.

The VIBS has also some another advantages as an accelerator for the ITER NB. A large ceramic insulator can be eliminated for the beam source. Having no surrounding structure as in the GIBS, the VIBS allows rapid pumping of residual gas molecules in the accelerator by increased conductance through the accelerator's grid supports. Hence, it is expected that the VIBS gives lower stripping losses of the negative ions in the accelerator than that in the original GIBS. A result of 3-dimensional Monte Carlo gas analysis shows that the stripping loss of the ions is 25% in the ITER VIBS at nominal operating pressure of 0.3Pa [7]. This stripping loss is about a half of the GIBS.

IMPROVEMENT OF 1 MV HOLDING CAPABILITY BY LARGE STRESS RING

By using the VIBS, we have succeeded in accelerating negative ion beams up to 971 keV in July 2001 [6]. However, the negative ion beam was not produced stably, and the current was limited to the low level of 5 A/m^2. Conditioning time to produce high current beam was too long, resulting in poor voltage holding capability of the insulator columns. Breakdown was always accompanied with a large amount of outgassing. From residual gas analysis, it was observed that the contaminant gas consisted of hydrocarbons [8]. In addition, careful observation of the FRP insulator columns revealed that traces due to the melting of epoxy runs along the inner surface of FRP. Figure 2 shows the photograph of FRP inner surface after the voltage holding test.

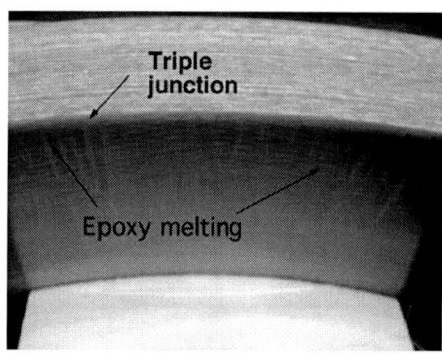

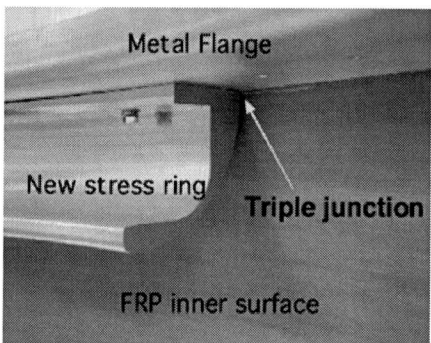

FIGURE 2. Damage of the FRP inner surface after the voltage holding test.

FIGURE 3. The photograph of the new large stress ring.

Melting of the epoxy is clearly seen in this photograph, and the meltings always start at the triple junction (interface of metal flange, FRP, and vacuum).

FRP is a dielectric material; therefore, electric field tends to concentrate at the triple junction during the high voltage operation. This high electric field might cause continuous micro discharge at the triple junction. The heat load on the surrounding FRP causes the melting of the epoxy contained in FRP and the release of hydrocarbons. High voltage breakdown is thought to occur during the outgas of hydrocarbons. Therefore, we think that the decrease of the electric field at the triple junction is essential to improve the voltage holding capability of the VIBS.

In order to suppress the concentration of electric field at the triple junction, a large stress ring was designed and installed as shown in Fig.3. By enlarging the size and curvature of the ring, electric field strength at the triple junction was lowered to 1.2 kV/mm from the original 3.6 kV/mm. A result of the voltage holding test with and without the stress ring in one stage of the FRP (rated voltage: 200 kV) is shown in Fig.4. With the new stress ring, flashover voltage reached the rated voltage of 200 kV within the first 2 hours, and a voltage of more than 300 kV was stably sustained after 8 hour conditioning. The conditioning time to sustain the rated voltage of 200 kV was drastically reduced by the new stress ring. As shown in Fig.4, the accelerator without the new ring could not reach 200 kV even after 8 hour conditioning.

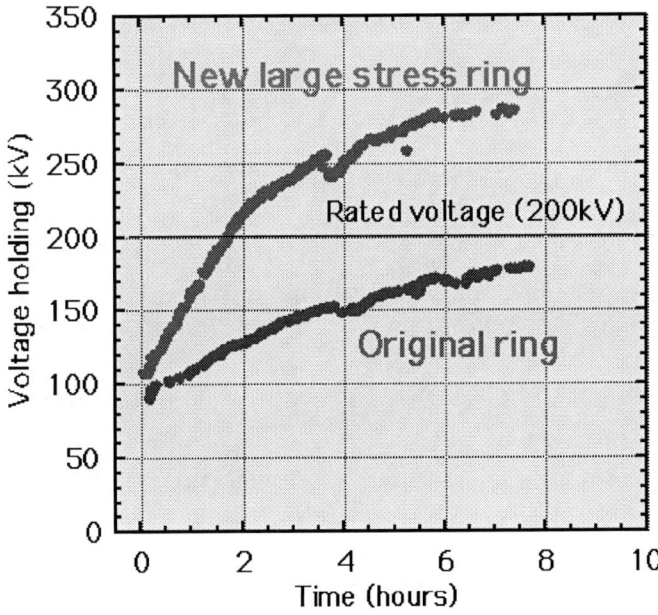

FIGURE 4. Result of voltage holding test with and without the new stress ring.

The improvement of the voltage holding was confirmed by lowering the electric field strength at the triple junction. The newly developed large stress rings were installed to all the five stages of FRP insulator columns. Figure 5 shows the

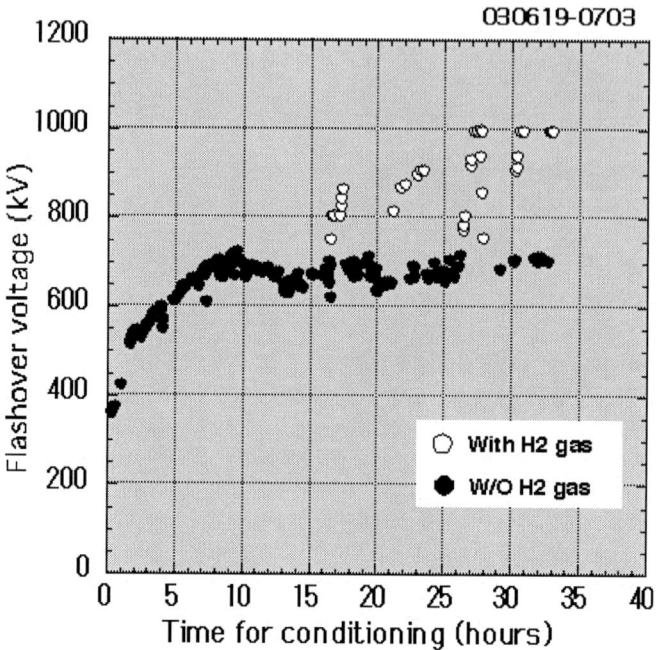

FIGURE 5. Conditioning history of VIBS with new large stress ring.

conditioning history of the VIBS equipped with the new large stress ring. Flashover voltage increases with increasing the conditioning time, however, it was saturated at 700 kV after the 10 hours of conditioning. Although each FRP insulator was confirmed to sustain the rated voltage of 200 kV, the VIBS of five stages sustained only 700 kV even after the 30 hours of conditioning. However, the drastic improvement of the flashover voltage was observed when H_2 gas was introduced into the FRP insulator stack to $0.5 - 1.0 \times 10^{-1}$ Pa (base pressure; 2.0×10^{-4} Pa), and the voltage holding reached 1 MV as shown in Fig.5. The H_2 gas is considered to prevent the creeping discharge along the FRP insulator surface. It was found that the VIBS could sustain 1 MV under the presence of H_2 in the range of $0.07 - 0.2$ Pa [9,10]. This pressure range corresponds to that of the accelerator during the operation, and it was confirmed that the VIBS can sustain 1 MV stably for more than 2 hours under the H_2 gas feed of this range. Conditioning without feeding H_2 gas is effective to attain the MV class high voltage holding with H_2 gas. We performed the conditioning of the VIBS for more than 10 test campaigns, and at every time, 30 hours conditioning before the H_2 feeding is needed to attain 1 MV. It was considered that gases absorbed in the FRP need to be removed to suppress discharges, and hence, it takes long time for conditioning.

VOLTAGE HOLDING DURING BEAM ACCELERATION

By installing the large stress ring, we have succeeded in sustaining stable 1MV operation. However, during beam acceleration, there are several issues to be clarified. 1) During the acceleration of the H⁻, secondary electrons will be produced due to the interception of the ions on the grid. These secondary electrons, as well as the stripped electrons, are accelerated with the H⁻ and produce X-rays by bremsstrahlung. These X-rays generate photoelectrons on the surface of the FRP, which might cause the breakdowns. 2) During the operation, Cs vapor is seeded into the source chamber to enhance the H⁻ production. The Cs has a low work function and a relatively high vapor pressure at low temperature, therefore, they might cause breakdowns if they leaked to the accelerator.

To clarify the above problems, negative ion acceleration test under the Cs seeding has been performed in VIBS. Figure 6 shows a history of beam acceleration at 900 keV under the Cs seeding. In this figure, beam current (closed circle) and the acceleration voltage (solid line) are shown as a function of the time. After the

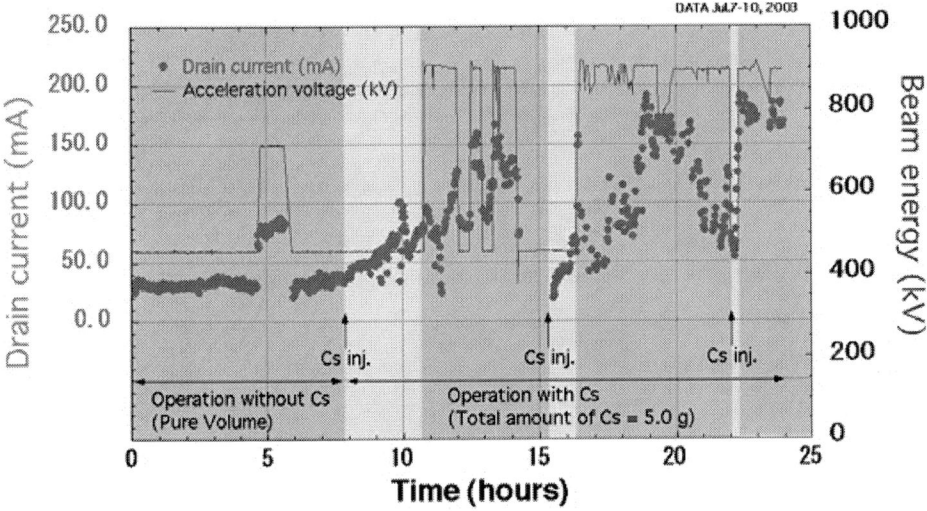

FIGURE 6. History of the beam acceleration test under the Cs seeding at 900 keV.

conditioning shots for several hours in pure volume operation, Cs was introduced into the beam source by heating the Cs oven to 470 °K. To prevent Cs oven power supply damage from high-voltage transients during high voltage breakdown, Cs was introduced to the source at the relatively low acceleration voltage of 450 kV. As Cs accumulates in the source, the negative ion current begins to increase as shown in Fig.6. After that, the Cs oven was turned off and the acceleration voltage was increased to 900 kV. The total amount of Cs introduced during the test campaign was 5.0 g, but the degradation of voltage holding was not observed. Pulse length of the beam was 0.5 sec and the interval between the shots was 60 sec. The beam acceleration of 900 keV, 0.1 – 0.2 A level was quite stable and repeated successfully for several hundreds shots. The breakdown due to photoelectrons generated by bremsstrahlung was not observed during the beam acceleration. It was confirmed that VIBS can sustain 1 MV stably under the condition of high current beam acceleration and Cs seeding operation.

ACCELERATION OF 100 A/m² NEGATIVE ION BEAM AT 800 keV

After the successful voltage holding test with beam acceleration, high current density H beams of 100 A/m² class were accelerated with seeding cesium. To enhance the H density under the low operating pressure, some minor modifications were made on KAMABOKO source: the filter magnet was strengthened from 460 to 745 Gcm, and the number of filaments were increased for higher arc power operation up to 40 kW. Although the plasma grid and extraction grid of the source has 49 apertures (7 x 7 lattice pattern) to extract the ions, the apertures were masked at PG and the ions were extracted only from 9 apertures (3 x 3) to limit extraction current. The acceleration energy of the beam and the pulse length was limited to 800 keV and 0.5 sec, respectively; this is because the beam target (copper tube with external-fin cooled by the water of 10 m/sec) cannot handle such a high power beam. In case of 800 keV and 100A/m² beam with the divergence of 7 mrad, surface heat flux at the beam target was estimated to be 6.4 kW/cm², and this is close to the critical heat flux (CHF) for the copper cooling tube [11].

Figure 7 shows the H beam current and the acceleration drain current as a function of the input arc power at the energy of 800 keV. The source operating pressure was 0.2 Pa and the extraction voltage was adjusted so as to obtain the maximum beam current at each arc power. The H current was measured by the calorimeter made of

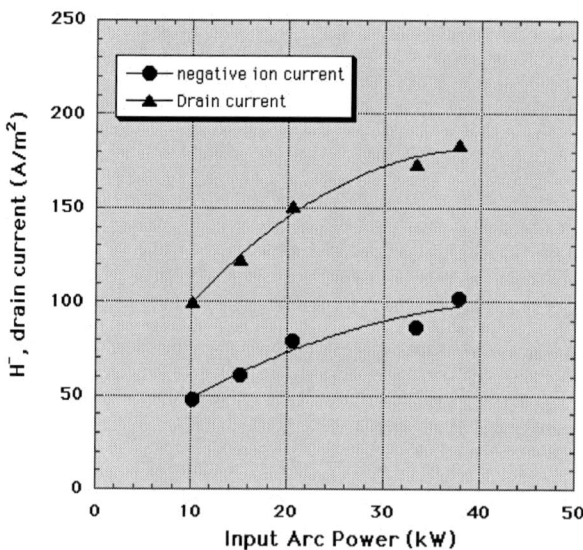

FIGURE 7. H beam current as a function of arc power.

copper, which was installed 2.0 m downstream from the beam source. The extracted H^- current increases with input arc power, and a H^- ion beam of 100 A/m^2 (total H^- current 140 mA) was stably accelerated (arc power 40 kW, extraction voltage 5.5 kV, source pressure 0.2 Pa). The current seems to saturate at higher arc power indicating that almost all ions near the extraction aperture are extracted (emission limit). We have already succeeded in extracting 300 A/m^2 H^- ion beams at KAMABOKO source in previous work (acceleration voltage 45 kV) [12], therefore, further increase of the current density is possible by tuning the cesium condition and arc discharge power input. For the higher power beam acceleration test, development of the new beam target using swirl tube technology is in progress.

CONCLUSIONS

In this paper, the present status of the vacuum insulated beam source for the ITER NB was presented. By using the newly developed large stress ring, voltage holding capability was drastically improved, and we have succeeded in sustaining 1 MV for more than 2 hours. No degradation of the voltage holding due to the beam acceleration and/or Cs seeding was observed and the H^- of 0.1- 0.2 A level at 900 keV was stably accelerated for several hundreds of shots. The maximum beam current density at present is 100 A/m^2 at 800keV (80 A/m^2 at 900 keV), which can be applicable to a practical NB system.

REFERENCES

1. T.Inoue et.al., JAERI-Tech 94-007.
2. Y. Fujiwara et al., JAERI-Research 99-071 (1999).
3. E. Hodgson et al., J. Nucl.Mater, 256-263, 1827-1830 (1998).
4. Y. Fujiwara et al., Fusion Eng. Design 55,1(2001).
5. T. Inoue et al., Rev. Sci. Instrum. 71(2), 744-746 (2000).
6. K.Watanabe et.al., Rev.Sci. Instrum.. 73(2), 1090-1092 (2002).
7. M.Hanada, et.al., Fusion Engineering and Design, 56-57, 505-509(2001).
8. T.Inoue et.al., Proc. 7th Production and Neutralization of Negative Ions and Beams, AIP Conference Proceedings 380, (1995) pp.397 – 405.
9. T.Inoue et.al., Fusion Engineering and Design 66-68, 597-602(2003).
10. M.Taniguchi et.al., Nuclear Fusion, 43, 665-669 (2004).
11. M.Araki et.al., Fusion Technol., 29, 519-528 (1996).
12. N.Miyamoto et.al. Proc.7th Production and Neutralization of Negative Ions and Beams, AIP Conference Proceedings 380, pp.300 – 306(1995).

Space Charge Lens for Focusing of Negative Ion Beams

I.A.Soloshenko, V.P.Goretsky, and A.M.Zavalov

Institute of Physics of NAS Ukraine, prosp. Nauki, 46, Kiev 03028, Ukraine

Abstract. A brief review of the results of experimental and theoretical studies of the space charge lens for focusing negative ion beam is presented. An idea of such lens was formulated earlier by the first two authors. For the present time, two versions of the lens are studied experimentally and calculated numerically. In the first version the focusing space charge is formed in the lens as a result of gas ionization by the negative ion beam itself; in the second version – as a result of gas ionization by both the beam ions and the electrons with energy of about 100 eV introduced from the special emitter placed in the lens volume. In both regimes focusing field values of about 100 V/cm are reached, which enable obtaining focal length of less than or equal to 20 cm. However, in the first case pressure working gas (argon, krypton, xenon) is about 10^{-3} Torr. A significant portion of negative ions is lost due to collisions with neutral particles at such a pressure. Introduction of additional ionizer enables significant lowering of the working gas pressure and avoiding losses of the negative ions. Calculations performed with the particle-in-cell method show good agreement with the experiment. The developed lens represents the simplest device - 3 electrodes with a voltage applied between them, which is one order of magnitude less than that at the ion source. Power consumed by the lens in the first regime is 2-3 orders of magnitude less than the beam power.

INTRODUCTION

In the injectors of neutral particles the beam of negative ions commonly passes a short distance to a charge-exchange target, where it is converted to a flow of neutral particles. In such cases attention is usually focused on forming a low-divergence beam at the exit of the ion extraction system. However, in a number of tasks of nuclear physics and technology it is necessary to use beams of small diameter, which should be transported at relatively longer distances. Lenses should be used for beam focusing in such cases.

The use of conventional "classic" lenses is not optimal for most cases. Electrostatic lenses are satisfactory only at the small beam current values, and a magnetic ones – only at small energies and masses of the ions. The only suitable "classic" device, which is sufficiently versatile and does not consume unreasonable amount of electric power is represented by magnetic quadrupole lens. There may be an alternative option for it represented by the space charge lens, which is less expensive and more versatile.

An idea of the space charge lens for focusing of the beam of negative ions was proposed by the two first authors in [1,2]. The space-charge lens principle is shown in Fig.1. Here the on-axis negative-ion beam potential dependence vs. neutral gas pressure is shown. The main peculiarity of the dependence is the existence of the point

CP763, *Production and Neutralization of Negative Ions and Beams*, edited by J. D. Sherman and Y. I. Belchenko
© 2005 American Institute of Physics 0-7354-0248-5/05/$22.50

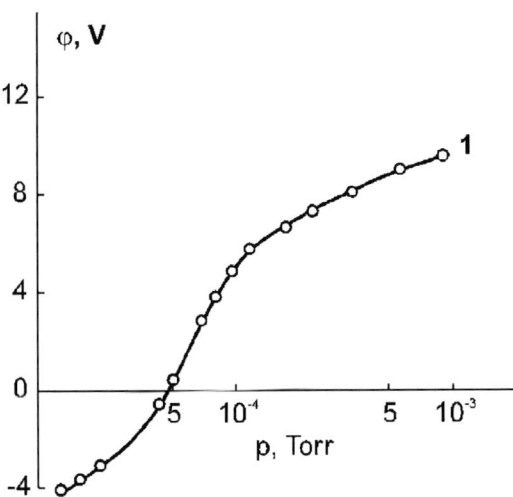

FIGURE 1. Dependence of potential vs air pressure. Beam current is 5 mA, beam energy is 15 keV. Note: potential in the beam center is considered with respect to grounded chamber walls.

with the zero beam potential. At this point space charge of the beam is completely compensated by positive ions formed as a result of the gas ionization by fast beam ions. Electron concentration formed by ionization is several orders of magnitude smaller due to high electron mobility. To the left of this point the beam space charge is undercompensated, and to the right it is overcompensated.

The last case is of interest for creation of the space-charge lens. In this regime the so called "gas" focusing of the beam is observed, however, this effect is small, because focusing fields are limited by the temperature of plasma electrons. As long as the pressure increases, density of positive ions n_i grows up, however, concentration of electrons n_e also grows up, so that the system is practically a quasi-neutral one. An idea for enhancing the lens focus power consists in removal of electrons from the beam by longitudinal electric field in regime of high pressure values, when $n_i \gg n_-$ (n_- is the concentration of negative beam ions). In the simplest case the lens is represented by three cylindrical electrodes placed sequentially along the axis. Outermost electrodes are grounded, and negative potential is applied to the middle one. As a result, electrons formed in the lens volume go onto the outermost electrodes along the axis. Positive ions formed as result of gas ionization go onto the middle electrode; however, due to their large inertia the regime is formed, at which $n_i \gg n_e$, and $n_i > n_-$. In the initial works [1,2] it has been demonstrated that by means of such device one could create focusing field in the beam of ~ 100 V/cm and to obtain the lens with focal length less than 20 cm. Drawback of such device consisted just in fact that even in mentioned heavy gases (argon, krypton, xenon) the focusing fields were created only at pressures of ~ 10^{-3} Torr. At such pressures a significant part of the beam might be lost due to electron stripping.

For excluding this negative effect it was proposed to increase the rate of gas ionization and thus to lower the working pressure in the lens. Background pressure reduction can be realized by introduction of 100 eV electrons into the lens volume, which could provide efficient gas ionization.

In the present work the results of experimental and numerical studies of the lens are given, both for the regime of gas ionization by the beam itself, and for the regime with the additional ionizer.

DESCRIPTION OF THE SETUP

The experimental setup is shown schematically in Fig. 2. The beam of hydrogen negative ions with energy of $\sim 10 \div 12$ keV and current of $10 \div 30$ mA was extracted through the slit with 0.5 mm × 15 mm dimensions from the surface-plasma source 1. Preliminary forming and bending of the beam was provided by 2 kG magnetic field produced by electromagnets 2. After passing through lens electrodes 3-4 the beam came to the collectors 7 and 8. Collector 7 with 10 cm diameter served for measurement of total beam current, and collector 8 with 2 cm diameter – for measurement of the beam compression degree. Distance from source emission slit to the first electrode of the lens was 20 cm. Distance from lens output plane to the collector was ~ 30 cm. The lens consisted of three electrodes: central one (3) composed of a cylinder with 15 cm diameter and 10 cm length. The outermost electrodes had 5 cm diameter, 1.5 cm length and were placed coaxially with central electrode with 0.5 cm gap. (In some experiments the central electrode was composed of a 5-cm diameter cylinder). Potential of the central electrode was varied in range from 0 V to –2000 V, and outermost electrodes were grounded. Gas was supplied either through the hole in central electrode, or through the emitter.

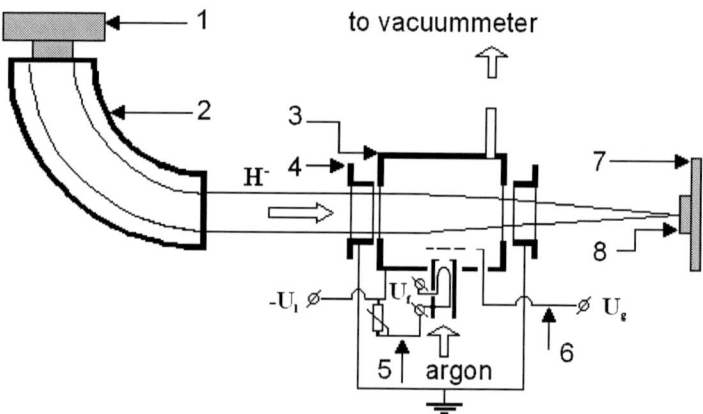

FIGURE 2. Schematic drawing of the experiment: 1 - pulsed source of hydrogen negative ions, 2 - beam bending magnet, 3 - lens cabinet, 4 - grounded entrance and exit electrodes of the lens, 5 - circuit of electron emitter filament, 6 - circuit of the grid for electron current control, 7 - collector of the beam current, 8 - collector of the beam current density. Gas is supplied through the electron emitter.

Gas pressure in the lens was varied in range of $1 \cdot 10^{-4} \div 1.5 \cdot 10^{-3}$ Torr, and pressure value in a region of the beam transportation was lower by one order of magnitude. Electron emitter was placed in a hole in the lens central electrode and consisted of two electrodes: 1) cathode, made of tungsten wire with 7 cm length, 0.3 mm diameter wound in a helix of 1 cm length, and 2) grid with 100 V bias with respect to the cathode.

Since working gas was supplied to the lens through the emitter of electrons, non-self-maintained discharge was ignited in a gap between the cathode and the grid. It enabled reduction of electron space charge limit, and to obtain electron current from the emitter into the lens volume of up to 200 mA.

EXPERIMENTAL RESULTS

Focusing effect of the space-charge lens is demonstrated in the most convincing way by Fig.3, which exhibits dependence of the beam compression coefficient (ratio of current value onto small collector with the lens on to that with the lens off) on the value of negative potential applied to central electrode of the lens. One can see that at low pressure, as it might be expected, applied negative potential does not result in the beam focusing. Only at pressure value higher than the certain critical one, it leads to the focusing effect, which increases with the pressure because of the increase of positive ions formation rate. The focusing effect dependence on lens negative potential

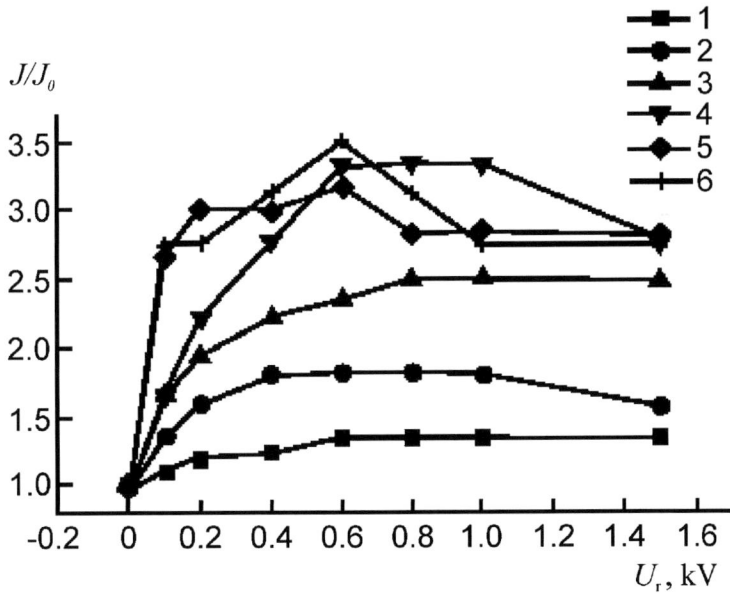

FIGURE 3. Dependence of hydrogen negative ion beam compression (J/J_0) on negative potential of retarding cylinder. Argon pressure in the lens: 1 - $3 \cdot 10^{-4}$; 2 - $7.6 \cdot 10^{-4}$; 3 - $1.5 \cdot 10^{-3}$; 4 – $2.2 \cdot 10^{-3}$; 5 – $3.6 \cdot 10^{-3}$; 6 - $6.4 \cdot 10^{-3}$ Torr. Diameter of lens central electrode 5 cm.

exists up to the certain point, where major portion of electrons is effectively removed from the lens volume onto the outermost electrodes, so that the space charge of electrons remaining in the lens is negligible. Particularly, for the lens with 5 cm diameter of central electrode the decrease of the potential value below $V \approx -300$ V does not cause important changes in focusing features of the lens.

Reported considerations were also confirmed by the dependencies of the beam compression coefficient on gas pressure (argon, krypton, xenon), presented in Fig.4. One can see that at increase of the atom mass (ion inertia), as it might be expected, optimum focusing is reached at lower pressure. Optimum compression degree does not depend on the gas type. The best focusing to ion collector is achieved at optimum pressure, so that further increase of focusing field will lead to overfocusing. Smaller value of compression degree (by factor of 4-6) is probably due to fact that the beam used in the experiments was not axi-symmetrical one. Negative influence might be also provided by the beam current oscillations due to plasma instability in the ion source.

One can easily estimate that at optimum current value at collector 8 (the beam is compressed to a maximum extent) lens focal length comprises ~ 20 cm. Focusing fields of ~ 100 V/cm are required for that purpose. Calculations show possibility of achieving such parameters if contribution of the electrons to space charge of the system is negligible. As one can see from the figures, optimum beam compression is reached at pressures of ~ 10^{-3} Torr. It is easy to demonstrate that at such pressure negative ion losses due to electron stripping already become important.

It was proposed to introduce an additional ionizer for lowering the working pressure value and for increasing the positive ion forming rate. Such ionizer was represented by electron flow with initial energy of ~ 100 eV. Results of the experiments with the additional ionizer are presented in Figs. 5a and 5b, where the lens focal distance normalized to the optimal value, achieved at high pressure, are

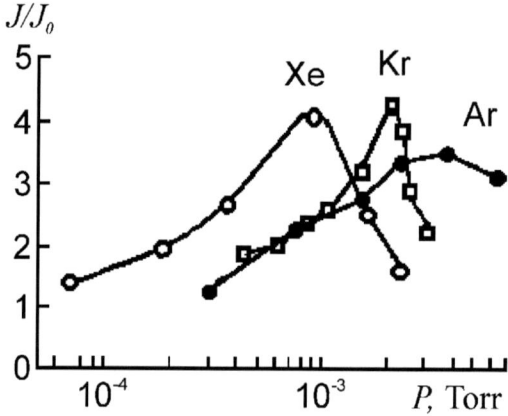

FIGURE 4. Dependencies of hydrogen negative ion beam compression (J/J_0) on gas pressure in the space charge lens for different gas (Ar, Xe, Kr). H⁻ beam current 15 mA; beam energy 10 keV. Diameter of central electrode of the lens 5 cm.

shown in dependence on electron current (5a) and gas pressure (5b). One can see that introduction of the electrons essentially increases the lens power, and that the electron effect increases with pressure decrease. Particularly, at emission current of 50 mA and at pressure P=2·10^{-4} Torr, the focal length is just twice longer than the optimum one. At such pressure ion losses in lens region are negligible. Thus, an efficient device is created for focusing of intense negative ion beams with the device cost being much smaller than that of currently existing lenses for ion beam focusing. Physical processes taking place in the lens will be considered at the discussion of results of numerical calculations.

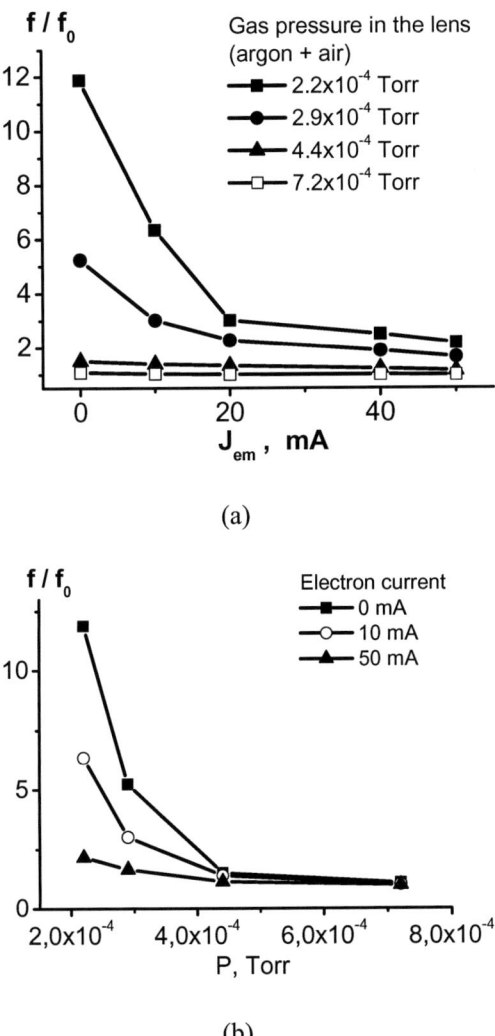

(a)

(b)

FIGURE 5. Dependencies of normalized value of lens focal length on electron current (a) and on gas pressure (b). The lens potential U_l = −1400 V. The H− ion beam current is 15 mA, energy is 12 keV.

NUMERICAL CALCULATIONS

Numerical calculations were performed by the "particle-in-cell" method. Geometry of the system used in the calculations is presented in Fig.6 and corresponds to the experimental one. A filament ring having 14 cm length was considered as electron emitter. The grid electrode with 2 cm width and 12 cm diameter was placed at 1 cm distance from the ring. Grid potential with respect to the ring was 0÷100 V; electron current was 0÷100 mA. The calculations were performed for potential difference values of -1500 V and –2000 V between central and outermost electrodes. Current of hydrogen negative ions was chosen to comprise 10 or 15 mA, and the beam energy – 12 or 15 keV. Production of positive ions in the lens volume was provided through Ar gas ionization by both the beam ions and electrons from the emitter. Electron production occurred due to both ionization processes, as well as electron stripping of the beam ions.

Initial energies of the electrons were calculated in the same way as in [3], and initial ion energies were assumed to be equal 1 eV with uniform angular distribution. Place of origin of each new particle in the lens volume was determined by random number generator. Coordinates of the new and already trapped particles were determined from motion equation which included field values produced by particle space charges and electrode potentials.

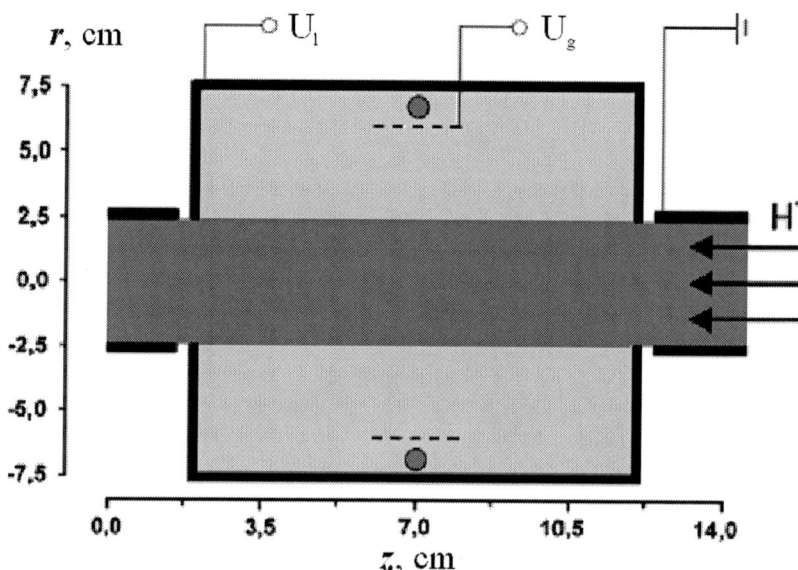

FIGURE 6. Model of the lens for numerical experiments. Schematic drawing of longitudinal cross section. Electrodes are shown by bold black line. Electron emitter is installed in central transverse plane (shown by circle near the electrode), and coaxial grid electrode with 2 cm width is placed closer to the axis.

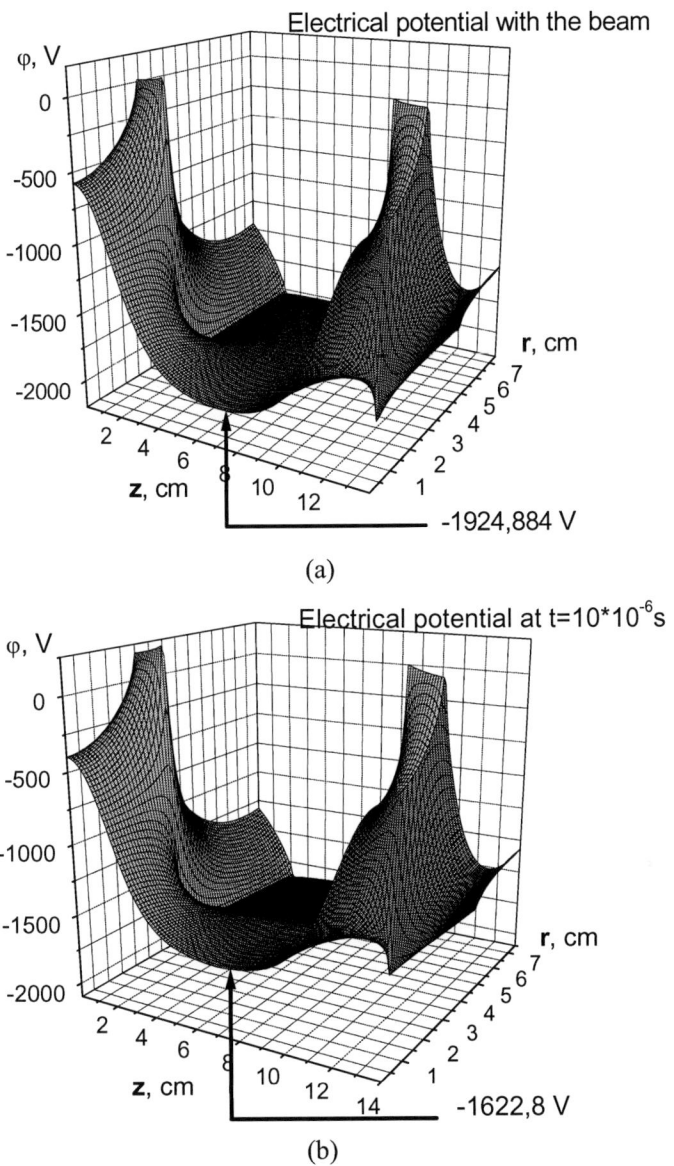

FIGURE 7. Distribution of potential in the lens with gas ionization by the beam. H⁻ negative ion beam with current 10 mA and energy 12 keV. a) Initial distribution of electric field potential formed by electrodes in the lens with H⁻ negative ion beam. b) Distribution of electric field potential in the lens in stationary regime at $t = 10 \cdot 10^{-6}$ s.

The time step used was 10^{-11} s. After 10^{-9} s time period the space charge distribution $\rho(z,r)$ for electrons, argon positive ions and the beam ions was calculated by the "cloud-in-cell" method taking in account the coordinates of all particles. On a base of calculated space charge distribution $\rho(z,r)$ the potential distribution $\varphi(z,r)$ was found for each point of spatial grid ($\Delta r=0.1$ cm, $\Delta z=0.1$ cm). Then introduction of the new particles was performed again, and calculation procedure was repeated. Stationary state was usually reached within a couple of microseconds, and calculated time never exceeded 10 μs.

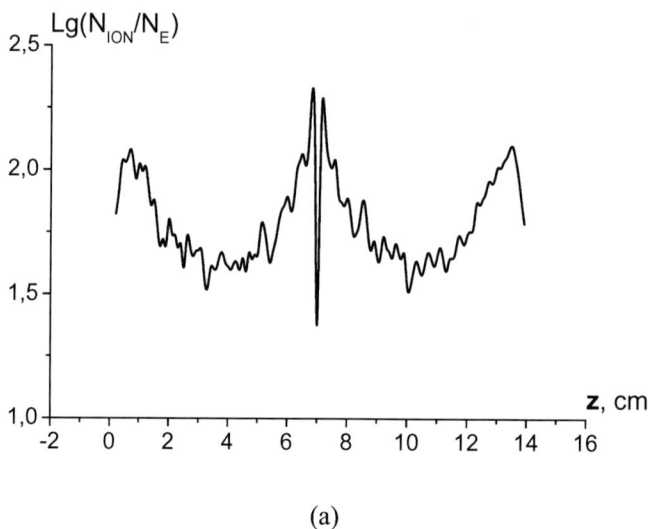

(a)

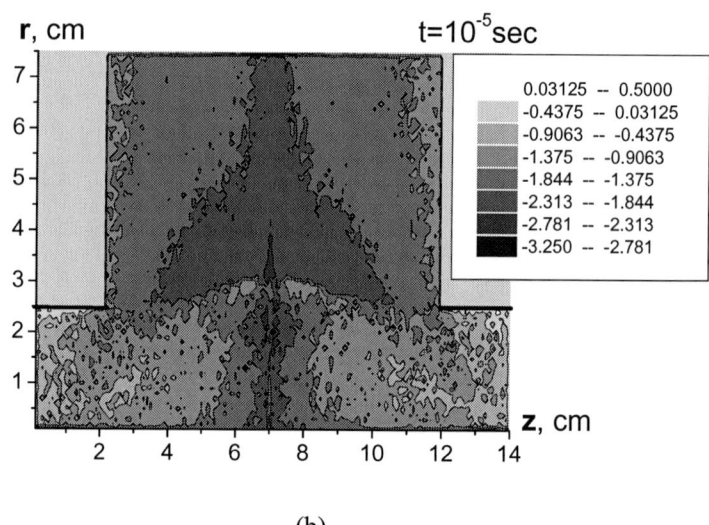

(b)

FIGURE 8. Distribution of ratio of concentration of positive ions to that of electrons at the lens axis (a) and in its volume (b). H⁻ negative ion beam current 10 mA, energy 12 keV. Ionization is provided by the beam.

Let us consider the results of calculations in absence of additional ionizer first. Fig.7a exhibits potential distribution in the lens at initial time, and Fig.7b – at stationary operation regime of the lens. One can see that potential at the lens center increased by about 300 V due to positive space charge of the ions. All formed ions left the lens volume onto the central electrode, and all electrons – onto the entrance and exit electrodes. Electron concentration at all points of the lens was at least one order of magnitude less than the concentration of positive ions. It is confirmed by Figs.8a and 8b, where the distribution of electron to ion density ratio is shown for lens axis and for its volume. Fig.9 represents trajectories of the beam ions in the lens and in drift region. One can see that the beam is focused, and focal length has 18 cm value, which is close to experimentally measured one.

We should note that calculated increase of the current, focused to collector has the value ~ 25, which is significantly higher than the experimental value. Such discrepancy, as it was mentioned above, may be due to non-ideal character of the beam used in the experiments, particularly, due to its asymmetry. The system behavior is more complex in presence of the additional ionizer. Let us consider in details the process of achieving the operation regime for the lens at electron emission current of 100 mA, energy of 100 eV and gas pressure of $7 \cdot 10^{-4}$ Torr. The temporal dependence of potential at the lens center is presented in Fig.10. At initial stage ($t=0 \div 0.5 \cdot 10^{-6}$ s) potential value in the lens decreases, because space charge of positive ions is not yet able to compensate space charge of the beam negative ions and the electrons from emitter. With a lapse of time the space charge of positive ions starts to be accumulated in the lens, and potential in the beam starts to grow. At this stage electrons from emitter are already able to perform ionization, and that contribution of electrons to the

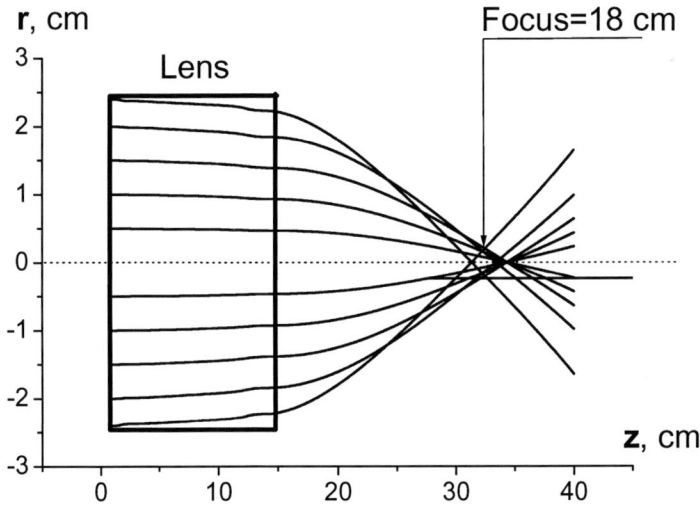

FIGURE 9. Trajectories of the beam particles inside the lens and outside it. H⁻ negative ion beam current 10 mA, energy 12 keV. Ionization is provided by the beam.

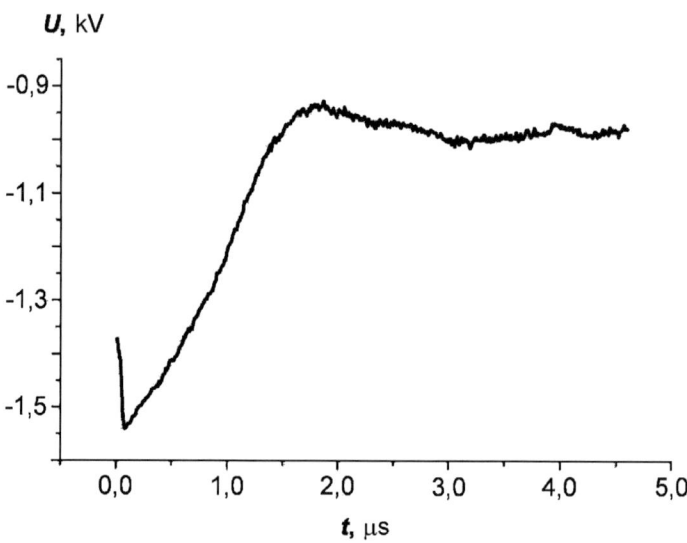

FIGURE 10. Dependence of potential at the lens center on time. H⁻ ion beam current 15 mA, energy 15 keV. Electron emitter current 100 mA.

ionization increases with the growth of potential at the lens center. This positive feedback leads to rapid growth of potential, so that the accumulation process is finalized in $0.5 \cdot 10^{-6}$ s by reaching stationary state, in which formation rates of the ions are exactly equal to corresponding leaving rates. While all positive ions leave onto central electrode of the lens, electrons formed in a process of gas ionization leave onto the entrance and exit electrodes, and electrons from emitter mainly leave onto the grid (just a small portion of them leaves onto the entrance and exit electrodes). The concentration of positive ions in the lens in stationary state exceeds concentration of electrons to a similar extent as that of the no additional ionizer case. However as it follows from Fig.11, this excess is not so different as in the case without additional ionizer. It should also be noted that radial distribution of potential in stationary regime is close to parabolic one, i.e. spherical aberrations are at their minimum, and focal length is 25 cm.

A set of numerical experiments was performed for different values of argon pressure and emission current in the present work. The results are presented in Fig.12 in form of ratios of the focal length value to its optimum one at different values of emitter current. One can see that, as well as in the experiment, an additional ionization of the gas by electron flux results in significant increase of the lens power, and that influence of the electrons increases with diminishing pressure and electron current growth. Calculated dependencies are close to experimentally obtained ones.

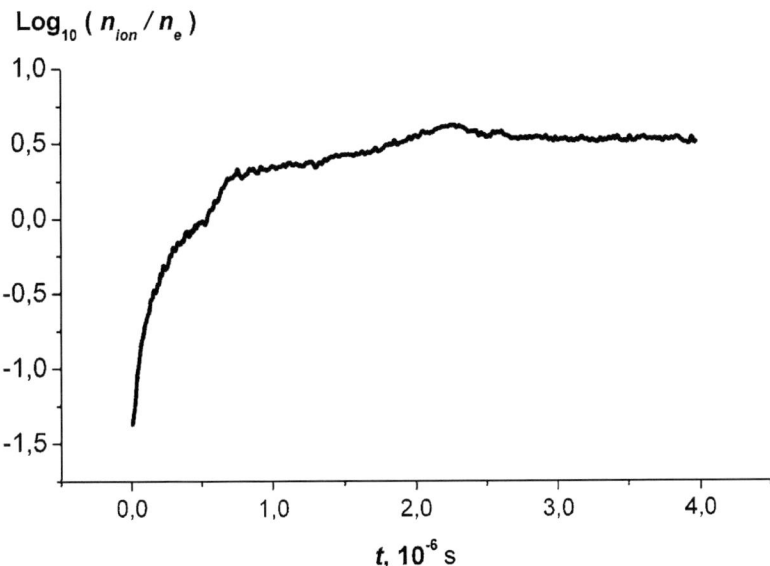

FIGURE 11. Time dependence for logarithm of ratio of densities of positive ions and electrons in the lens. H⁻ ion beam current 15 mA, energy 15 keV. Electron emitter current 100 mA.

.

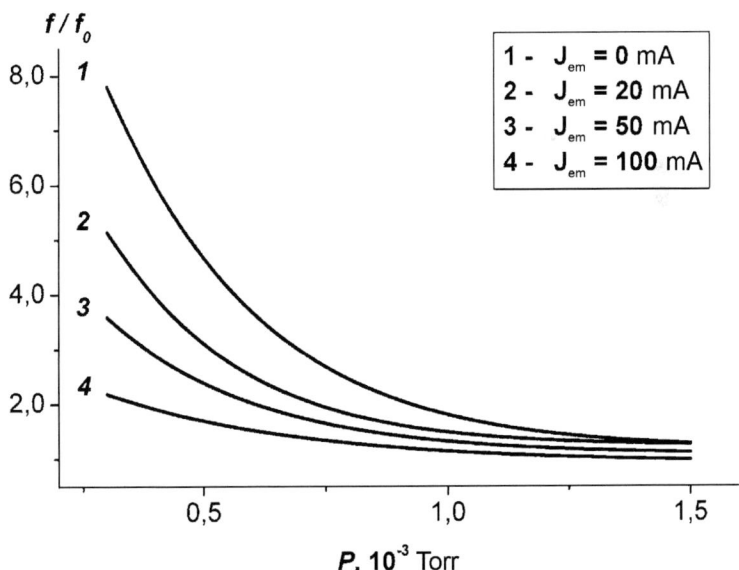

FIGURE 12. Dependence of relative focal length of the lens on pressure for different values of emission current. Here $f_0 = 20.0$ cm.

BRIEF CONCLUSIONS

The present experimental and numerical investigations have resulted in a creation of a simple and efficient space charge lens for focusing negative ion beams. The focusing is provided by a potential, applied to central electrode of the lens, with value being one order of magnitude less than the ion source potential. Power loss in the lens is as well significantly less than that to be consumed at the focusing by means of known lenses, particularly quadrupoles. Introduction of additional gas ionization by an electron flow enabled realization of the beam focusing at sufficiently low value of working gas pressure.

ACKNOWLEDGMENTS

This work is accomplished with the support of Foundation for Fundamental Researches at the Ministry of Education and Science of Ukraine (Grant # F7/297-2001).

REFERENCES

1. Goretski, V.P., Soloshenko, I.A., and Shchedrin, A.I., *Plasma Phys. Rep.* **27**(4), 335 (2001).
2. Zavalov, A.M., Gorshkov, V.N., Goretski, V.P., and Soloshenko, I.A., *Plasma Phys. Rep.* **29**(6), 480 (2003).
3. Golovinskii, P.M., and Shchedrin, A.I., *Journ. Tekh. Fiz.* **59**(2), 51-56 (1989), in Russian.

Sub Microsecond Notching of a Negative Hydrogen Beam at Low Energy Utilizing a Magnetron Ion Source with a Split Extractor

Douglas P. Moehs

Fermi National Accelerator Laboratory[a], P.O. 500, Batavia IL 60510 USA

Abstract. A technique for sub-microsecond beam notching is being developed at 20 keV utilizing a Magnetron ion source with a slit extraction system and a split extractor. Each half of the extractor is treated as part of a 50 ohm transmission line which can be pulsed at ±700 volts creating a 1400 volt gradient. This system along with the associated electronics is electrically floated on top of a pulsed extraction voltage. A beam reduction of 95% has been observed at the end of the Fermilab 400 MeV Linac and 35% notching has recently been achieved in the Booster.

INTRODUCTION

Expanding neutrino research at Fermilab has created a demand for 8 GeV proton intensities greater than the existing Proton Source (Linac and Booster) has ever provided [1]. Increases in proton intensity are presently limited by component activation associated with proton losses at high energies. Several projects have been initiated to address this problem including realignment, new magnets, and beam collimation in the Booster. As part of this effort, beam notching at low energies is being developed specifically to reduce the losses associated with beam injection and extraction from one accelerator to the next.

Effort to cleanly notch low energy beams has persisted, despite complications associated with space charge [2], due to the low beam rigidity and the relatively low cost associated with fast kilovolt pulsed-power supplies. At the SNS, a pulsed electric quadrupole in the 65 keV LEBT is being used to create chopped beams [3]. At KEK a modulated surface-plasma converter source has modulated beams with 70 ns rise and fall times [4]. Furthermore, a traveling wave chopper utilizing special timing for a 35 keV beam was presented at the LINAC 2002 conference [5]. The split extractor technique presented here is complementary to these efforts. It is hoped that low energy space charge problems will be avoided by notching in the extraction region where the beam is expected to be space charge limited, due to the strong electric field of the extractor. The high gas pressure inherent to this region should also facilitate fast space charge recovery.

---

[a] Work Supported by University Research Associates Inc., contract number DE-AC02-76CH03000

CP763, *Production and Neutralization of Negative Ions and Beams*, edited by J. D. Sherman and Y. I. Belchenko
© 2005 American Institute of Physics 0-7354-0248-5/05/$22.50

EXPERIMENTAL SETUP

A Magnetron ion source with a slit extraction system is used at Fermilab for the production of H minus ions [6]. Ions are extracted from the source at 20 keV, magnetically bent through 90 degrees and then further accelerated through a 750 kV column. In order to deflect the ion beam without adding additional electrodes to the cramped source region the extractor has been split down the middle. Figure 1 shows the Fermilab magnetron ion source and the split extractor. The 30 kV standoff and electrical feed through, on the left, replaced an observation window. Glass beads are now used to insulate the cables inside the vacuum instead of the Kapton tubing, shown in figure 1, which would turn greenish-black after several month of use and start to conduct along its exposed surfaces.

FIGURE 1. A picture of the split extractor used to extract H minus ions from the magnetron. The extractor components are mounted on 2.54 cm ceramic standoffs and are aligned by eye with the anode slit. The Kapton tubing shown here has been replaced by glass beads in recent experiments.

The tip of the extractor and the anode slit are aligned by eye and separated by approximately 2 mm using ceramic standoffs as mounting posts. The extractor slit width is also set to 2 mm using a feeler gauge. The anode slit is currently 0.7 x 10 mm.

Beam notching is accomplished by treating each half of the extractor as part of a 50 ohm transmission line which can be pulsed at ±700 volts creating a 1400 volt gradient across the split extractor. Electrically the extraction electrodes appear as a 10 cm break in the transmission line. A 20 ns electrical pulse in standard 50 ohm coax has about a 4 m wave front so the extractor component of the transmission line should be invisible to the pulse. To verify this assumption, the reflected power was measured using TTL pulses and time domain reflectromitry. A reflected power of 5-10% was measured. This result was strongly coupled to the connection quality at the extraction electrodes. The polarity of the voltage gradient is significant and is selected in order to optimize beam notch.

The ±700 volt high-voltage (HV) pulser, the 50 ohm loads and the vacuum feed through are electrically floated on top of the pulsed extraction voltage, typically 20 kV. The gate for the HV pulser is provided by a fiber optic network producing TTL pulses of the desired width. At least 3 Booster buckets, each 26 ns in width, are required to fire the Booster extraction magnet so the notch should be at least 78 ns wide.

The effective beam transition time can be computed by summing the electrical transition time of the HV pulser, the ion transit time through the extractor, and the curved path through the 90 degree bending magnet. The beam dynamics for a pair of deflection plates is described with more detail in reference [7]. An in house, dual-polarity high-voltage FET power supply was built for this application. Occasional sparking of the extraction system does not appear to be a problem for this supply. The electrical rise times are around 25 ns while the 55 ns fall times are slower than expected. Due to path length differences in the magnet, ions extracted from the source simultaneously will exit the magnet at different times. A SIMION model of the magnet suggests that 20 keV ions are dispersed in time by roughly 10 ns. Adding the above times to the 5 ns ion transit time through the extractor, 9.5 mm wide along the beam path, gives effective beam fall and rise times of 40 ns and 70 ns respectively.

ACCELERATOR RESULTS AND DISCUSSION

Figure 2 shows a beam trace from a Pearson current transformer with a 50 ns response time in the 750 keV beam transport line. In this case, 8 notches were created to match the number of turns being injected into the Booster. The burst frequency is set to match the Booster

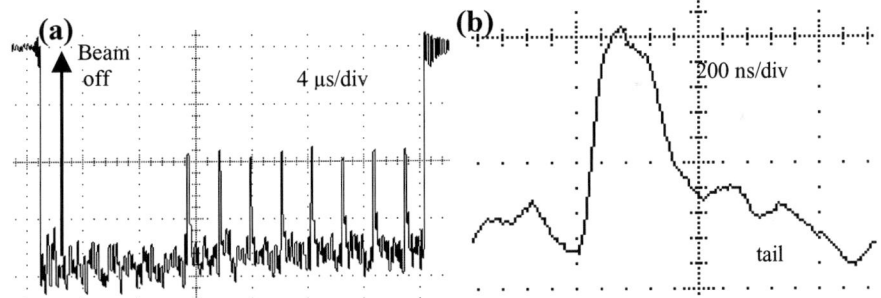

FIGURE 2. (a) A 20 mA beam of H minus with 50% notching observed at 750 keV using a Pearson current transformer. **(b)** The rise and fall times of one notch is magnified in figure b.

injection frequency of 454 kHz. The applied voltage gradient at this time was 1200 volts, limited by resistors in the HV supply. In figure 2, the notches are 100 ns wide with approximately 40 ns and 70 ns rise and fall times respectively, as expected. The droop in beam intensity over the notching period, in this case about 1.5 mA, is associated with an extended tail in the beam recovery time of each notch. The cause of this tail is not yet understood. Theories being investigated include electrode charging, meniscus disruption and recovery, and beam space charge recovery.

The low energy end of the Fermilab Linac [8] operates at 200 MHz. Once the H minus ions are bunched and captured in the Linac the beam structure should be retained. A Beam Position Monitor (BPM) was used to measure one notch at 10 MeV. The raw BPM signal as observed by a fast oscilloscope is shown in figure 3. The beam current is proportional to ½ the signal strength so the notch depth at 10 MeV is 50% as expected.

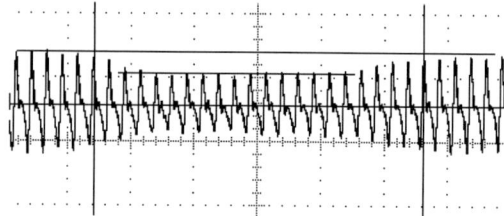

FIGURE 3. The raw BPM signal of a single notch measured at 10 MeV. Vertical and horizontal lines are added to guide the eye. The RF frequency is 200 Mhz and the horizontal scale is 20 ns/div.

Figure 4(a) shows the first 4 turns of charge being injected into the Booster as measured by a toroid. The notch appears to grow with each injection. Figure 4(b) shows the signal from a phase detector in the Booster and compares notched and un-notched beam at 400 MeV, 2.4 μs into the machine cycle, just before the Booster extraction magnet would fire creating the traditional Booster extraction gap. In turn 4, right side of figure 4a, the notch appears to be about 50%, however in figure 4b the beam reduction is only 35%. During charge injection into the Booster the Booster RF is turned down allowing the 200 MHz linac structure to fade prior to rebunching at 38 MHz for acceleration to 8 GeV. Thus the notch is adversely affected by the longitudinal space charge forces and in this case a reduction of 15% in notching efficiency was observed.

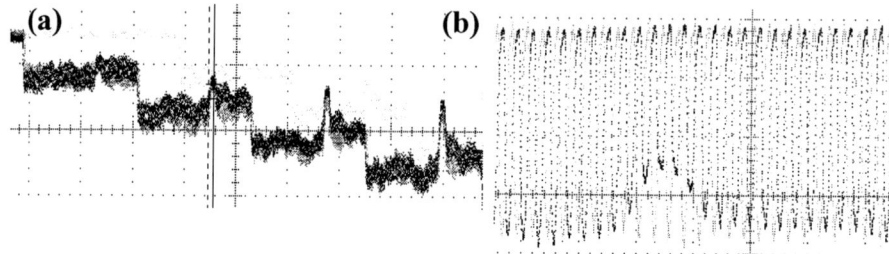

FIGURE 4. **(a)** Toroid signal of the first 4 turns of charge being injected into the Booster with the notch present on the left of each step. **(b)** Phase detector signal showing the Booster notch at 400 MeV, before acceleration. The lighter trace shows an un-notch Booster cycle.

In this experiment, only a trigger was available from the Booster. Thus the 100 ns pulse train, shown in figure 2, was created by a pulse generator in burst mode. The frequency of the pulse generator was adjusted to optimize the notch efficiency. Better timing and a cleaner notch at 750 keV should help improve the notching efficiency in the Booster. Once implemented, the notch will be timed in reducing the amount of beam kicked out at 4 MeV.

An attempt has also been made to reduce the losses at the Linac switching magnet, where beam is sent to the Booster. A 150 ns notch was created and timed in using BPM's near the switching magnet at 400 MeV. Under normal operating conditions, a reduced activation rate of 25% was observed at the front end of the switching magnet. In this experiment, a beam reduction of 95% was measured at 400 MeV, during the notch, while the reduction at 750 keV was only 50%. This suggests that under certain conditions ions may continue to fall out of the notches as they propagate down the Linac. Efforts to scrape the beam in the 750 keV line have begun in hopes of improving the low energy notch.

ACKNOWLEDGEMENTS

The author would like to thank James Wendt for his expert modifications of the ion source, Raymond Hren and Brian Stanzil for their ready help with cabling and electronics, Chris Jesnsen, for developing the FET HV pulser, James Lackey for help with TTL time domain reflectromitry, Milorad Popovic for accelerator assistance, Raymond Tomlin for his assistance with the BPM systems, William Pellico and Richard Meadowcroft for providing and fixing the fiber optic network and triggers, and Charles Schmidt for stimulating discussion and positive encouragement!

REFERENCES

1. Fermilab Proton Committee Report, www.fnal.gov/directorate/program_planning/studies/ProtonReport.pdf (2003).
2. J.G. Alessi, *Rev. Sci. Instrum,* **61**, 1 (1990).
3. S. Nath, J. Billen, J. Stovall, H. Takeda, L. Young, Proceedings of LINAC 2002, p. 130, Korea (2002).
4. K. Shinto, A. Takagi, K, Ikegami, and Y. Mori, *Rev. Sci. Instrum.* **71**, 696 (2000).
5. A. Novikov-Borodin, LINAC Conference Proceedings, Korea, 2002, pp. 115.
6. D. Moehs, "Studies on a Magnetron" in *Production and neutralization of Negative Ions and Beams,* edited by M. Stockli, AIP Conference Proceedings 639, Gif-sur-Yvette, France, 2002, pp. 115.
7. M. Stockli, US Particle Accelerator School Lecture notes: "Ion Beam Chopping", Williamsburg VA (2004).
8. M. Popovic, L. Allen and C. W. Schmidt., Particle Accelerator Conference Proceedings, 1995, pp. 917.

Mathematical Formulation and Numerical Modelling of the Extraction of H⁻ ions

Reinard Becker

Institut für Angewandte Physik der Johann Wolfgang Goethe – Universität, Fach 180
D-60054 Frankfurt/M, Germany

Abstract. In the past, ion extraction of volume produced H⁻ ions has either been simulated with programs for the extraction of positive ions [1] or by programs, which wrongly claimed to have a "real" and "genuine" option for H⁻ ions [2]. Although "reasonable" results have been obtained in both ways, the mathematical formulation of the physics at the plasma sheath is wrong, and the modelling of electrodes near the sheath [3] must fail. In this paper a self consistent formulation of the extraction problem for H⁻ ions is presented, which takes into account any number of positive ions, like fast or thermal protons, thermal cesium and molecular ions like H_2^+ and H_3^+, which all are essential for the generation of H⁻ ions in the plasma volume. Equally important is the porting of this formulation to a simulation program and the verification of experimental results. This has lead to the development of the program nIGUN©.

INTRODUCTION

The present theory is the result of several steps of development: In 1997 the virtual cathode behaviour of protons, entering the plasma sheath and being reflected by the field for H⁻ extraction has been pointed out [4], however, without considering the role of thermal positive ions, like molecules and cesium ions in the case of cesium seeding. This feature has been added in 2002 [5] for one kind of thermal ions. In 2003 the attempt has been made, to include the whole sheath from the plasma potential to the extraction in one theory [6]. This, however, is too comprehensive, because the important effect of changing the electron to H⁻ fraction by a bias of the plasma electrode [7] cannot be treated in a linear sheath model. The present formulation therefore is an extension of the 2002 theory, allowing for any number and kind of positive ions in front of the plasma electrode, like protons coming from the plasma potential and protons, molecules, and cesium ions with thermal energy, being created in the neutralized vicinity of the plasma electrode. Since H⁻ are assumed to be thermal there, it will not matter, if these will be produced by surface or by volume processes. Other work on the extraction of H⁻ ions is reviewed in ref. [6]. PBGUNS [2] uses wrong expressions for the electron space charge and ignores fast positive ions as well as additional thermal ones, like cesium and molecules.

CP763, *Production and Neutralization of Negative Ions and Beams*, edited by J. D. Sherman and Y. I. Belchenko
© 2005 American Institute of Physics 0-7354-0248-5/05/$22.50

THE INVERTED SHEATH

In contrast to positive ion extraction, where the potential fall in the sheath is continued by the acceleration field of the extraction potential [8], ion sources for H⁻ production are more complex: The natural potential fall of the plasma sheath needs to be reversed for the extraction of negative ions (and electrons). In general this is provided by a transverse magnetic field (filter) in front of the plasma electrode, which forces fast electrons to follow the flux lines, while slow ones may have enough collisions to move across them by ambipolar diffusion. This localized changes of the electron velocity distribution and of the electron density will cause a fall of potential towards the plasma electrode, favouring the migration of H⁻ ions into this region [9]. According to the axial potential model shown in Fig. 1, we can formulate the space charge term for each kind of particle in the extraction region, which is the region of interest for this paper:

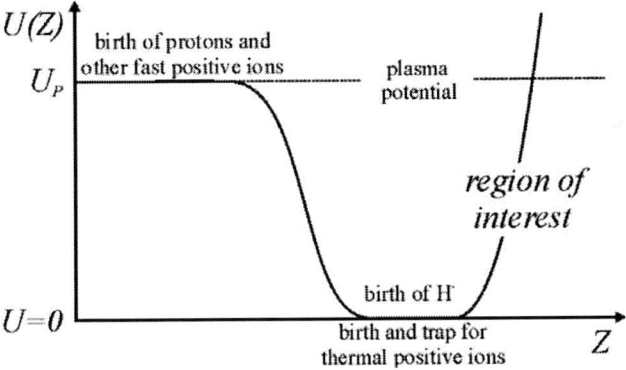

FIGURE 1. Axial potential model for the definition of space charges in the extraction region

The density of electrons with energy U_e will be reduced by acceleration in the same way as the H⁻ density:

$$n_e(U) = \frac{n_e(0)}{\sqrt{1 + \dfrac{U}{U_e}}}, \qquad n_-(U) = \frac{n_-(0)}{\sqrt{1 + \dfrac{U}{U_-}}} \tag{1}$$

the density of fast protons (and other fast ions) is dying out by the virtual cathode process [4]:

$$n_P(U) = n_P(0)\left(1 - \frac{U}{U_P}\right) \tag{2}$$

Other positive ions are considered to be thermal and trapped between the extraction field and the plasma, hence will obey a Boltzmann distribution:

$$n_i(U) = n_i(0) \exp\left\{-\frac{U}{U_i}\right\} \tag{3}$$

This problem now has 2 unknowns for each charged particle, the density and the energy (either directed or thermal). While the energies must be known to obtain solutions, the densities can be expressed by the definition of the electron-to-H⁻-current (Γ)

$$\frac{n_e}{n_-} = \Gamma\sqrt{\frac{m_e U_-}{M_- U_e}} \tag{4}$$

and by the condition of quasineutrality at $U = 0$

$$\frac{n_P}{n_-} = 1 + \frac{n_e}{n_-} - \frac{n_1}{n_-}\sum\frac{n_i}{n_1} \tag{5}$$

We then calculate from the balance of charged particle currents at the plasma electrode with wall potential U_w the relation of the density of the first kind of positive thermal ions to the density of H⁻ ions:

$$\frac{n_1}{n_-} = \frac{1 + \Gamma - \left(1 + \Gamma\sqrt{\dfrac{m_e U_-}{M_- U_e}}\right)\left(1 - \dfrac{U_w}{U_P}\right)^{\frac{3}{2}}\sqrt{\dfrac{\pi M_- U_P}{M_P U_-}}}{\sum\dfrac{n_i}{n_1}\left\{\sqrt{\dfrac{M_- U_i}{M_i U_-}}\exp\left[-\dfrac{U_w}{U_i}\right] - \left(1 - \dfrac{U_w}{U_P}\right)^{\frac{3}{2}}\sqrt{\dfrac{\pi M_- U_P}{M_P U_-}}\right\}} \tag{6}$$

PARAMETER RELATIONS OF SOLUTIONS

By solving eq. 6 for assumed values of the directed or thermal ion energies and for choosing $M_1 = 1$ or $= 133$, relations will be obtained for the parameters of solutions either without or with cesium seeding. The assumption of a negative wall potential, as used in a former presentation [5] has been dropped, because the potential model (Fig. 1) and the associated space charge terms (eq. 1-3) then will break down. In Fig. 2, the relation of parameters is shown with cesium seeding for a directed proton energy of 10 times the H⁻, cesium, and electron temperature. A positive wall potential always will reduce the electron to H⁻ ratio, as observed in experiments, and the cesium ion density always must be lower than the H⁻ density. Lower values of Γ exist with cesium, as observed in experiments, too.

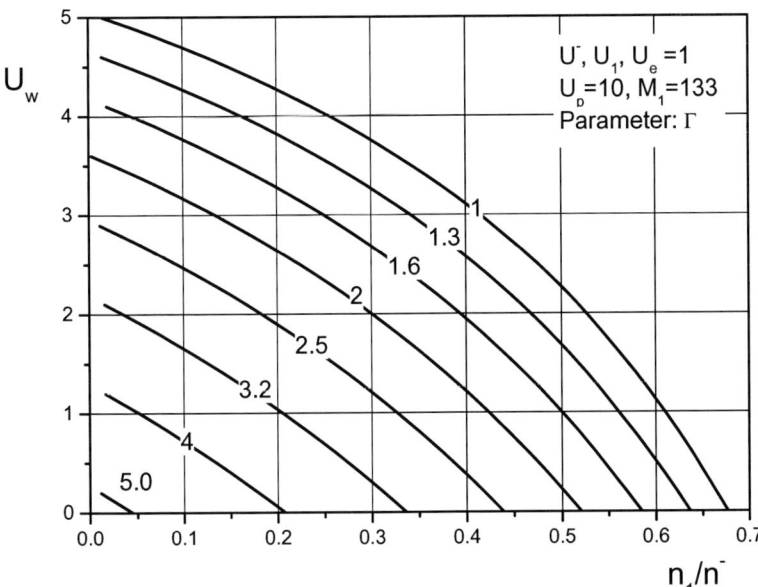

FIGURE 2. Relation of parameters for solutions of the inverted sheath with cesium seeding

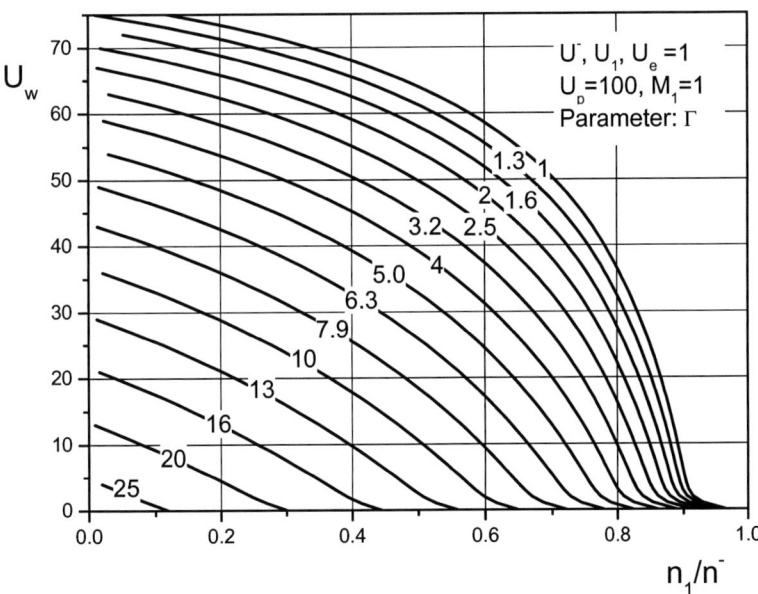

FIGURE 3. Relation of parameters for solutions of the inverted sheath without cesium seeding

IMPLEMENTATION INTO nIGUN©

The 2D simulation program for positive ion extraction, IGUN© [8] has been modified to simulate negative ion extraction on the basis of the theory presented in this paper and given the name nIGUN©. The evaluation of eq. 6 is part of the input procedure. For the "real" solution, however, the densities of the different particles must be known absolutely. This is achieved by the integration of Poisson's equation with the space charge terms eq. 1-3, resulting in eq. 7:

$$
\frac{\varepsilon_o U'^2}{2e} = n_- \left\{ \begin{array}{c} 2\frac{n_e}{n_-}U_e\left[\sqrt{1+\frac{U}{U_e}}-1\right] + 2U_-\left[\sqrt{1+\frac{U}{U_-}}-1\right] \\ -\frac{n_P}{n_-}\left(U-\frac{U^2}{2U_P}\right) + \frac{n_1}{n_-}\sum\frac{n_i}{n_1}U_i\left(\exp\left[-\frac{U}{U_i}\right]-1\right) \end{array} \right\}
\tag{7}
$$

For any combination of the potential and the field strength in the extraction region the absolute H⁻ density can be determined from eq. 7. The other densities are following then from eq. 4-6. It has been found that this determination of densities is done reasonably at $U=U_P$, where the space charge of fast positive ions has disappeared. As a result, the densities of a cesium seeded plasma are shown in fig. 4 from the quasineutral plasma (on the left) towards the extraction region (right side) together with the increase of the potential function for a planar diode of 20 mesh thickness.

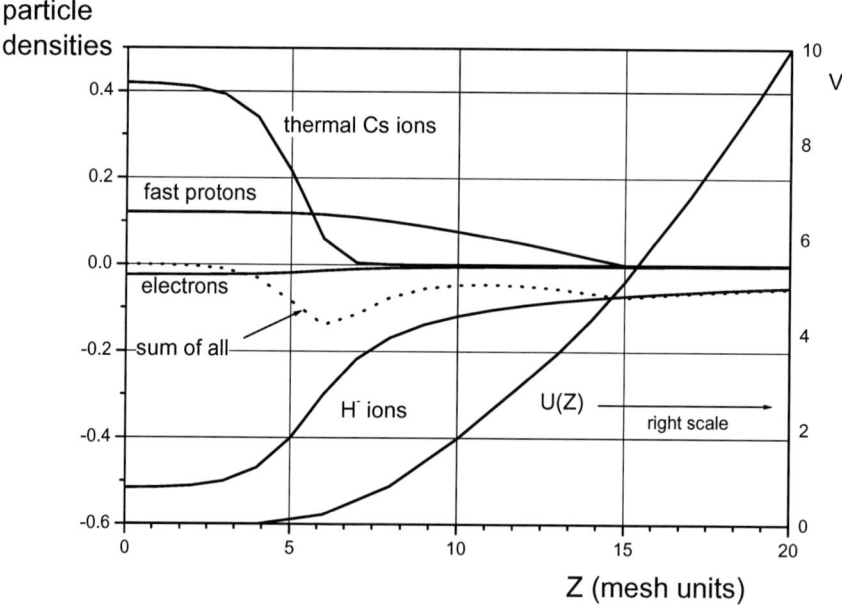

FIGURE 4. Self-consistent densities of electrons, H⁻ ions, fast protons and thermal cesium ions in an inverted plasma sheath according to the presented theory and its implementation into nIGUN©

THE 3/2 POWER LAW

Poisson's equation for charged particles, being accelerated in a diode from zero velocity, reads:

$$\Delta U = -\frac{\rho}{\varepsilon_o} = \frac{j}{\varepsilon_o \sqrt{\dfrac{2e}{M} U}} \qquad (8)$$

For H⁻ extraction, the expression for the space charge term looks quite different:

$$\Delta U = \frac{e}{\varepsilon_o} \left\{ \frac{n_e}{\sqrt{1+\dfrac{U}{U_e}}} + \frac{n_-}{\sqrt{1+\dfrac{U}{U_-}}} - n_P\left(1-\frac{U}{U_P}\right) - n_1 \sum \frac{n_i}{n_1} \exp\left[-\frac{U}{U_i}\right] \right\} \qquad (9)$$

If the energy is sufficiently high, however, the contributions from positive ions disappears and the electron and H⁻ term may be approximated by:

$$\Delta U \approx \frac{j_{H^-}}{\varepsilon_o \sqrt{\dfrac{2e}{M} U}} \left(1+\frac{\Gamma}{42}\right) \qquad (10)$$

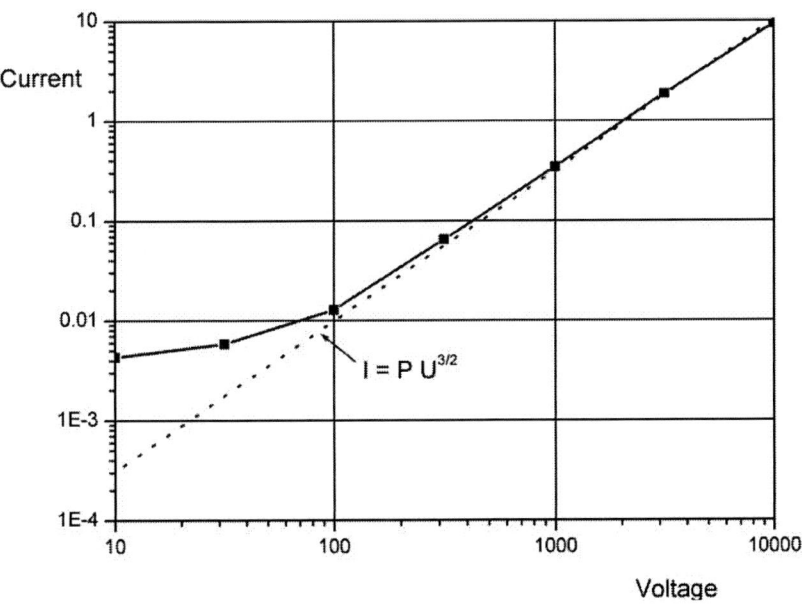

FIGURE 5. Simulations with nIGUN© for a plane diode compared with the voltage$^{3/2}$-law.

This shows, that for high voltages there will be approximately a 3/2 power law, but the current will be an effective one, just as described by Leitner et al. [1], when using a computer program for positive ion extraction in order to simulate the extraction of negative ions. It should be emphasized, that this asymptotic appearance of the 3/2 power law may be useful only for the prediction of the current in a planar diode, but other questions, like the shape of the plasma electrode in the vicinity of the beam boundary need a correct consideration of the potential function there. It has been found [6], that an angle of 45° between the beam boundary and the plasma boundary will provide an aberration free beam boundary, in similar way as the famous Pierce angle of 62.5° for solid emitters.

CONCLUSIONS

The presented theory for H- extraction is self-consistent and can take into account any number of fast and of thermal positive ions. The axial potential function starts with vanishing field strength at the potential of the wall electrode (if there is no bias on the PE). This model accepts birth of H- ions in the volume around the extraction aperture as well as on the wall electrode. In agreement with measurements, the proton energy is much lower with added cesium than without and lower values for the electron to H - current are found for cesiated plasmas than for cesium free ones. The mathematical formulations have been implemented into a new program, called nIGUN$^{\copyright}$ [10], for the simulation of the extraction of H⁻ ions.

ACKNOWLEDGEMENTS

The author wants to thank Wulf Kunkel, Ka-Ngo Leung, Jens Peters, Marthe Bacal, and Klaus Volk for sharing their experimental experience over many years.
This work has been supported by contract EU-project HPRI-CT-2001-50021.

REFERENCES

1. M.A. Leitner, D.C.Wutte, K.N.Leung, *Nucl. Instrum. and Meth.*, **A 427**, 242 (1999)
2. J.E. Boers, http://thunderbirdsimulations.com/T
3. R.F. Welton et al., *Rev. Sci. Instrum.* **73**, 1013 (2002)
4. R. Becker, K.N.Leung, W.Kunkel, *Rev. Sci. Instrum.*, **69**, 1107 (1998)
5. R. Becker, *AIP conference Proceedings* **639**, 82 (2002)
6. R. Becker, *Rev. Sci. Instrum.* **75**, 1687 (2003)
7. K.N. Leung, S.R. Walther, and W.B. Kunkel, *Phys. Rev. Lett.* **62,7**, 764 (1989)
8. R. Becker, W.B.Herrmannsfeldt, *Rev. Sci.Instrum.* **63**, 2756 (1993)
9. M. Bacal, J.Bruneteau, P.Devynck, *Rev. Sci. Instrum.*, **59,10**, 2152 (1988)
10. http://www.egun-igun.com

H⁻ AND D⁻ SOURCES

ECR-Driven Multicusp Volume H⁻ Ion Source

M. Bacal, A.A. Ivanov Jr., C. Rouillé, P. Svarnas

Laboratoire LPTP, Ecole Polytechnique,UMR 7648 du CNRS, 91128 Palaiseau, France

S. Béchu, J. Pelletier

Laboratoire EPM, ENSHMG, BP 95
38402 Saint Martin d'Hères Cedex, France

Abstract. We studied the negative ion current extracted from the plasma created by seven elementary ECR sources, operating at 2.45 GHz, placed in the magnetic multipole chamber "Camembert III". We varied the pressure from 1 to 4 mTorr, with a maximum power of 1 kW and studied the plasma created in this system by measuring the various plasma parameters, including the density and temperature of the negative hydrogen ions. We found that the electron temperature is optimal for negative hydrogen ion production at 9.5 cm from the ECR sources. The tantalum-covered wall surface pollution reduces the extracted negative ion current and enhances the electron current. Tantalum evaporation has a positive effect. The use of a grid and of a collar in front of the plasma electrode did not lead to any enhancement of the extracted negative ion current.

Keywords: Negative ion source, Electron cyclotron resonance, Hydrogen
PACS: 29.25.Ni, 52.70.-m, 52.80.Pi

INTRODUCTION

The main advantage of microwave-driven negative ion sources over the filament discharge ones is the absence of short life components (the filaments). Several attempts were made to use an ECR microwave discharge for negative ion production [1-8]. In these earlier works the microwaves were injected using a waveguide, followed by a window, in a magnetized discharge chamber, where ECR took place. In some works [1-6] a magnetic filter was used to separate this "driver" chamber, containing fast electrons, from the extraction region. Fukumasa and Matsumori [4] reported that in the ECR source the electron temperature does not decrease across the magnetic filter as much as in filament discharge plasma. Gobin *et al* [7, 8] assumed that the electromagnetic wave non absorbed by the plasma was heating the electrons in the extraction region, behind the magnetic filter. They replaced the magnetic filter by a negatively biased stainless steel grid to stop the microwaves and the energetic electrons from penetrating into the extraction region [7, 8]. This led to an enhancement of the extracted negative ion current.

In this work we installed a two-dimensional network of elementary ECR plasma sources [9] on the upper flange of the negative ion source Camembert-III [10] conserving the original multicusp confinement system of the source. We investigated

CP763, *Production and Neutralization of Negative Ions and Beams*, edited by J. D. Sherman and Y. I. Belchenko
© 2005 American Institute of Physics 0-7354-0248-5/05/$22.50

the parameters of the hydrogen plasma produced in this system using probes and laser photodetachment, and the extracted negative ion and electron currents, for a total cw microwave power (2.45 GHz) up to 1 kW. The first results obtained with this source in June-July 2003 were reported in [11] and can be summarized as follows:

1. A quasi-maxwellian, low temperature, electron distribution was found, without use of additional magnetic filter.
2. A bi-Maxwellian negative ion distribution, with lower temperatures compared to the filament discharge.
3. The effect of increasing the microwave power in the range 0.5 to 1 kW is beneficial.
4. Similar dependence of extracted currents on the plasma electrode bias, to that observed with the filament discharge. A particularly low ratio $Ie/\bar{I}=5$ was observed.
5. Important effect of magnetic multipole chamber.

In this paper we report the long-term (of the order of six months) time evolution of the negative ion density and extracted current. In the first experiments the wall was covered with tantalum, deposited in earlier filament discharge experiments. With time the wall was polluted and we found a reduction of the negative ion extracted current. Evaporation of fresh tantalum from a filament led to the recovery of negative ion characteristics.

We studied the influence of the distance (D) from the ECR sources to the plasma electrode on the extracted negative ion and the electron currents, and from the ECR sources to the probe (d) on the plasma characteristics (electron temperature and density, negative ion density and temperature).

We also tested the effect of introducing a collar, with and without an additional magnetic filter. Peters [12] reported an enhancement of the negative ion current in a pure hydrogen (non-cesiated) rf ion source when a collar was used. Earlier Leung et al [13] found a reduction of the electron current only, when the collar was added in a pure hydrogen filament discharge, but also in an rf discharge.

Following the positive result obtained in another ECR source [7, 8] we also introduced a stainless steel grid, which could be biased negatively, in front of the plasma electrode.

EXPERIMENTAL SETUP

Plasma Source.

We installed on the top flange of the stainless-steel chamber of Camembert III a two-dimensional network of seven elementary independent ECR plasma sources, shown in Figure 1. These sources are described by Lacoste et al (see Fig. 1 in [9]). Briefly, each plasma source is made of two main parts, a permanent annular magnet with azimutal symmetry around its magnetization axis, and a microwave applicator constituted by a coaxial line, parallel to the magnetization vector. The inner conductor of the coaxial line penetrates inside the annular magnet. Each magnet is completely encapsulated in a stainless steel envelope and is water-cooled. These sources operate at 2.45 GHz microwave frequency. Thus the magnetic field intensity required for the ECR condition is 875 G. The maximum microwave power acceptable by a single source is 200 W. Figure 2 in [9] shows the configuration of the magnetic field for a samarium-cobalt magnet 30 mm long, with an outer diameter of 20 mm and an inner diameter of 6 mm. The plasma is produced by the electrons accelerated in the region of ECR coupling by the microwave electric field applied via the coaxial line. The fast electrons oscillate between the two mirrors in front of the opposite poles of the magnet and drift azimuthally around the magnet. The plasma produced by the inelastic collisions of these fast electrons diffuses away from the magnet. The network of seven elementary ECR sources consists of six sources with the magnets of the same magnetization, while the central source has the opposite one.

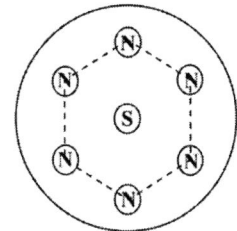

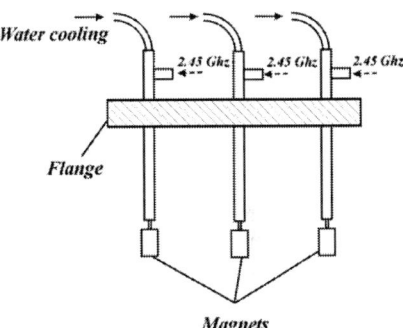

FIGURE 1. Top-down and side view scheme of the network of seven ECR sources implemented on the upper flange of Camembert III.

This network is operated from a single microwave power source of 1.2 kW (2.45 GHz) by dividing its microwave power into seven equal parts. The cylindrical stainless-steel sidewall of Camembert III, described in [10] is 44 cm in diameter and 45 cm long. It contains sixteen columns of samarium-cobalt magnets with the north and south poles alternatively facing the plasma. The second end of the chamber is bounded in part by the stainless-steel plasma electrode of the extractor (10 cm in diameter), which contains an extraction hole of 0.8 cm in diameter, and in part by a water cooled annular copper plate, connected to the sidewall (which is grounded). The extractor has been described elsewhere [14]. The neighboring plasma is magnetized and due to this and to a small positive bias of the plasma electrode, large densities of volume produced negative ions concentrate in this region [14].

In the present arrangement, the source contains three distinct regions: (i) a driver region, located near the network of seven elementary ECR sources and possibly on the perimeter of the device, in the strong multicusp magnetic field); (ii) an extraction region, which extends over the central, field-free region; (iii) a weakly magnetized region with high n^-/n_e, bounded by the plasma electrode with the extraction opening. The magnetic filtering effect is provided by the magnetic field of the elementary ECR sources and possibly by the multicusp magnetic field near the wall, which confine the fast electrons.

Diagnostics.

The plasma parameters (electron density and temperature) were measured using a microcomputer-controlled electrostatic probe (0.5 mm diam and 15 mm long) made of tungsten. The probe was located on the axis of the chamber, at a distance d from the lower end of the central elementary ECR source. The distance D between the plasma electrode and the lower end of the ECR sources was varied from 19.5 to 24.5 cm by moving the ECR sources. The probe was located at a fixed distance of 15 cm from the plasma electrode. Thus the distance (d) between the probe and the lower end of the central ECR source was varied from 4.5 cm to 9.5 cm. The H^- ion density, n^-, was measured by the photodetachment technique, reviewed recently in [15]. The H^- negative ion temperature was measured using the two-laser photodetachment technique [15]. The negative ion temperature kT^- is determined from the negative ion density recovery curve after the negative ions have been destroyed by photodetachment in a small cylindrical region. The present fitting technique allows to determine the respective temperatures and fractions of two negative ion populations [16, 17].

RESULTS

Plasma Characteristics.

The electron temperature and density and the negative ion density (see Fig. 2) were measured in the pressure range from 1 to 3 mTorr for a total applied microwave power of 1 kW (*i.e.* 140 W/antenna) and for two distances from the ECR source to the probe (4.5 and 9.5 cm). Note that the electron temperature for P > 1.5 mTorr at the distance of 9.5 cm remains in the optimum range for H^- ion production, *i.e.* Te < 1 eV. Correspondingly, the negative ion density is highest for this, larger, distance. The electron density goes up linearly with hydrogen pressure and attains, at 3 mTorr and 1 kW, 3.4×10^{10} cm^{-3}.

The negative ion density also linearly increases with pressure and is not much affected by wall effects. Note that the electron temperature obtained in Camembert III is significantly lower than the one we measured in a chamber without multipolar confinement, in which it was from 1.6 eV to 2.3 eV for a pressure of 3 mTorr and power from 100 to 150 Watt/antenna. The electron density measured in Camembert III at 3 mTorr is by a factor of two higher than in the chamber without multipolar confinement, for 150 W/antenna.

The influence of multipolar confinement on plasma is discussed in details in [18]. The main multicusp magnetic field effect is the trapping of "primary" electrons – that is, the trapping in the multicusp field of the energetic electrons creating the plasma by neutral molecule ionization. The lifetime of these energetic electrons considerably increases by being trapped in magnetic field compared to field-free lifetime. The plasma itself (that is, the ions and low-temperature electrons with energies ~1eV) actually is not confined well enough by the multicusp magnetic field. The increase of

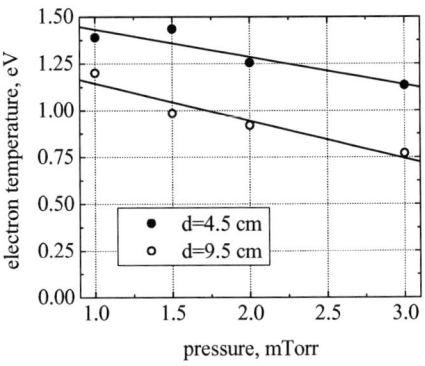

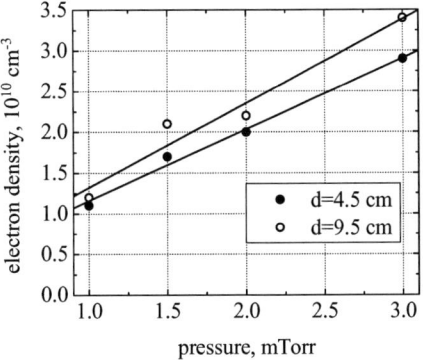

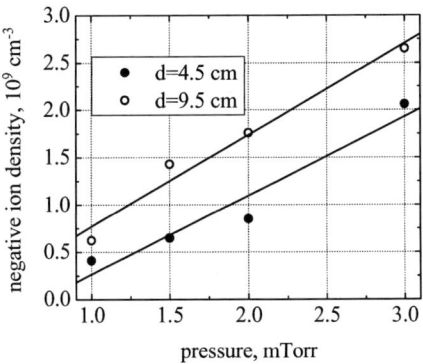

FIGURE 2. Dependence of the electron temperature, electron density and negative ion density on hydrogen pressure, for two values of the distance (d) between the central minisource and the probe.

the plasma density due to multipolar configuration occurs essentially because of the higher ionization rate provided by an increased lifetime of energetic electrons situated at the periphery of the chamber. The multicusp field also traps energetic electrons present in the central part of the device, several mechanisms of free electron trapping are proposed in [18]. So, the multipolar field selectively traps energetic electrons, thus the average electron energy in volume decreases.

In our system the plasma is created by ECR discharge, so the first mechanism limiting the transport of energetic electrons in volume is of the same origin as the ECR heating – the magnetic field of elementary ECR sources confines the hot electrons. The network of multipolar elementary ECR plasma sources forms an additional multicusp configuration in the chamber, and the fast electrons are formed already inside this region, so they are well confined. The energetic electrons who succeeded however to penetrate into the central, field-free part of the source, encounter the second barrier – the multicusp field of Camembert III chamber which rapidly traps these free energetic electrons. This is the reason of the observed decrease of temperature compared to the experiments with elementary ECR sources but without multicusp configuration.

The negative ion temperature dependence on pressure is presented in Fig. 3, for two distances from the ECR sources. The average value T_0 (from one-temperature fit of the recovery curve) is shown for both distances d. We found that the negative ion population contained two groups with

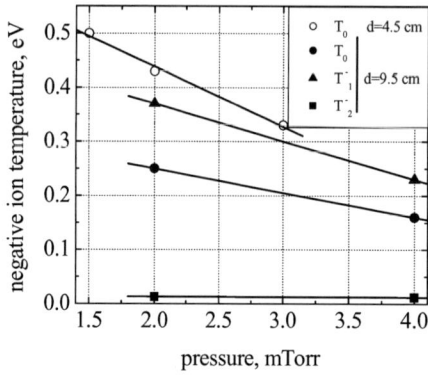

FIGURE 3. Dependence on hydrogen pressure of negative ion temperature (average value) at two distances d from the ECR sources. Two values found for the two H⁻ populations are shown for the distance d=9.5 cm only. Applied power 1 kW.

different temperatures (T^-_1, T^-_2) at the longer distance d = 9.5 cm. This was true for only one pressure (3 mTorr) at the shorter distance d=4.5 cm. Note that the negative ion temperatures found here are lower than those found in a filament discharge (see Ref. 17 and 19).

The population fractions corresponding to the two temperatures obtained from the two-temperature fit for d = 9.5 cm, are reported in Ref.11.

Extracted Negative Ion And Electron Currents.

The extracted negative ion and electron current dependence on the plasma electrode bias was measured for four pressures in the range 1 to 3 mTorr (see Figure 4). The results in Fig. 4 correspond to the maximum distance from the ECR sources to the plasma electrode (D=24.5 cm). The extraction and acceleration voltages were both 2kV.

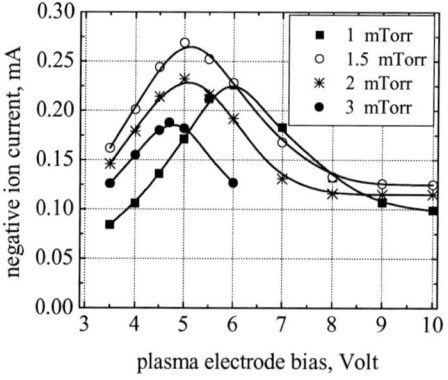 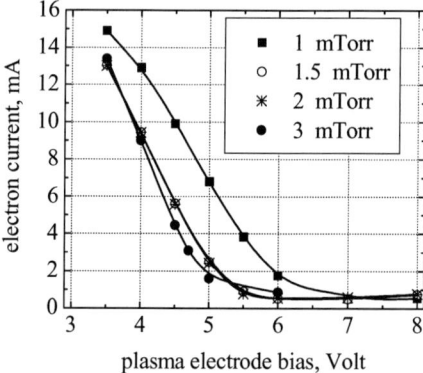

FIGURE 4. Effect of hydrogen pressure on the extracted negative ion and electron currents. Maximum distance between ECR sources and plasma electrode (D = 24.5 cm). Good wall conditions (72 hours after the second tantalum evaporation).

The negative ion current goes through an optimum for a positive plasma electrode bias. The value of this bias, which is close to the plasma potential, is maximum for the lowest pressure and goes down with the pressure increase. The maximum negative ion currents are obtained for 1.5 mTorr. With D−19.5 cm the maximum negative ion current is 10% higher, as can be seen in Fig. 5, where the extracted negative ion currents are shown for the two extreme values of D studied. The optimum value of the plasma electrode bias is lower with the larger distance D. The corresponding extracted electron currents are also shown in Figures 4 and 5. Note that the extracted electron current is higher when the distance D is lower.

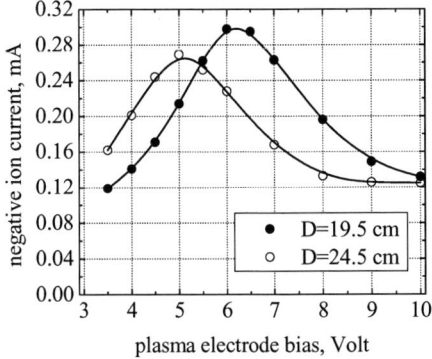 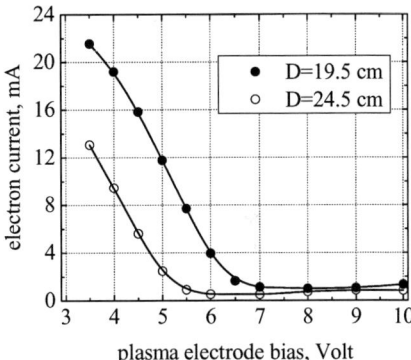

Figure 5. Effect of distance (D) from ECR minisources to the Plasma Electrode. Two distances are compared : 19.5 cm and 24.5 cm. Experiment effected under optimum pressure conditions (1.5 mTorr) with fresh tantalum deposited on the wall.

Effect Of The Wall Surface State.

The study of the source during the first six months, during which there were two longer interruptions of the pumping, indicated a continuous deterioration of the source characteristics. At optimum plasma electrode bias, a reduction of the negative ion current up to 20%, and a corresponding increase of the extracted electron current by a factor five were observed. We suspected that this change was related to the modification of the wall surface. While at the beginning of our study the wall was covered with tantalum, resulting from tantalum filament evaporation, the wall surface may have been polluted during the source operation and the interruptions of the pumping.

In order to verify this assumption, we installed a tantalum filament which could deposit tantalum on a part of the wall. Figure 6 shows the change of the extracted currents after two tantalum depositions. Note the very low ratio, $I_e/I^- = 2.5$, obtained after the second tantalum deposition for the plasma electrode bias of +6 V. These observations confirm those reported in Ref. 19 relative to the effect of tantalum deposition on the wall in a filament discharge.

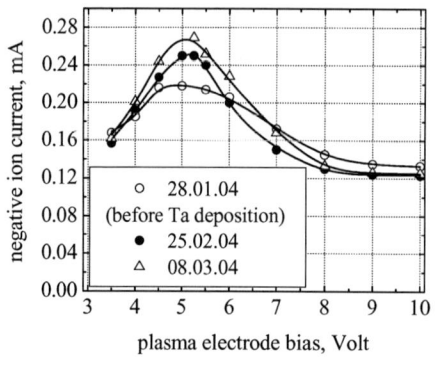

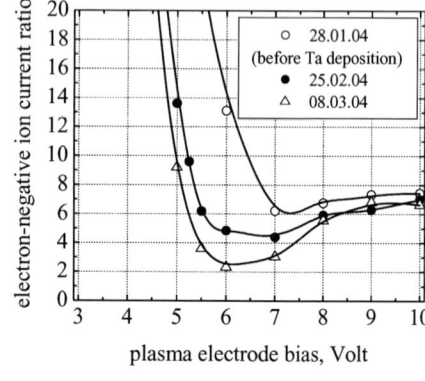

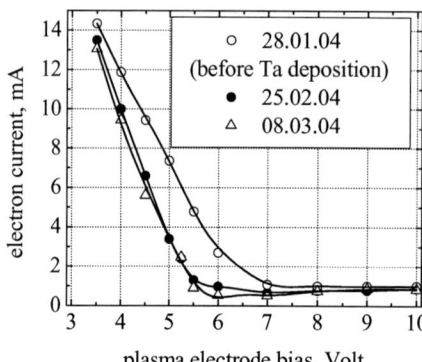

Figure 6. Evolution of the extracted currents and their ratio with depositing tantalum. Hydrogen pressure: 1.5 mTorr. The distance between the ECR minisources and the plasma electrode is 24.5 cm. The data of 280104 were taken before tantalum deposition.

Effect Of Negative Plasma Electrode Bias.

There was much controversy on whether a positive or a negative plasma electrode bias was more favourable for enhancing the negative ion current. In Ref. 14 as well as in earlier reports (Refs. 20, 21) it was shown that the negative ion current went through a maximum at a positive plasma electrode bias, approximately equal to the plasma potential. Here we explored the operation of our source with negative (with respect to the plasma potential) plasma electrode bias (see Fig. 7, curve without collar/without MF)). Note that there is a non-negligible negative ion current extracted even with negative bias on the plasma electrode, which is due to the effect of the positive extraction voltage. However there is no enhancement of the negative ion current compared to the value extracted at the optimum PE bias. The extracted electron current is maximum in the negative bias range. Thus there is no advantage for our source operation with negative plasma electrode bias.

This observation led us to consider two different modes of populating with negative ions the plasma in front of the plasma electrode.

1. In the case of negative with respect to the plasma potential PE bias the electrons oscillate in front of the plasma electrode, but are not collected by it. They produce locally negative ions, which are extracted.

2. In the case of PE bias close to the plasma potential, the electrons are collected by the plasma electrode, and the mechanism described in Ref. 14 and analyzed theoretically in Ref. 22 works : the negative ion density is enhanced in front of the plasma electrode when this region is depleted in electrons due to the positive bias and small transverse magnetic field by bringing there negative ions from the main plasma volume. The second mechanism leads under certain conditions to higher extracted negative ion currents than the first one and indeed to lower extracted electron currents.

Effect Of A Collar.

Following the positive results obtained by Peters [12] when using a collar in an rf source operated in pure hydrogen, we studied the effect of such a collar in our source.

We fixed the collar on the plasma electrode, which led to a reduction of the extraction opening diameter from 8 to 7 mm. The collar potential is imposed by the plasma electrode bias. The collar is made of copper, its inner wall surface is covered with a tantalum foil. Its inner diameter is 18 mm and its inner length 16 mm. The experiments were performed at 2 mTorr, with the distance D=24.5 cm.

As shown elsewhere [14] a weak transverse magnetic field (maximum 20 Gauss) is present in the plasma in front of the plasma electrode and is due to a leak from two magnets located in the extraction electrode. When the plasma electrode is biased positive this weak magnetic field leads to the increase of the negative ion density and the reduction of the electron density in front of the extraction opening [14]. Our first experiments with the collar were effected in the presence of this weak magnetic field only, without adding any additional magnetic filter (MF). The result is shown on Figure 7 by the curve labeled 'with collar/without MF'. It can be compared to the curve with the label 'without collar/without MF'. Note that the presence of the collar reduces the negative ion current approximately three times, much more than the factor

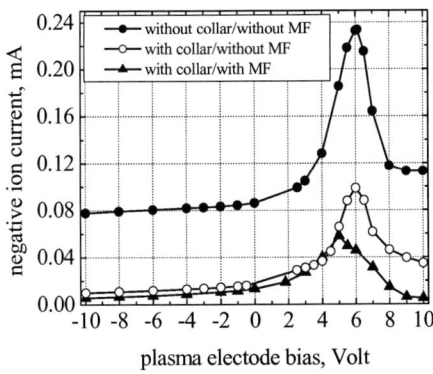 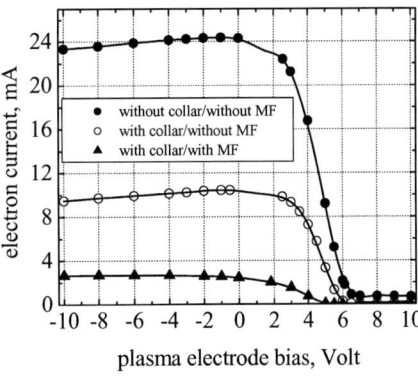

FIGURE 7. Illustration of the effects of introducing a collar and applying a negative plasma electrode bias. Hydrogen pressure 2 mTorr.

49/64=0.76 due to the reduction in the extraction opening. The beneficial effect is the reduction of the electron current.

In the next experiment we introduced an additional magnetic field produced by two small magnets, which was aligned with the existing one. The total maximum transverse field is 200 Gauss. The result is shown on Figure 7 by the curve labeled 'with collar-with MF'. One can note a further reduction of the negative ion current, as well as that of the extracted electron current. The experiments were conducted at positive and negative plasma electrode bias. As can be noted from Figure 7, there is no enhancement of the negative ion current with increasing the negative bias of the plasma electrode.

We should note that in our case the collar is introduced in a plasma region which does not contain considerable densities of fast electrons, which was not the case in [12]. Our results are similar to those reported by Leung *et al* [13] in pure hydrogen, in the sense that the effect of the collar was to reduce the extracted electron current, without any increase of the negative ion current. In our case, the negative ion current is also reduced by the introduction of the collar.

Effect Of A Stainless Steel Grid.

Following the report by Gobin *et al* [7] that a negatively biased stainless steel grid placed in front of the plasma electrode could enhance the negative ion production in an ECR source, we introduced a large stainless steel grid (0.55 mm wire diameter, 2 mm gap) in Camembert III, at 5 cm from the plasma electrode. No improvement was obtained by the presence of the grid and its negative bias. More exactly, no currents were extracted in the presence of the grid. This is a demonstration of the fact that in our device the microwaves are completely absorbed before reaching the grid and no energetic electrons are produced near this grid.

CONCLUSION

The elementary ECR sources appear to be useful components for a negative ion source. The studied network of seven sources produced at a distance of about 10 cm from them a plasma with optimum electron temperature for negative ion production, not much affected by wall effects. The extractor location should be optimized in order to take advantage of the properties of this plasma.

We can note that the pressure affects differently the plasma characteristics, measured near the ECR sources at distance d, and the extracted currents, measured at larger distances D, where d = D – 15 cm. The negative ion density increases monotonously with pressure, up to 4 mTorr, while the extracted current measured 15 cm farther, attains a maximum at 1.5 mTorr.

It seams that we deal with two different regimes at distance range 5 – 10 cm (where probe measurements are made) and at distance range 19.5 – 24.5 cm (where extracted currents are studied).

In the first regime we observe direct negative ion formation in the plasma flowing from the ECR sources, with the characteristic effect of electron temperature, enhancement of negative ion density with gas density and weak wall effect.

In the second regime a gas pillow interferes with this process (by H^- detachment and mutual neutralization), limiting the optimum pressure to 1.5 mTorr and leading to a strong wall effect.

The study of the extracted negative ion and electron currents in a wide range of plasma electrode bias, including bias negative with respect to plasma potential, indicates than negative ions can be extracted with negative (with respect to plasma potential) PE bias, but their current under these conditions is lower than that extracted at the optimum bias value, which is near the plasma potential.

The use of a collar and of a stainless steel grid in front of the plasma electrode did not lead to any positive result.

ACKNOWLEDGEMENT

The support of European Community (Contract No. HPRI-CT-2001-50021) is gratefully acknowledged.

REFERENCES

1. Hellbloom, G., Jacquot, C., *N.I.M.* **A243**, 255 (1986).
2. Mozjetchkov, M., Takanashi, T., Oka, Y., Tsumori, K., Osakabe, M., Kaneko, O., Takeiri, Y., Kuroda, T., *Rev. Sci. Instrum.* **69**, 971-973 (1998).
3. Hashimoto, K., Asano, S., *Fusion Eng. Design* **26**, 495 (1995).
4. Fukumasa, O., Matsumori, M., *Rev. Sci. Instrum.* **71**, 935 (2000).
5. Ciubotariu, C.I., Thèse de Docteur en Sciences de l'Université Paris XI Orsay No. 4793 (1997).
6. Tanaka, M., Amemiya, K., *Rev. Sci. Instrum.* **71**, 1125-1127 (2000).
7. Gobin, C.I., Delferrière, O., Ferdinand, R., Harrault, F., Benmeziane, K., Gousset, G., Sherman, J.D., *Rev. Sci. Instrum.* **75**, 1741 (2004).
8. Benmeziane, K., Thèse de Docteur en Sciences de l'Université Paris XI Orsay No. 7545 (2004).
9. Lacoste, A., Lagarde, T., Béchu, S., Arnal, Y., Pelletier, J., *Plasma Sources Sci. Technol.* **11**, 407-412 (2002).
10. Courteille, C., Bruneteau, A.M., Bacal, M., *Rev. Sci. Instrum.* **66**, 2533 (1995).
11. Ivanov Jr., A.A., Rouillé, C., Bacal, M., Arnal, Y., Béchu, S., Pelletier, J., *Rev. Sci. Instrum.* **75**, 1750-1753 (2004).
12. Peters, J., *Rev. Sci. Instrum.* **69**, 992 (1998).
13. Leung, K.N., Hauck, C.A., Kunkel, W.B., Walter, S.R., *Rev. Sci. Instrum.* **61**, 1110 (1990).
14. Bacal, M., Bruneteau, J., Devynck, P., *Rev. Sci. Instrum.* **59**, 10 (1988).
15. Bacal, M., *Rev. Sci. Instrum.* **71**, 3981- 4006 (2000).
16. Bacal, M., Ivanov Jr., A.A., Rouillé, C., Nishiura, M., Sasao, M., Proceedings of the 30th EPS Conference on Controlled Fusion and Plasma Physics, St. Petersbourg, Russia (2003).
17. Ivanov Jr., A.A., *Rev. Sci. Instrum.* **75**, 1754-1756 (2004).
18. *Microwave Excited Plasmas, Plasma Technology*, vol. 4, Chap. 10-12, edited by M. Moisan and J. Pelletier (Elsevier, Amsterdam, 1992).
19. Bacal, M., Ivanov Jr., A.A., Glass-Maujean, M., Matsumoto, Y., Nishiura, M., Sasao, M., Wada, M., *Rev. Sci. Instrum.* **75**, 1699 (2004).
20. Leung, K.N., Ehlers, K.W., Bacal, M., *Rev. Sci. Instrum.* **54**, 56 (1983).
21. Bacal, M., Hillion, F., *Rev. Sci. Instrum.* **56**, 2274 (1985).
22. Sakurabayashi, T., Hatayama, A., Bacal, M., *Rev. Sci. Instrum.* **75**, 1770 (2004).

Finding the Optimum Frequency and the H⁻ Distribution in the HERA RF- Volume Ion Source

J. Peters

DESY, Hamburg, Germany

Abstract. The HERA RF-Volume Source is the only source available that delivers routinely an H⁻ current of 40 mA without Cs. The dependency of the quality of the H⁻ beam on the frequency was investigated. A frequency range of 1.65 – 9 Mhz was scanned and the emittance was measured for several H⁻ currents up to 40 mA. The production mechanism for H⁻ ions in this type of source is still under discussion. Laser photodetachment measurements have been started at DESY in order to measure the H⁻ distribution in the source. The measurements have also been done under extraction conditions at high voltage. The results of the measurements with and without extraction are a basis for the development of a theory for the transition between plasma and vacuum (sheath), a cornerstone for beam transport programs. Knowledge of the H⁻ distribution and where they are produced makes further source improvements possible.

INTRODUCTION

The first RF driven volume source [1] was built in 1991 at LBL with an RF antenna immersed in the plasma. It became the prototype for the SSC [2] source which was later cesiated. The amount of Cs needed for this type of source is so low that it is possible to use Cs dispensers mounted on a collar around the plasma aperture. The source installed at SNS [3] is similar with a high duty cycle of 6%.

The DESY source [4] was modeled on the SSC source. It has a low duty cycle (0.12 %). 80 mA were reached with a tantalum collar [5] and the same internal antenna type which was used for SSC and LBL designs. It turned out that these antennas contained potassium (K) which works similar than Cs.

For reliability reasons the RF coupling coil was placed behind a ceramic outside of the plasma [6] (Fig. 1. and 2.). More than 50 mA were recently reached with this source. The lifetime of the system is more than a year compared to 500 – 700h for a filament system. The RF volume source works best with a starter system. At DESY a tandem source is used for this purpose.

CP763, *Production and Neutralization of Negative Ions and Beams,* edited by J. D. Sherman and Y. I. Belchenko
© 2005 American Institute of Physics 0-7354-0248-5/05/$22.50

FIGURE 1. The discharge chamber of the DESY volume source with an external RF coupling outside of the plasma.

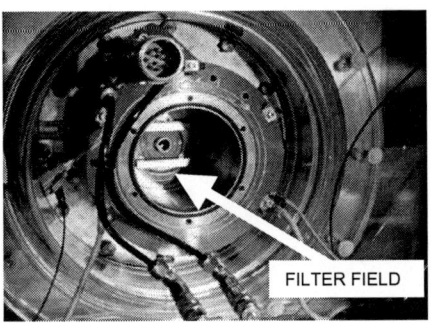

FILTER FIELD

FIGURE 2. Filter field magnets, collar, and plasma aperture of the DESY volume source.

FINDING THE OPTIMUM FREQUENCY FOR AN RF VOLUME SOURCE

Accelerator users need a maximum H^- current and a low beam emittance. For a low emittance, the H^- have to be delivered at a low plasma temperature. According to the volume process a high H^*_2 production is necessary for a maximum H^- current. There are at least two reasons for frequency dependence:

1.) The optimum energy level for producing H^*_2 has to be reached in half a period. When the time is too short (frequency too high) it is possible to increase the input power but this increases the plasma temperature.

2.) The plasma penetration is different due to the frequency dependent skin effect.

Determining the Parameters of the RF Coupling Coil

The magnetic field H and the electric field E inside of a coil (see Fig. 3.) can be

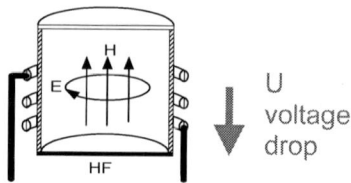

U
voltage
drop

FIGURE 3. Schematic of RF source, showing RF discharge chamber, coil , coil voltage drop and fields.

expressed with the following [7] :

$$H = I_{RF} \frac{N_{COIL}}{\sqrt{l_{COIL}^2 + 4R_{COIL}}} \qquad (1)$$

$$E = \frac{1}{2} r\omega B_0 \sin(\omega \cdot t) \qquad (2)$$

where l_{COIL} and R_{COIL} are the length and the radius of the coil. N_{COIL} is the number of coil windings. I_{RF} is the current delivered to the source by the transmitter and the transforming network. The electric field E depends on the magnetic induction $B_0 = \mu$ H and r the radial distance from the center of the coil. The voltage drop over the coil depends on the frequency ω and the inductance L. The dependence is given by:

$$U = I_{RF} \omega L \qquad (3)$$

$$L \approx N_{COIL}^2 \qquad (4)$$

The electric field produced by this drop can be reduced by using special winding techniques. A high field will disturb the plasma. It is important to note that the H field depends linearly but the voltage drop quadratically on the number of windings

Coupling of the Transmitter Power to the Plasma

The cooperation between DESY and Frankfurt University made it possible to bring a transmitter tunable with four coils and a capacitor in the range between 1-13 MHz to DESY. A coupling box with a variable transformer and a tunable capacitor (Cres) was built by DESY. Cres forms together with the coil inductance a series resonance circuit (Fig. 4.). The power delivered to the source is influenced by the skin effect, parasitic capacitances and the proximity effect. Power measurements were done with a dual directional coupler. Measurements directly at the coil turned out to be very difficult due to the nonlinear characteristic of the source load and the phase sensitivity of the parameters.

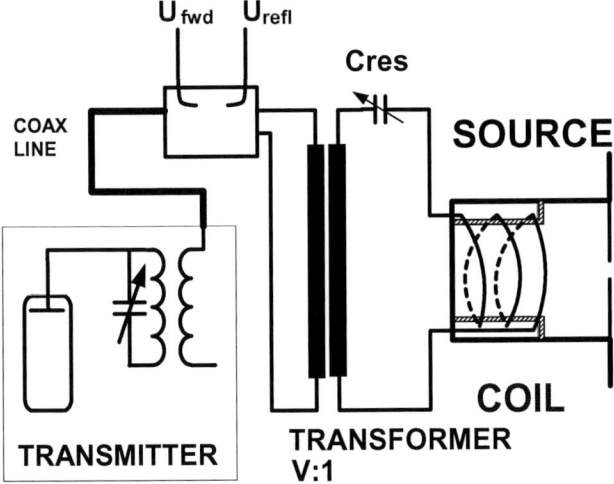

FIGURE 4. Coupling of the RF transmitter to the source.

H⁻ Current vs. RF Power for Different Frequencies

Fig. 5. shows the frequency and RF power dependence of the H⁻ current. It turns out that with the set up used a frequency range between 2 MHz and less than 4 MHz demonstrates the best power efficiency. For higher frequencies the plasma penetration is reduced due to the skin effect, the voltage drop over the coil increases, parasitic capacitances and the proximity effect become more effective. In the low frequency range the voltage drop and the electrical field is reduced. These two effects can be compensated partly by a higher number of windings.

Unfortunately it is not possible to compare these measurements with those of the SNU [8] (13.56 MHz) and the two ECR type sources at SACLAY [9] and Yamaguchi University [10] both at about 2.5 GHz, because their RF power and the currents are much lower.

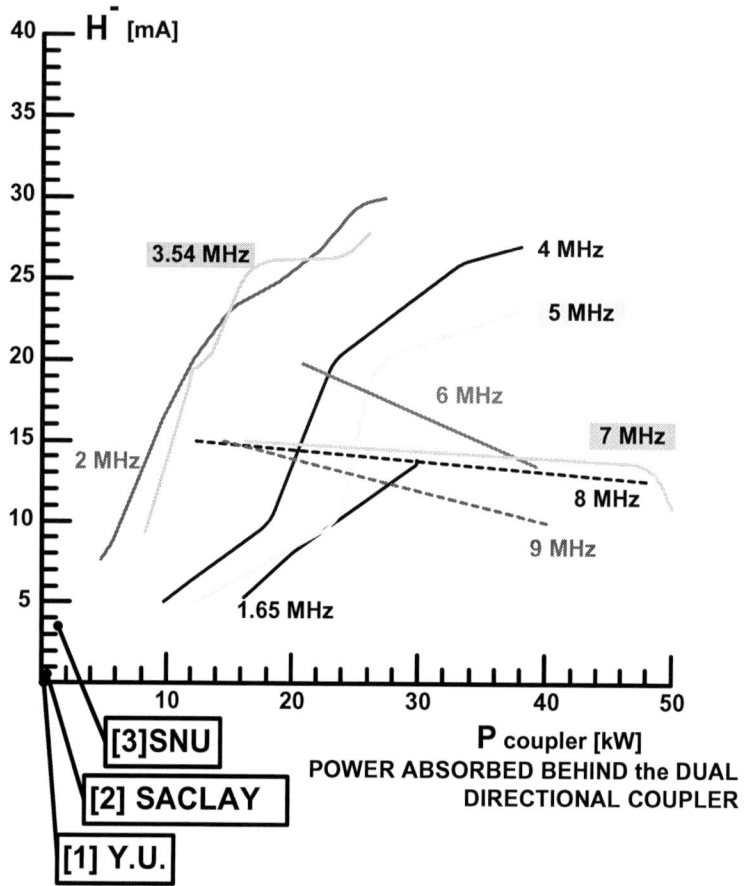

FIGURE 5. Frequency and RF power dependence of the H - current.

Measurement of the Emittance vs. Frequency

Fig. 6. shows the frequency and H current dependence of the normalized 90% rms emittance. The dependence on the H current was used for comparison because this is a parameter which can be measured easily. The curves demonstrate almost the same characteristics. This might be due to a slight change in the position of the beam. Only the 4 MHz curve seems to be different. Note that it is difficult to keep all parameters during the long emittance measurements constant. Especially the gas pressure and the tuning of the series resonance seem to be very sensitive.

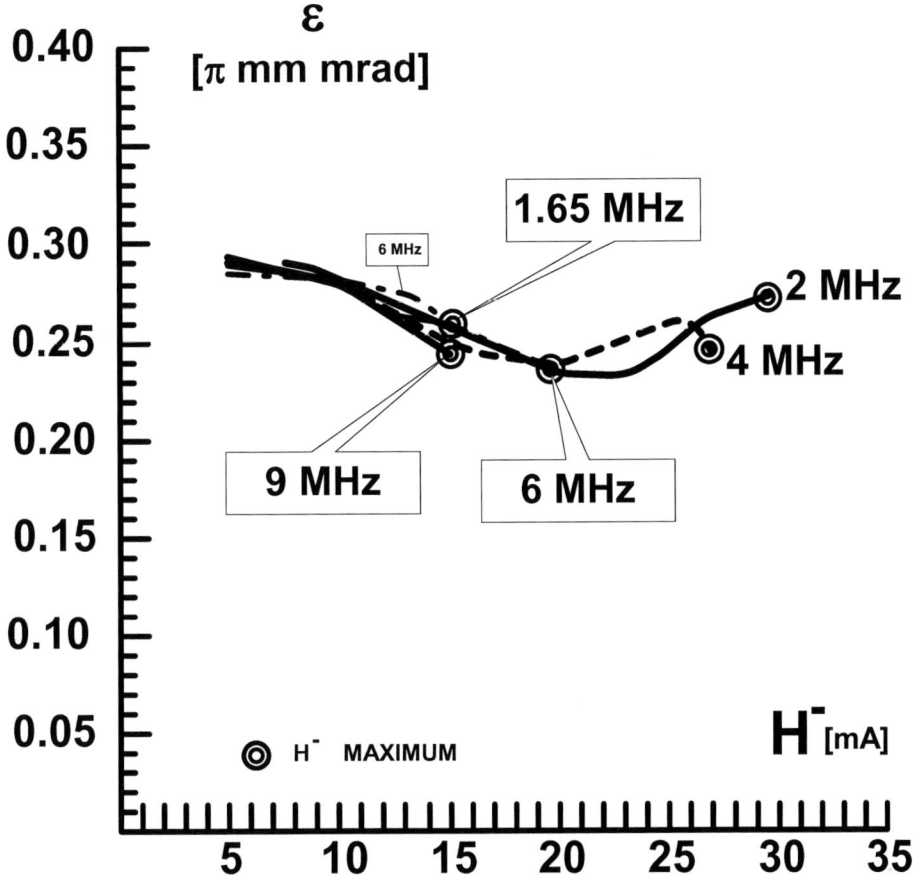

FIGURE 6. Frequency and H⁻ current dependence of the rms emittance.

H⁻ Current vs. RF Power for Different Frequencies and Numbers of Windings

Coils with 6.5, 20 and 40 windings were used in order to measure the frequency dependence of the H⁻ current (see Fig. 7.). It was not possible to get a stable performance of the source for all coils at higher frequencies.

A power of 26 kW was reached for all coils at the frequencies 1.65 MHz and 2 MHz (Fig. 8). This made it possible to plot the frequency dependence of the H⁻ current with the number of windings as parameter and the winding dependence with two frequencies as parameter. Fig. 9. demonstrates the improvement of performance at lower frequencies with the number of windings (H~N_{COIL}). It was not possible to reach the 6.5 wdg. values. In Fig. 10. the reduction of the H⁻ current due to the voltage drop $U \sim N_{COIL}^2$ is demonstrated. It reaches values of $U \sim 18$ kV. In further experiments a shielding of the electrical field will be applied.

Adding up all improvements and using a high power transmitter we were able to reach a peak H^- current of 61 mA with an average of 58 mA (see Fig. 11). The Rf power was about 35 kW.

FIGURE 7. RF coupling coils with 6.5 and 40 windings

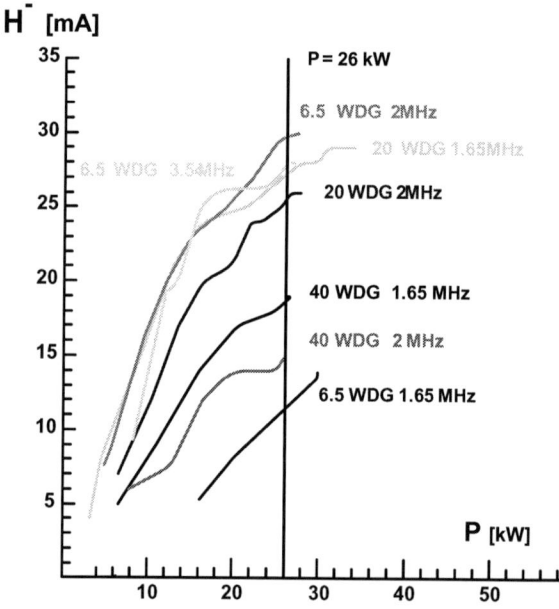

FIGURE 8. H^- current dependence on RF power and the number of windings.

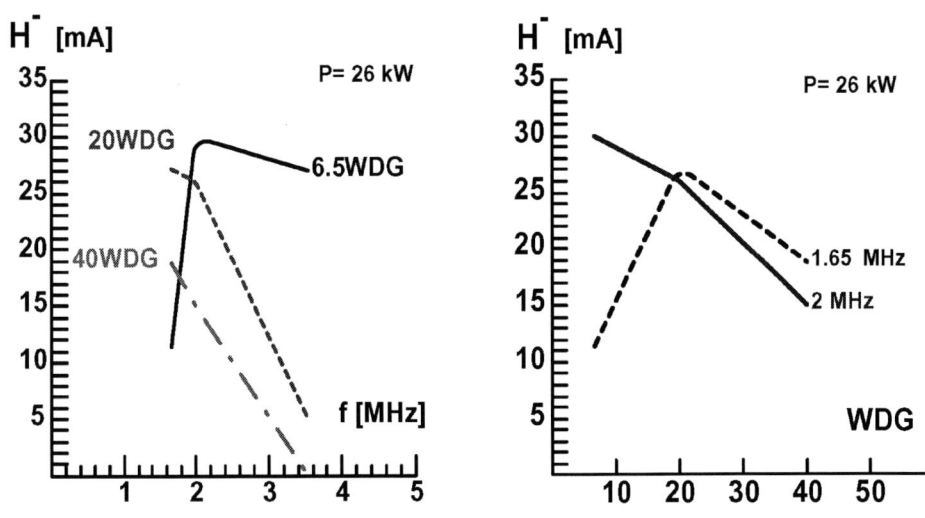

FIGURE 9.   **FIGURE 10.**

| H - current dependence on frequency with the number of windings as parameter at 26 kW. | H ¯ current dependence on the number of windings with the frequency as parameter at 26 kW. |

PEAK 61 mA H¯

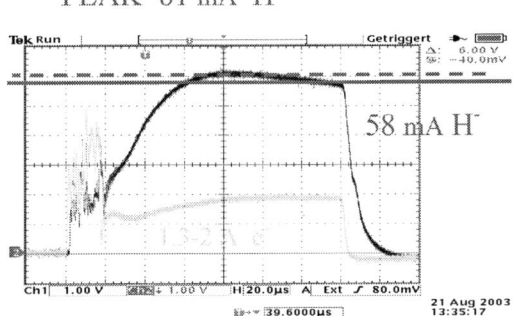

FIGURE 11. H ¯ current at the source exit toroid and the electron current.

COOLING SECTION BETWEEN DISCHARGE CHAMBER AND FILTER FIELD

Cooling sections with the length of 0.5, 1, 1.5 and 2 cm were added between the discharge chamber and the filter field. The usual multi cusp field was kept. It turned out that the brightness=I_{H^-} /εy can be increased with insertions for H⁻ currents higher than 40 mA. However the maximum of the H⁻ current is reduced.

PHOTODETACHMENT MEASUREMENTS

Photodetachment measurements in order to measure the density of negative ions have been done as early as 1969 [11]. First density measurements of H⁻ are reported in 1979 [12]. In the HERA source a modification of a technique with a cylindrical metal probe (Langmuir probe) aligned parallel to a laser axis [13] was used. The details of the source are given in several papers [14], [15] ,[16].

Measurement Set Up

The power applied at the RF coil of the HERA H – source is pulsed. With a positive bias (U_{LP}) of 5V close to saturation one draws an electron current to the tip of the probe. This current (I_{LP}) is measured with a toroid (see Fig.12). The current signal is shown in Fig.13a. The affinity of the electron attached to the hydrogen atom is with 0.75eV very low [11]. By photodetachment H⁻ + h ν = H + e- an increase in electron density is produced.

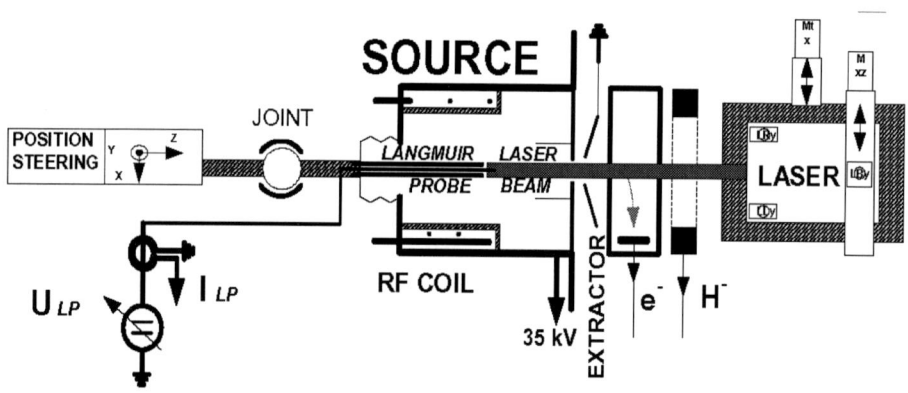

FIGURE 12. HERA RF source with mounted langmuir probe on the back side and a laser beam shooting through the extractor hole.

To clearly interpret the signals it has to be made sure that no other photon processes like photoionisation take place. Fig. 13b shows the increase in electrons detected when a 9 nsec, 1064 nm puls of a Nd: YAG laser travels on the axis of the source. The maximum pulse energy of our laser was 650 mJ per pulse. The 8 mm ∅ laser beam was compressed to 3mm ∅ with an optical system.

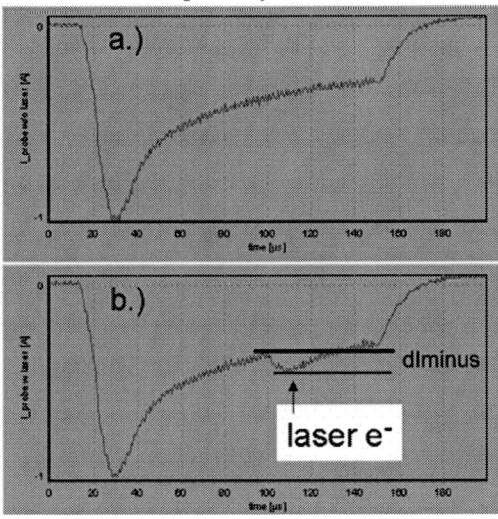

FIGURE 13. Current pulse of the Langmuir probe (I_{LP}) without laser beam (a) and with photo detached electrons (b).

The beam was dumped on the probe or on a ceramic dump surrounding the end of the probe. A set up where the laser beam is dumped outside of the source is given in [17]. The probe tip and inner conductor are a molybdenum wire of 0.4 mm ∅. The center wire is shielded by a metal tube which is completely isolated from the plasma with sealed ceramic tubes (see Fig. 14).

Due to the environment there is noise on the signals in Fig. 13. By averaging over many pulses it was possible to get very stable values.

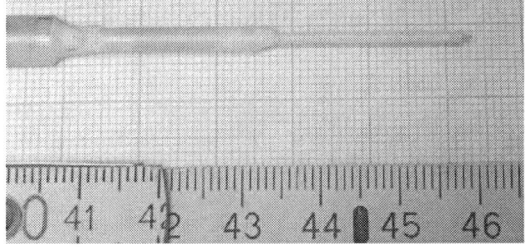

FIGURE 14. Ceramic shielded probe with tip.

We tried different ways [18] to detect the increase in the number of electrons due to the laser beam. First the signal without laser beam was subtracted from the signal with laser beam. It turned out that detecting the difference between maximum and the minimum in the flat part of the same pulse before the laser starts delivers the same results.The laser beam was fixed on axis of the source. The probe tip was moved step by step in rectangular planes perpendicular to the beam axis (see Fig. 12). The x and y movement is done by turning the probe in a joint the z motion by pulling the probe in and out. The movements in x, y and z are done with a three table system. A long bellow is used for transforming the movements into the vacuum.The size of the planes over which measurements were done varied from 4mm x 4mm in the collar area to 10mm x 10mm in the RF coil range and had in the final part of the source a size of 6mm x 6mm. All measurements were done with a 0.5 mm step size. The z planes were measured in 5mm steps.

Results

Fig. 15 gives a 3D sample presentation of typical H^- intensities measured on selected rectangular areas along the source. It was possible to associate measurement patterns to different zones of the source. The H – intensity (Nh-) is proportional to the additional electron current delivered due to the laser pulse . The Nh- measurements were taken along the central axis of the source in arbitrary units. Measurements were taken at 0 V and at 10 kV extractor gap voltage.

Strong Cusp Magnet Field Area

In the range between z=65mm and 95mm a strong multicusp field is present in the source. Only a small maximum was detected in an area of about 1.5mm by 1.5mm which is a factor 2 smaller than the laser beam diameter. Significant peaks with a Nh-value around 1 were measured.

RF Field Area

Between z=100 mm and 130-140 mm a strong RF field is applied to the plasma. The cusp field is reduced and the dipole filter field starts only at the end of the range. One finds now laser produced electrons in a circle with 10mm diameter. A plateau is formed and in addition there are spikes. The spikes become more numerous when the acceleration voltage is applied. Apparently the freed electrons are accelerated by the RF field and this movement is modulated by the acceleration voltage. In this range a maximum Nh- value of 2.28 was found.

Filter Field Area (in front of the collar)

A dipole field is applied which has a maximum of about 20 mT at z= 157 mm and goes to zero in the middle of the acceleration gap. The plateau becomes lower which could be due to a reduced RF field. Just in front of the collar the size of the measurement plane was reduced in order to avoid damage. Here the H $^-$ densities were less than half the maximum values.

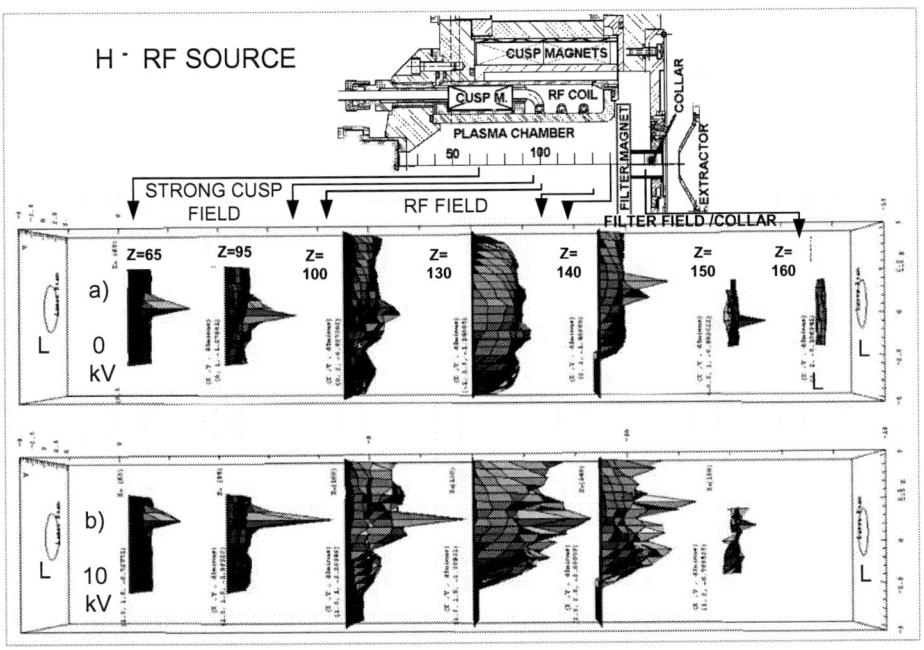

FIGURE 15. H⁻ intensities along the source axis without extractor gap voltage (a.) and with 10 kV applied (b.). L marks the laser beam diameter.

Collar Area

In this region only one measurement was done without acceleration voltage due to sparking which occurred under high voltage. With the 6.5 mm plasma aperture the plasma density is here reduced. The H⁻ density (Nh⁻) is only about 5% of the maximum values detected. In case of an applied acceleration field there would be a competition between the field from the probe tip and the accelerating field.

Acceleration Voltage (On/Off)

Contrary to expectations a large intensity change was found when an extraction voltage of 10 kV was applied. This difference is most obvious in the RF field area. The voltage was applied in the usual operation mode with extractor at ground and the source at high voltage. The mechanic for the probe was grounded and isolated from the probe and source bucket.

Alignment

In the strong cusp magnet field area the position of the maximum intensity varies only 0.5 mm, the step size. In the RF field area the variation is bigger and with HV applied many peaks occur. In case of the filter field range without HV the adjustment was lost due to a power failure.

Photodetachment and H⁻ Generation Uncertainty

The biased tip of the Langmuir probe collects the photodetached electrons which are seen as a local maximum within about 5 μsec (see Fig. 13). If there is a plasma potential with a gradient present or if there is an RF field there will be an uncertainty. Electrons detached in the 9 nsec laser light channel flash can be driven to the probe or away from the probe dependent on the plasma gradient. A contribution to the dIminus maximum will depend on distance of the electrons from the tip and their speed. A local maximum of the plasma potential will collect additional electrons from all sides a gradient plasma mainly from lower side. A complicated plasma potential surface can result in very different pictures of the measured dIminus.

The field gradient due to the RF field has a similar effect. Dependent on the RF frequency the electrons will be moved several times back and fore before it hits the tip. As seen in Fig. 15 & 16 a transversal component is added to the picture in the area of high RF levels.

In a low density plasma there is a strong effect of the extraction voltage on the plasma potential. Measurements of the HERA source showed only a small influence of the extraction voltage but a big one of the collar potential [16].

Uncertain is also where the measured H⁻ are produced if there is a plasma gradient or a RF field. They are also moved by these fields and within the pulse of 100 μsec there is plenty of time to redistribute.

The RF source, four distributions of photodetached electrons and the plasma potential are shown in Fig. 16. The photodetached electrons are concentrated where the plasma potential has its maximum. Photodetaced electrons and H⁻ are moved to this spot by the voltage gradient.

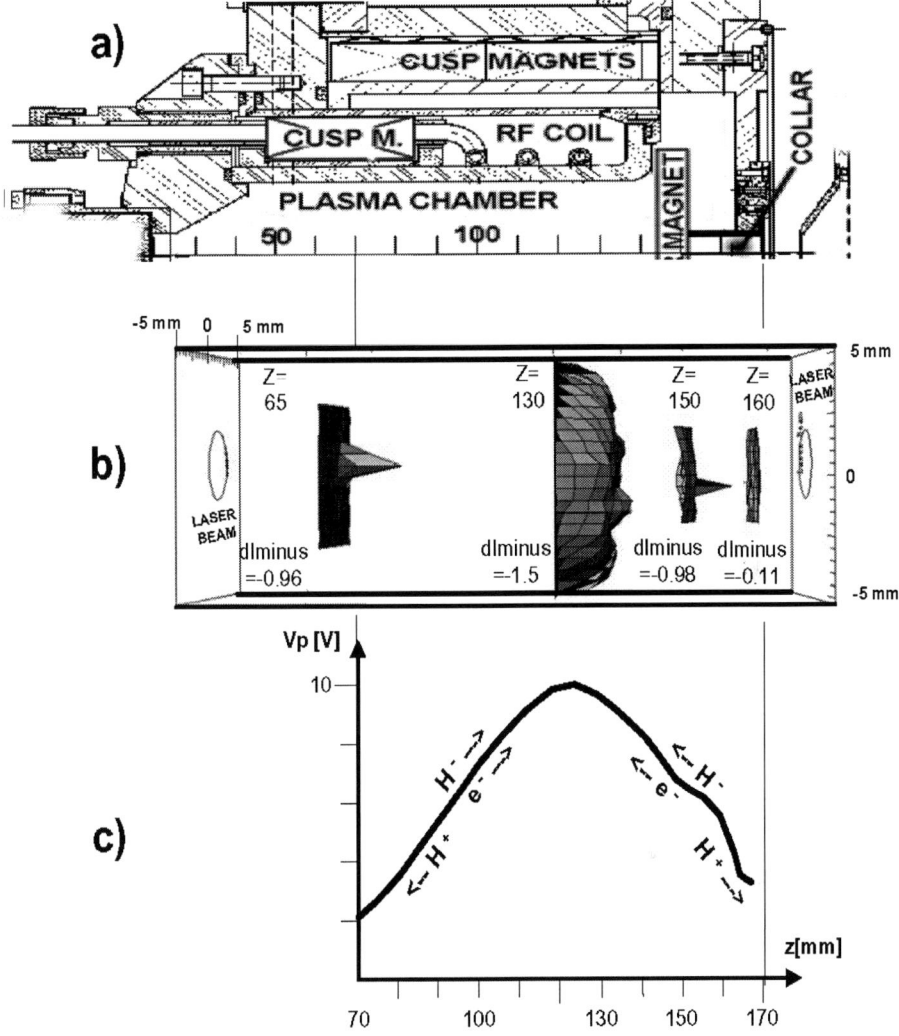

FIGURE 16. a) RF source b) Photodetached electrons c) Plasma potential . The photodetachment electrons are concentrated where the plasma potential has its maximum. photodetaced electrons and H- are moved to this spot by the voltage gradient.

CONCLUSIONS

Photodetachment measurements of the H⁻ ions show that the extraction voltage has an unexpected effect on the H⁻ distribution in the plasma. The photodetached electrons and the H⁻ are strongly influenced by the different magnetic fields and the RF which is coupled into the source plasma which results in the shown plasma potential. A 20mm shift of the plasma potential maximum to the entrance of the collar should improve the

H $^-$ distribution and lead to a higher extracted H $^-$ current. Care has to be taken that the electrons which attach to the vibrationally excited hydrogen molecules keep their temperature. It will be important to study the H $^-$ distribution not only on axis but in the whole source. The magnetic and RF fields together with the laser light should be carefully measured. Varying the filter field and observing the change in the plasma transition (sheath) will be of special interest.

ACKNOWLEGMENTS

The author would like to thank H. Klein, U. Ratzinger and K. Volk for helpful discussions and providing a tunable transmitter.Many thanks are due to M. Bacal for help and advice with this work.The author is grateful for the contribution of the following colleagues at DESY: I.Hansen, H.Sahling and R.Subke. The thesis (Diplomarbeit) of C. Sehnke [18] was an important basis for our studies. The author also wishes to thank H. Weise and the technical groups at DESY for their support and M.Lomperski of DESY for helpful suggestions to the wording of the article. The support of EEC (Contract HPRI-CT-2001-50021) is gratefully acknowledged.

REFERENCES

1. K.N.Leung, G.J.DeVries, W.F.DiVergilio and R.W.Hamm, Rev.Sci.Instrum.62(1),100(1991).
2. K. Saadatmand, et al., Rev.Sci.Instrum. 67(3), March 1996, pp. 1318-1320.
3. R.Keller, et al., AIP Conf. Proceedings 639, 2002, pp. 47-60.
4. J.Peters, Proceedings of the XVIII International Linear Accelerator Conference (August 1996), pp. 199-201.
5. J.Peters, Rev.Sci. Instrum. 69 (2),(1998), pp. 992-994.
6. J.Peters, Proceedings of the XVIV International Linear Accelerator Conference (August 1998), pp.1031-1035.
7. H. Löb and J. Freisinger, Lectures Plasmaphysics, Giessen 1994, pp. 71-72.
8. I.S.Hong, AIP Conf. Proceedings 639, 2002, pp. 61-66.
9. R.Gobin, Rev.Sci.Instrum. 75, No. 5, May 2004, pp.1729-1731.
10. O.Fukumasa, Rev. Sci. Instrum., Vol.71, No. 2, February 2000, pp.935-938.
11. J.Taillet, C. R. Acad. Sci., Ser. B 269, 52 (1969).
12. M.Bacal and G. W. Hamilton, Phys. Rev. Lett. 42, 1538 (1979).
13. M.Bacal, Rev. Sci. Instrum., Vol. 71, No. 11, November 2000.
14. J.Peters, Rev. Sci. Instrum., Vol. 75, No. 5, May 2004, pp.1709-1713.
15. J.Peters, Proceedings of the XXI International LINAC Conference, Gyeongju, Korea, August 19-23, 2002.
16. J.Peters, Ph.D. thesis, Universität Frankfurt,2001.
17. Y.Matsumoto et al, Rev. Sci. Instrum., Vol. 73, No. 2, February 2002, pp.952-954.
18. C.Sehnke, Entwurf u. Implementierung eines modularen Sofwaresystems zur Plasmadiagnostik mittels Langmuir-Kurven u. Photodetachment, Diplomarbeit, FH Wedel, Februar 2004

H⁻ Source with the Volume-Plasma Formation of Ions

Yuri V. Kursanov *, Petr A. Litvinov, Vladimir A. Baturin

*Institute of Applied Physics, National Academy of Science of Ukraine,
58 Petropavlovskaya St. Sumy, 40030 Ukraine, E-mail: baturin@ipflab.sumy.ua
* Sukhumi Institute of Physics & Technology, Abkhazia*

Abstract. In the presented work results of experimental researches of two versions of H- ions source with axial-symmetric and slot-hole geometry of ions extraction are submitted. Formation of ions occurs in a volume of hydrogen plasma (without additives of cesium) due to two-step dissociative attachment of thermalized electrons to vibrationally excited molecules. These sources were executed in such a manner that processes of vibrational excitation and dissociative attachments of electrons are divided in space. It allows optimizing these elementary processes separately. In these sources ions are extracted from paraxial plasma along a divergent magnetic field. Due to this fact the suppression of electrons occurs.

INTRODUCTION

In the presented work results of experimental researches of two versions of H⁻ ions source with axial-symmetric and slot-hole geometry of ions extraction is submitted. Formation of ions occurs in a volume of hydrogen plasma (without additives of cesium) due to two-step dissociative attachment of thermal electrons to vibrationally excited molecules H_2 [1]. The cross-section of this process quickly grows up to a significant size $(>10^{-17}$ cm$^2)$ with growth of oscillatory quantum number at electron energy of several eV.

Optimization of source conditions for vibrational excitation of molecules and for the subsequent formation of negative ions is realized due to creation of discharge system that generates in the emission chamber two areas of plasma - peripheral, with rather big fraction of fast electrons and paraxial with cold electrons.

SOURCE DESIGN

In Fig. 1 the design of a source with axial-symmetric geometry of formation of ion beam is schematically shown [2]. Its magnetic system is designed on the basis of permanent Sm-Co$_5$ magnets which create in an interpolar gap magnetic field $B_z = 0,09$ - 0,12 T. The discharge chamber 5 consisting of the cathode 1 and anode 2 represents inverse gas magnetron geometry, which works on the basis of glow discharge in crossed ExH fields. The electromagnetic valve 7 supplies gas magnetron with working gas. Under submission of a pulse of a voltage onto electrodes of the discharge chamber, plasma of a tubular configuration is generated. Under the certain conditions

CP763, *Production and Neutralization of Negative Ions and Beams*, edited by J. D. Sherman and Y. I. Belchenko
© 2005 American Institute of Physics 0-7354-0248-5/05/$22.50

[3] plasma penetrates through an annular slot into the emission chamber 6 and reaches the emission electrode 3. For stable current transmission, the additional electrode 4 is placed near a face part of magnetron, on which voltage higher than anode potential is applied.

Due to formation of double layer before a narrow annular slot at magnetron output, fast electrons are delivered to the area of its volume. In peripheral plasma conditions favorable for vibrational excitation of molecules are created. Internal paraxial plasma, formed by diffusion of peripheral tubular plasma across a magnetic field, will contain the vibrationally excited molecules and the enriched fraction of slow electrons, while the fraction of fast electrons does not penetrate here because of the action of magnetic filter. Thus, in the internal plasma there are necessary conditions for effective realization of a finishing phase of two-step process of formation of negative ions.

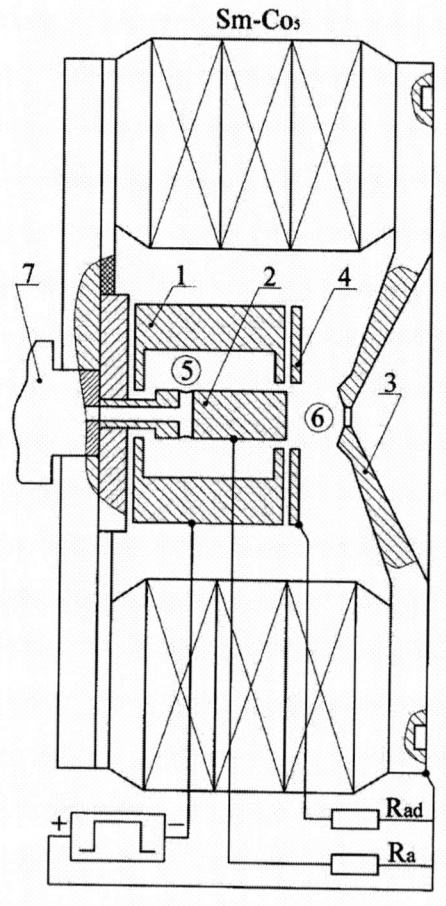

FIGURE 1. Source with axial-symmetric geometry.

EXPERIMENTAL RESULTS

Extraction of ions from a source is made from paraxial zone of the emission chamber. Suppression of accompanying electrons occurs due to their moving along a magnetic field onto the emission electrode serving as the source anode. Negative ions are practically not affected by the influence of a magnetic field and at observance of a condition $\lambda_i > d$, they participate in emission of ion beam. Here λ_i - mean free path of a negative ion, d - distance from a place of its formation to the emission aperture.

In Fig. 2 dependences of an emission current of ions H- (curve 1) and a current of accompanying electrons (curve 2) are given as a function of discharge current. Each point on this curve is given at the optimized source pressure. Extraction voltage was varied to keep the perveance constant with the discharge current increase.

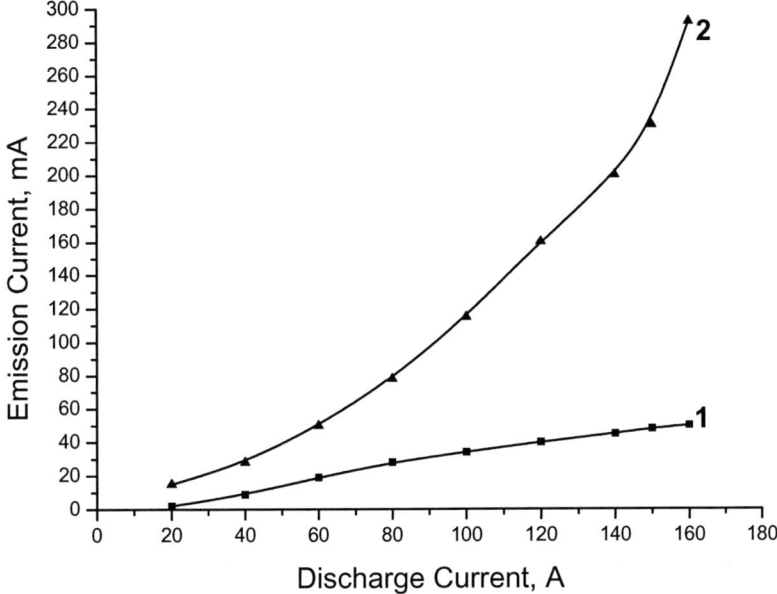

FIGURE 2. Dependence of negative ions current (curve 1) and accompanying electrons current (curve 2) on discharge current.

At discharge current I_{dc} = 160A and U = 98 kV ion current I_i = 50 mA is received with emission density of ion current J $\approx$ 0,22 A/cm^2 and an electron current of I_e = 292 mA. The dependences show, that at the discharge current > 140 A the ion current curve is no longer linear, and the curve of electron current grows more abruptly. It is possible to explain such behavior by development of fluctuations in plasma with the discharge current increase and by violation of stability of its tubular structure, and also by decrease of Sm-Co$_5$ magnetic strength as a result of thermal flux from hot emission electrode with which they are in good thermal contact. The basic parameters received for this source, are given below in Table 1.

TABLE 1. The basic source parameters.

Ion beam current I_i	~ 50 mA
Ratio I_e/I_i	$\sim 3 - 6$
Ion beam energy	$- 10 - 100$ keV
Emission current density	$\sim 0,22$ A/cm^2
Impurity ions	$\sim 3\ \%$
Pulse duration	$\sim (10^{-4} - 10^{-3})$ s
Repetition rate of pulses	$\sim (1 - 10)$ s^{-1}
Life time of source	$\sim (1\text{-}2)\ 10^6$ pulses

The lifetime of the ion source was limited by the electromagnetic gas valve. It is possible to increase the H- ion current from source (without introduction of additional processes responsible for their kinetics in plasma) by expansion of volume of negative ions generation and by the corresponding increase of the emission aperture area. Increase of the emission aperture area is limited by allowable values of phase characteristics of ion beam and by aberrations of ion-optical system.

In the next ion source version, shown in Fig. 3, expansion of area of H$^-$ ions generation is carried out by formation of internal paraxial plasma in the form of the cylinder of the height d and of the oval cross section extended along the emission slot.

The area of the emission slot is 2x15mm^2 [4]. As in the emission area of the source concentration of negative ions is maximal in the paraxial zone, and concentration of electrons is maximal on periphery, then the output of negative ions and accompanying

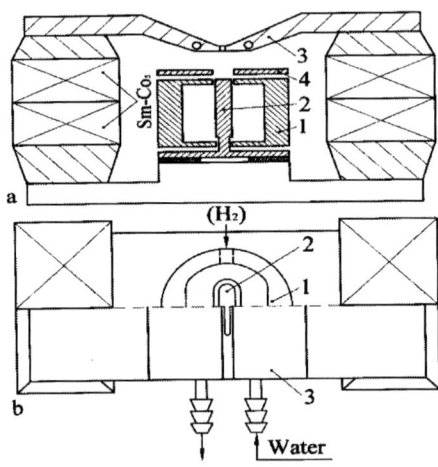

Figure 3. Ion source with slot-hole geometry.

electrons will depend on value Δr, which is equal to the distance between the perimeter of an emission slot and a projection of the magnetron anode end onto the emission electrode. It is necessary to optimize this Δr value for achieving the acceptable ratio I_e/I_i. Experiments showed that with a real instability of peripheral plasma and under non-ideal coaxiality of the discharge chamber, value Δr in this source should be about ~ (2 - 3) mm.

Dependence of negative ion beam current as a function of discharge current is shown in Fig. 4. H⁻ ions current obtained has a value of ~ 60 mA for this source modification. The instability of the oval form of peripheral plasma begins to appear with the further increase of discharge current.

Dependence of I_e/I_i ratio as a function of discharge current at optimum values of pressure in the source is represented in Fig. 5. The dependence has a nonlinear character. In this ion source version current of accompanying electrons is higher than in the source with axial-symmetric geometry. It can be explained by the fact, that the appearance of plasma inhomogeneity along the azimuthal direction becomes more probable with the oval form of tubular plasma. Rotation of these inhomogeneities in the crossed ExH fields results in rotary oscillations and, correspondingly, to emission of fast electrons to the area of emission slot.

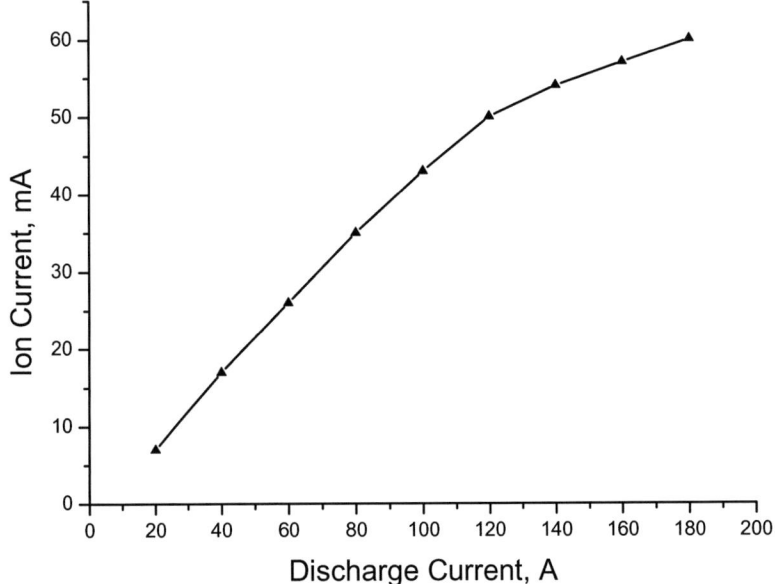

FIGURE 4. Dependence of H- ion current on the discharge current.

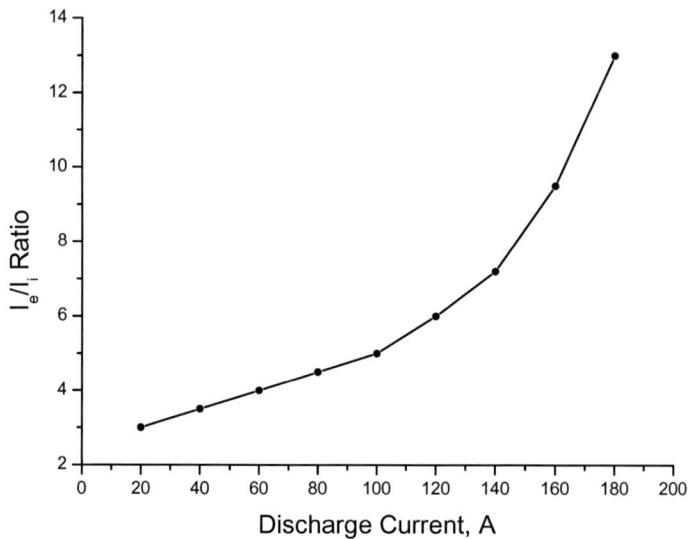

FIGURE 5. Dependence of I_e/I_i ratio as a function of discharge current.

DISCUSSION

The ion sources described were in operation for a long time in the structure of RFQ accelerator of the Sukhumi Institute of Physics & Technology. We propose to obtain further increase of negative ions current in the source by introduction of additional active mechanisms, which will allow to increase the concentration of slow electrons and vibrationally exited molecules. The concentration of slow electrons can be increased by slowing down fast electrons of peripheral plasma in the emission chamber using reflective electrode. The concentration of vibration-exited molecules can be increased by placing metal hydride in the end face of electrode 2 (Fig.1). This material desorbs vibration-exited molecules of hydrogen under certain conditions [5].

REFERENCES

1. Hiskes J.R. and Karo A.M., *J. Appl. Phys.*, **56**, 1927 (1984).
2. P.A. Litvinov, *Vopr. Atom. Nauki Tekhnol. Ser. Nucl. Phys.* 4,5 (31,32), 48 (1997) -in Russian.
3. P.A. Litvinov, V.A. Baturin, *Flash chamber of a quasi-continuous volume source of negative ions* (in these proceedings).
4. P. Litvinov, V. Baturin, *Nucl. Instr. and Meth. in Phys. Res. B.* **171**, 573 (2000).
5. Shmalko Yu.F., Lototsky M.V., Klochko Ye.V. and Solovey V.V., *J. Alloys and Compounds*, **231**, 856 (1995)

Flash Chamber of a Quasi-Continuous Volume Source of Negative Ions

Petr A. Litvinov, Vladimir A. Baturin

Institute of Applied Physics, National Academy of Science of Ukraine,
58 Petropavlovskaya St. Sumy, 40030, Ukraine, E-mail: baturin@ipflab.sumy.ua

Abstract. The design of plasma generator with cold cathode for negative ion source is described, and results of experimental studies are submitted. The generator consists of three chambers connected in series with the dosed gas leaking between them. The electrode system of the first chamber (with high pressure) and of the second one (with low pressure) represents two-chambered inverse gas magnetron. The second chamber connects via an annular slot with the third (emission) chamber, in which plasma of the tubular form is generated. In the third chamber plasma is divided into two areas (peripheral and paraxial) where necessary conditions for effective formation of negative ions are created. The submitted generator will allow optimizing a design of negative ions source.

INTRODUCTION

The direct current E x H glow discharges are attractive for generation of low-temperature plasma due to stability of discharge burning, simplified operation and an advanced reliability of the discharge cell at current density on the cathode up to tens of A/cm^2 and voltage on the discharge in some hundreds of volt. Their working pressure is in the range of 10-10^{-1} Pa. For use of glow discharges in the dc ion sources it is necessary to decrease gas pressure in the discharge chamber. This fact is especially important in the sources of negative ions, where negative ions are collected from all volume of their generation, therefore the mean free path of a negative ion in the emission area should be large. For minimization of negative ion neutralization it is necessary to reduce the pressure in front and behind of the emission aperture, where the velocity of negative ions is small and cross section of their neutralization in collisions with gas is large.

Decrease of gas feed into discharge system with cold cathode leads to sharp increase of ignition and burning voltage of the glow discharge. The inverse gas magnetron is one of the gas discharge devices, known at the present time, which operates at the lowest possible gas pressure. It is possible to decrease gas pressure to the level excluding a noticeable neutralization of negative ions in the pulsed negative ions source with the inverse gas magnetron, by the pulsed gas [1]. Gas flow dynamics in the discharge space and various velocity of neutral and charged particles is important in this case. There is an opportunity to decrease an ignition voltage and to reduce gas pressure for the stationary low pressure discharges. It can be obtained due to auxiliary discharge at high pressure, which delivers plasma into the discharge cell

CP763, *Production and Neutralization of Negative Ions and Beams*, edited by J. D. Sherman and Y. I. Belchenko
© 2005 American Institute of Physics 0-7354-0248-5/05/$22.50

with low gas pressure. The auxiliary discharge works as a plasma cathode, which supplies a low pressure discharge with the primary electrons.

The results of experimental researches on creation of three-chambered discharge system with cold cathode, which generates stationary cold plasma at low pressure in the output chamber are given in this work.

THE SOURCE ELECTRODE SYSTEM AND PHYSICAL PROCESSES IN THE GAS-DISCHARGE GAP

The gas discharge system (Fig.1) represents an axisymmetrical construction, consisting of three chambers with the dosed gas leak between them. The copper anode (1 and 2) and copper cathode (3 and 4) are made of two parts. Part 3 of the cathode and part 1 of the anode create the first discharge chamber 5 (high pressure), and part 4 of the cathode and part 2 of the anode create the second discharge chamber 6 (low pressure). Chamber 5 is connected with chamber 6 by convergent ring channel 7, and the discharge chamber 6 is connected with the third emission chamber 8 with the help of another convergent ring channel 9. Due to gas flow resistance of the convergent ring channels, the decrease of pressure in each following chamber is obtained. An additional decrease of working gas pressure due to its expansion to the large volume and due to radial differential pumping is obtained in the emission chamber 8. Sm-CO$_5$ magnets (14 and 15) is established between the emission electrode 11 and a magnetic pole 12. The framework 13 of the discharge system,

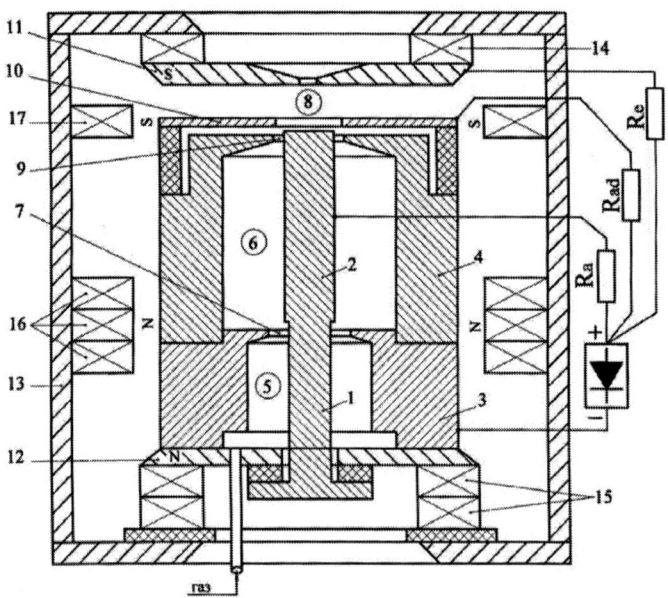

FIGURE 1. Schematic diagram of three-chambered discharge system. R$_a$ = 680 Ω, R$_{ad}$ = 630 Ω, R$_e$ = 13 Ω.

which is a component of magnet system, has a form of six-sided box. The ~ 35 mm gaps (not shown in Fig.1) between the longitudinal sides of the box provide the radial pumping of inter-electrode space. High-pressure chamber is supplied with the working gas through the bottom tube (Fig.1).

The electrode system, forming discharge chambers 5 and 6, represents two-chambered inverse gas magnetron, which works as E x H glow discharge. Both discharge chambers are connected to one power supply, as shown in Fig. 1. The voltages on the electrodes were distributed with the help of the resistive divider.

The efficiency of low pressure discharge is mainly limited by a lifetime of electrons. In this construction of the inverse magnetron, both magnetic and electrostatic trapping of fast electrons can be realized. The superposition of a longitudinal magnetic field, created by magnets 14 and 15, increases a lifetime of fast electrons, emitted by a cylindrical surface of the cathode and collected to the central anode. Flat sides of the magnetron chambers, having a cathode potential, provide electrons oscillations along the magnetic field. As a result, a lifetime of fast electrons in the magnetron chambers is large and they undergo numerous collisions with gas and have enough time to make sufficient number ionizations for maintenance of the independent discharge, before they get on a surface of the anode.

Physical processes in the two-chambered magnetron discharge with compressed plasma are rather difficult, therefore a creation of mathematical model, allowing to carry out engineering calculation of the construction is difficult, because of the complexity and multi-parametrical dependence of phenomena taking place here.

Computer modeling of the discharge chamber potential distribution predicts the electrode shapes and their potentials values This computer design was optimized experimentally later.

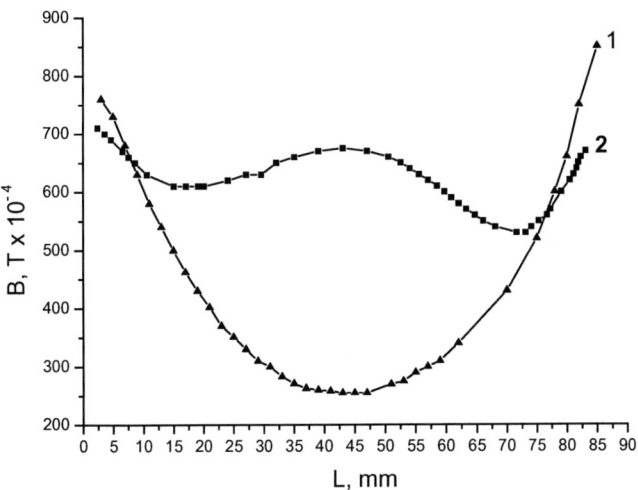

FIGURE 2. Axial distribution of magnetic field B_z in the interpolar gap of plasma generator. 1 - without adjusting magnets. 2 - with adjusting magnets.

EXPERIMENTS AND DISCUSSIONS

It is difficult to obtain a homogeneous magnetic field along all the length in discharge chamber of the real ion source. In the presented construction, the basic permanent magnets 14 and 15, which excite a magnetic field, are located at the bottom and top part of the chamber. The distribution of magnetic field produced in the interpolar space by this magnets is shown on Fig.2 (curve 1). It is visible, that the intensity of a field along the discharge cell is non-homogeneous.

To reduce this non-homogeneity of a field, the corrective magnets 16 and 17 were installed on each of six sides of the magnet box 13 (see Fig.1). After optimization of the corrective magnets position, the following distribution of magnetic field along of the discharge chamber axis was obtained (curve 2 in Fig.2). All experimental results were received with the corrected distribution of the magnetic field.

Dependence of ignition voltage U_b of magnetron as a function of gas pressure P_e in the emission chamber is shown at Fig. 3 by curve (gas flow is proportional to the pressure P_e). The maximal voltage applied was 800 V. At this voltage (800V) the discharge ignites when pressure is $P_e \approx 1,7 \cdot 10^{-3}$ Pa, and it ignites at lower pressure with the gas flow increase. Experiment has shown, that in a range of gas pressure $P_e = (1,7 - 3) \cdot 10^{-3}$ Pa the discharge glows only in the first chamber, and at $P_e \geq 3 \cdot 10^{-3}$ Pa the discharge begins to glows in the second chamber, at ignition voltage $U_b = 560$ V (as it follows from the diagram).

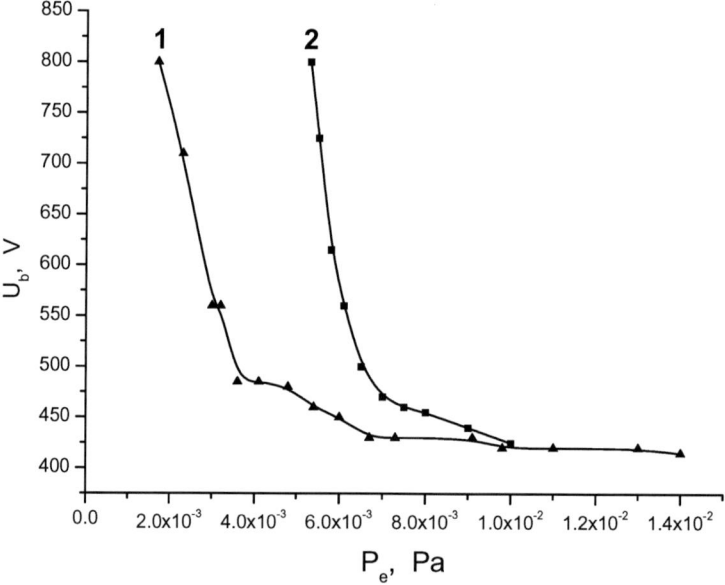

FIGURE 3. Dependencies of discharge ignition voltage for two-chambered (1) and one-chamber (2) gas magnetron on pressure in the emission chamber.

A similar dependence for the single-chamber gas magnetron is shown in the same figure (curve 2) for comparison. In this case, chamber 5 was filled with a copper insert. The outside surface of this insert adjoined tightly to the internal surface of the cathode 3, and a ring channel with width $\Delta r \sim 0,8$ mm was left between the anode 1 and the internal surface of insert for gas delivery into the chamber 6. The discharge ignition voltage in the chamber 6 was 560 V, and it occurs at pressure $P_e \approx 6\cdot10^{-3}$ Pa, two times higher as compared with that of the two-chamber magnetron.

For maintenance of a significant pressure difference between the second and third discharge chambers, and also for decrease the yield of the particles, sputtered from the cathode, it is necessary to reduce a width Δr of the annular slot. Its minimal width is limited by a thickness of a space charge layer, which prevents penetration of plasma into the emission chamber [2]. The thickness d of this layer can be evaluated from the Child law:

$$ d = \frac{2}{3}\left(\frac{\varepsilon_0 \left(\frac{2q}{M}\right)^{1/2} U^{3/2}}{j_i} \right)^{1/2} , \qquad (1) $$

where j_i - ions current density on the cathode; q - ion charge; M - ion mass; U - voltage across the layer; ε_0 - dielectric constant. Ions current density on the cathode can be evaluated from the Bohm law:

$$ j_i = 0.4en\left(\frac{2kT_e}{M} \right)^{1/2} . \qquad (2) $$

At $n = 10^{12}$ cm^{-3}, electron temperature $T_e = 5$ eV, oxygen ions current on the cathode $\sim 49,5$ mA/cm^2.Substituting this value of ion current density to the formula (1) and believing, that practically all the voltage drop $U = 350$ V is concentrated in the cathode sheath, we shall receive $d = 1,3$ mm. Taking into account that, the pre-sheath also adjoins to a space charge layer [3], then a minimal width of the annular slot can be chosen $\sim 1,5$ mm.

The maximal value Δr will be limited by allowable gas pressure in the emission chamber. At the fixed value of gas flow, the pressure in the emission chamber can be regulated due to change of the annular slot width 9 and also due to radial pumping of the channel. This pressure should be small enough to minimize the losses of negative ions by their neutralization on the drift way.

The neutralization losses of negative ions can be determined from the expression $n\sigma_{-0} l_e \approx 10^{-2}$. As a cross section of neutralization under ion energies in several tens eV is $\sigma_{-0} \approx 5\cdot10^{-16}$ cm^2, then at a choice of the length of the emission chamber $l_e = 1$ cm, the pressure should be $P_e \leq 8\cdot10^{-2}$ Pa. The experimental studies have shown that this condition $l_e = 1$ cm is carried out in all range of gas flow at $\Delta r \leq 2$ mm.

Physically it is clear, that the formation of an electron avalanche in the emission chamber begins when the electric field intensity between plasma and emission electrode will exceed the critical value $E > E_{kr}$. The value E_{kr} will depend on the initial density of electrons, penetrating through a compression aperture. Besides, for the transition of a discharge onto the emission electrode, a gas density in the chamber

239

should also exceed some minimum. For decreasing of this terminal pressure, it is important to have an increased gas density in the initial area of the emission chamber, where, under the certain conditions, plasma will penetrate from the gas magnetron and initiate the development of an electron avalanche.

The first experiment on a stretch of the discharge compression to the emission electrode was carried out at the following characteristics of the gap: $\Delta r = 1{,}5$ mm; $P_e = (5{\cdot}10^{-3} - 8{\cdot}10^{-2})$ Pa; $l_e = 1$ cm. The discharge current I_{dc} varied in a range $(0{,}1 - 3)$ A. The experiment has shown, that a stretch of the discharge was extremely unstable and it occurred spontaneously only at the maximal values P_e and I_{dc}. It is obvious, that at the chosen values of the gap, to carry out a stretch of the discharge was possible only due to increasing of intensity of field E between plasma, having a potential close to a potential of the anode, and emission electrode. It could be made either by applying to the gap of a higher voltage from an additional power supply, or by installing the additional electrode 10 into the emission chamber (see Fig.1), which would have a positive potential with respect to the anode 2.

An additional electrode was installed at distance $\Delta l = 1$mm from a face part of the cathode 4, and was connected to a positive pole of the power supply through the resistor $R_{ad} = 630$ W at first. Other gap parameters were unchanged. Thus, a stable stretch of the discharge began at the magnetron discharge current $I_{dc} = (0{,}5{-}0{,}7)$ A. At decreased value of Δl, a sustaining current of the supporting charge at which a stable current transmission occurred on the emission electrode, decreased too. At $\Delta l = 0{,}4$ mm the sustaining current was $\sim 0{,}25$ A.

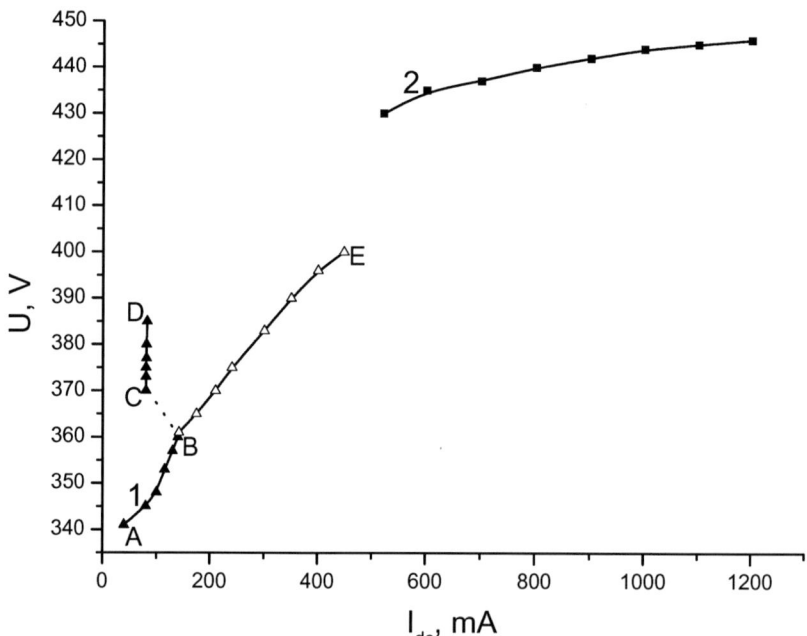

FIGURE 4. Volt-ampere characteristic of discharges in magnetron (1) and in the emission chamber (2) at $P_e = 1{,}5{\cdot}10^{-2}$ Pa.

Fig. 4 shows a volt-ampere characteristic of the magnetron discharge (curve 1) and of the discharge in the emission chamber (curve 2) at the following characteristics of the gap: $\Delta r = 1,5$ mm; $P_e = 1,5 \cdot 10^{-2}$ Pa; $\Delta l = 0,4$ mm; $l_e = 0,7$ cm. After the discharge initiation, a voltage between cathode and anode falls from value U_b to the value designated by a point A on curve 1. Further a volt-ampere characteristic corresponds to AB segment of this curve. At obtaining of the certain current (point B) a discharge initiation occurs in the emission chamber, and then the dependence of a current in the anode circuit from a voltage corresponds to CD segment. When the emission electrode is disconnected, then a volt-ampere characteristic corresponds to ABE curve. The change of gas pressure in the emission chamber results in the point B shift: point B is shifted to larger currents with pressure decrease, and to smaller current with pressure increase. The dependence of a current in the circuit of the emission electrode vs voltage (after voltage breakdown in the gap) is shown by curve 2.

The carried out experiment also has shown, that after discharge ignition in the gap, the pressure P_e can be decreased down to $\sim 8 \cdot 10^{-3}$ Pa at the same discharge current and with the same stability of discharge burning. In other words, a discharge ignition occurs at a little bit higher pressure than its stable burning.

CONCLUSION

The work of three-chambered discharge system can be presented as follows. At first, a discharge initiation occurs in the chamber of high pressure 5. The formed plasma, at sufficient width of the annular slot 7, penetrates along a magnetic field into the chamber 6. The flow of penetrating plasma promotes an ignition of the discharge in it and also a formation of plasma at lower pressure, than in the chamber 5. Further, under the certain conditions, plasma penetrates from gas magnetron into the emission chamber through the annular slot and initiates the development of an electronic avalanche in it. Plasma having a tubular configuration is generated in the emission chamber. An accurate circular trace on the emission electrode around the emission aperture evidences it.

Formation of a double layer before the narrow annular slot 9 and fast electrons delivery into the area of local increase of gas pressure, produced by gas escape from the magnetron region, creates the favorable conditions for the vibration excitation of molecules in this peripheral plasma.

As a result of peripheral plasma diffusion across a magnetic field, a plasma formation in the paraxial area takes place. The paraxial plasma, formed in such a way, along with the vibrationally excited molecules, will also contain the increased concentration of slow electrons, as far as fast electrons fraction does not penetrate into this area due to action "of the magnetic filter" [4]. Thus, the necessary conditions for negative ions formation by effective process of dissociative attachment of slow electrons to vibrationally excited molecules appear in the internal plasma [5].

REFERENCES

1. P.A. Litvinov, *Vopr. Atom. Nauki and Tekhnol. Ser. Nucl. Phys..* **4,5** (31,32), 48 (1997) -in Russian.
2. V.A. Kagadey, A. V. Kozyrev, I.V. Osipov, D.I. Proskurovskiy, *Journ.of Tech.Phys.* **71**, 22. (2001) -in Russian.
3. Alekseyev B.V., Kotelnikov V.A. *Probe method of plasma diagnostic.* M.: Energoatomizdat, 1988. P.240. (in Russian).
4. Holmes A. J. T., *Rev. Sci. Instrum..* **53**, 1517 (1982)
5. Lee Y., Gough R. A., Leung K. N., et al., *J. Vac. Sci. Technol.* **B**16 (6), 3367 (1998).

Advances in the Ion Source Research and Development Program at ISIS

D. C. Faircloth, J. W. G. Thomason, R. Sidlow, M. O. Whitehead

CCLRC, RAL, ISIS, Didcot, Oxon, OX11 0QX, UK

Abstract. This paper covers the advances in the ion source research and development Program at ISIS over the last 2 years. The work is a combination of theoretical finite element analysis calculations and experiments conducted on a purpose built development rig. The broad development goals are higher beam current with longer pulse length. A Finite Element Analysis (FEA) model is used here to understand the steady state and dynamic thermal behavior of the source, and to investigate the design changes necessary to offset the extra heating. Electromagnetic FEA modeling of the extraction region of the ISIS H⁻ ion source has suggested that the present set up of extraction electrode and 90° sector magnet is sub-optimal, with the result that the beam profile is asymmetric, the beam is strongly divergent in the horizontal plane and there is severe aberration in the focusing in the vertical plane. The FEA model of the beam optics has demonstrated that relatively simple changes to the system should produce a dramatic improvement in performance. The theoretical and experimental results are compared here.

INTRODUCTION

The ion source research and development program at Rutherford can be broadly split into 3 sections: Infrastructure, Thermal and Electromagnetic. Each of these sections involves the model, design and experiment cycle.

The practical work is carried on the Ion Source Development Rig (ISDR)[1] to allow testing without affecting ISIS itself.

Infrastructure changes to the ISDR include the up-rating of power supplies and flange and support redesign. Thermal work covers thermal finite element modeling of the source, allowing many different load and cooling scenarios to be tested. Electromagnetic modeling allows different extraction and beam transport electrode geometries to be studied and optimized.

INFRASTRUCTURE

The configuration of the ion source and magnet flange for the ISDR has now been changed so that the ion source assembly can be loaded from the top, rather than the back of the magnet flange, as shown in Figure 1. A schematic of the top loading ion source and extended magnet flange is shown in Figure 2.

CP763, *Production and Neutralization of Negative Ions and Beams*, edited by J. D. Sherman and Y. I. Belchenko
© 2005 American Institute of Physics 0-7354-0248-5/05/$22.50

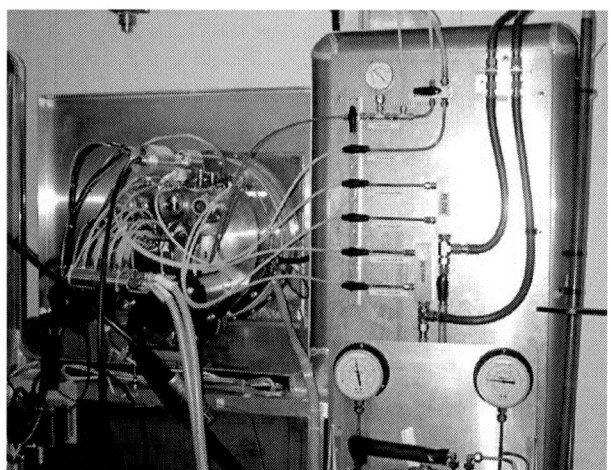

FIGURE 1. The top loading ion source mounting flange.

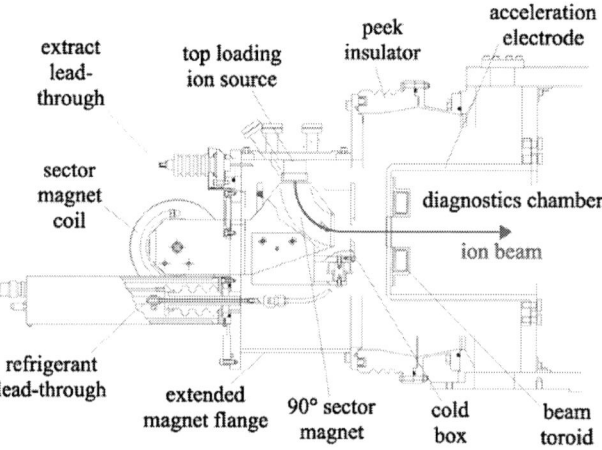

FIGURE 2. Schematic of the top loading ion source.

This allows greater flexibility for future source developments, where additional space for scaling of source components or more aggressive cooling strategies can be provided by inserting a spacer ring between the ion source flange and the magnet flange. The penalty for this innovation, however, is that the source has had to be moved back by about 200 mm from its original position. The emittance scanners on the ISDR were designed for use with the ISIS RFQ test stand[2], and were only required to scan over ~30 mm to cover the maximum extent of the Low Energy Beam Transport beampipe. This means that, with the distance from the cold box front plate to the emittance scanners now increased to 685 mm (56 mm of acceleration gap and 629 mm of drift), the extent of the divergent ion beam would be larger than the range of the scanners. This problem has been addressed by modifying the scanners so that two or more separate scans can be taken and then merged to create a single scan over a range of ~78 mm.

The high voltage platform has been extended provide space to fit a new extract voltage power supply. The new power supply will allow an increase of the extract voltage from 17 to 25kV and an increase in pulse length up to 2.5ms. An additional 3-phase isolating transformer has also been installed to power the new extract voltage power supply.

Work is currently underway to provide a separately excitable penning field to allow the effect of penning field to be studied.

THERMAL

A thorough understanding of the thermal characteristics of the ISIS ion source is essential if operation is to be extended to the higher duty factors, whilst maintaining an optimal regime for H⁻ ion production and source lifetime.

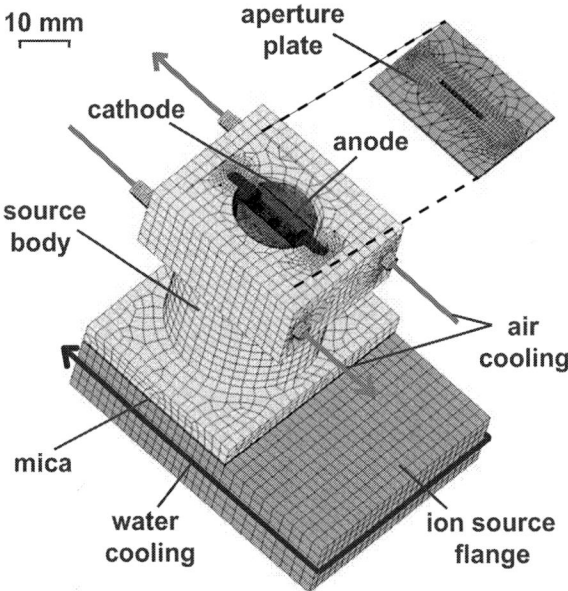

FIGURE 3. ALGOR thermal model of the ISIS ion source.

Figure 3 shows the model of the ion source. The source is of the Penning type, comprising a molybdenum anode and cathode between which a low pressure hydrogen arc is struck. Hydrogen and caesium are fed into the arc via holes in the anode; these can be more clearly seen in Figure 4. The anode and cathode are housed in the stainless steel source body. The anode is thermally and electrically connected to the body, whereas the cathode is isolated from the body by means of a ceramic spacer. The whole assembly is bolted to a flange, separated by a thin layer of mica to provide electrical isolation for the cathode.

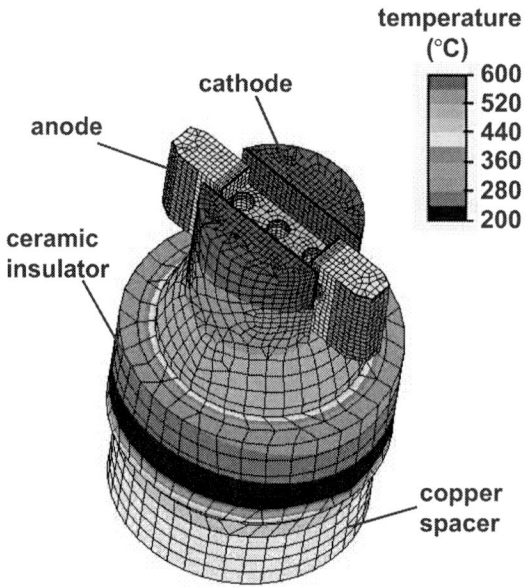

FIGURE 4. Component temperatures from the steady state model.

Source cooling is provided by two systems illustrated in Figure 3: air cooling via two pipes in the source body nearest the electrodes and water cooling via a channel cut into the ion source flange. Air flows along one pipe and is then returned down the other as shown in Figure 3. Air is used because of the safety hazards involved with having water close to the caesiated ion source.

The ions are extracted through the slit in the aperture plate.

ALGOR[3] FEA software has been used for thermal modeling. The details of the model and its validation have been discussed previously[4].

The typical ISIS ion source operating conditions are a 4 kW, 0.5 ms, 50 Hz arc. An assumption is made that all the electrical power as measured in the external circuit goes into heating the electrode surfaces exposed to the discharge. The arc is bounded on all sides by sections of the cathode, anode and aperture plate.

When all the parameters that correspond to normal operation of the ISIS source are applied to the model the temperatures obtained are very close to the temperatures measured in the actual source (Table 1). This provides validation that the model is realistic.

To obtain a steady state solution the average power densities over the 50 Hz cycle are applied to the electrode surfaces.

TABLE 1. Theoretical thermocouple, electrode surface and actual thermocouple temperatures.

Location	From Model			Actual ISIS
	Thermocouple	Surface	Difference	
Anode	456°C	496°C	40°C	400-600°C
Cathode	501°C	585°C	84°C	440-530°C
Source Body	416°C	441°C	25°C	390-460°C

In normal operation the source temperatures are monitored using three thermocouples: Cathode, Anode and Source Body. All these thermocouples are positioned some distance from the electrode surfaces exposed to the arc plasma so they do not give actual surface temperatures. The realistic model of the source allows this difference between measured and surface temperature to be calculated, Table 1. The difference between these values depends on the distance between the measurement point and the electrode surface. It is greatest for the cathode because the measurement point is at the very base of the cathode.

The electrode surface temperatures are an important factor when considering the performance of the ion source as these surfaces play an important role in the plasma physics of the arc; the caesiation of the surface is temperature dependent for example.

The aim of the modeling work is to find out what is required to maintain the electrode surfaces at the temperatures in the current source whilst increasing the duty cycle.

The duty cycle is doubled to 1 ms and the cooling represented by a Heat Transfer Coefficient (HTC) in the head and flange increased. Figure 5 shows the results.

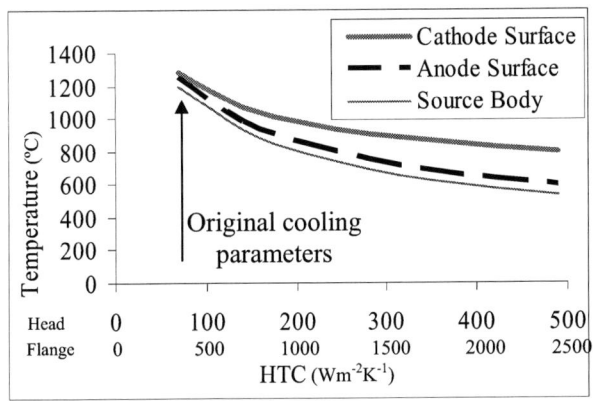

FIGURE 5. 1 ms duty source steady state temperatures for increased cooling.

Without increasing the cooling, the source temperatures approximately double. As the water and air flow rates are increased the anode and source body temperatures come down together, however the cathode surface temperature decreases more slowly. This is because the cathode is thermally isolated from the cooling systems by the layer of mica and the ceramic insulator. For a 1 ms duty cycle there is no combination of coolant flow rates that will produce the original surface operating temperatures.

Mica is a very poor thermal conductor. To improve the cooling to the cathode the layer of mica is removed from the model. With the mica removed the steady state surface temperatures shown in Table 1 can be reached easily for a 1 ms duty cycle.

The mica is present to provide electrical isolation of the cathode from the flange, so in practice a thin layer of material with good electrical insulation properties and high thermal conductivity (such as aluminum nitride) will have to be used.

Removal of the mica will cause problems with source start up. It will be difficult to heat the source up to reach operating temperature. Modifications being considered include a heating element to pre-heat the cathode.

The steady state calculations only calculate the average electrode surface temperature. Using the steady state solution as a starting point it is possible to run a transient study. This allows the peak surface temperatures reached at the end of the arc on period to be calculated. To ensure accurate results the elements of the FEA model near the electrode surfaces are made very thin (10^{-5} m) in the direction of heat flux.

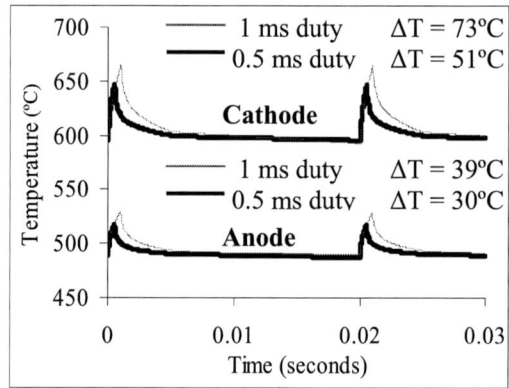

FIGURE 6. Anode and cathode surface temperatures for 0.5 ms and 1 ms duty cycles for the normal sized ion source calculated from the transient model.

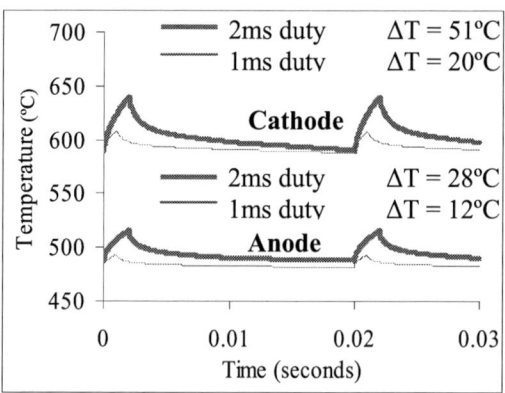

FIGURE 7. Anode and cathode surface temperatures for 1 ms and 2 ms duty cycles for the double sized ion source calculated from the transient model.

Figure 6 shows how the peak temperatures vary though the cycle for the 0.5 ms and 1 ms duties. During the on period there is a rapid increase in the surface temperature of the materials directly exposed to the plasma; this temperature then decays away as the arc energy dissipates into the thermal mass of the material. The peak source body temperature does not vary because it is not directly in contact with the plasma. In a similar way there is no detectable change in temperature at the thermocouple measurement points.

The exact implications of this electrode surface temperature rise during the on period are poorly understood for a surface Penning ion source. The temperature rise is clearly larger for the 1 ms duty cycle, but it is not known how this will affect the ion source operation. All that is known is that the existing source operates very well. Work is currently underway to test the ion source with a 1 ms duty cycle on the ISDR at RAL.

The electrode surface temperature rise during the on pulse cannot be mitigated with additional cooling because the energy does not have time to conduct away from the surface. The surface temperature rise is therefore mainly dependent on the power density applied by the plasma to the electrode surfaces and the length of time it is applied. The total arc power and length of time cannot be changed, therefore the electrode surface area must be increased to decrease the power density.

All linear dimensions in the model are doubled and the simulation repeated, keeping the instantaneous power at 4 kW as before. Scaling the ion source dimensions has been successfully implemented by previous researchers[5] at Los Alamos National Laboratory. Figure 7 shows the transient results. For a 1 ms duty cycle the surface temperature rises are significantly reduced.

In the double sized source, the duty cycle can be further increased to 2 ms and the surface temperature rises are very similar to the normal ion source for a 500 μs duty. This confirms that the surface temperature rise is largely dependant on surface power density and time. (4X duty balanced with 4X increase in surface area).

ELECTROMAGNETIC

A recent paper[6] has described MAFIA[7] modeling of the extraction region of the ISIS H- ion source[8,9]. This demonstrated that optimization of the beam optics should result in a significant improvement in the measured emittance of the source. The design incorporates new pole pieces for the 90° sector magnet, and a 'maximag' magnet steel tube (internal diameter 30 mm, wall thickness 5 mm) extending from 3 mm in front of the 90° plane to flush with the cold box exit. Together these should deal with fringe fields of the 90° sector magnet more effectively. In addition two new extraction geometries were specified: one a terminated version of the standard ISIS extraction geometry and the other a Pierce geometry[10] these are shown in Figure 8. All of these new components have now been manufactured and tested on the ISDR.

FIGURE 8. Extraction electrodes and aperture plates for the ISIS standard geometry (centre), terminated standard geometry (left) and Pierce geometry (right).

Emittance scans were taken for four combinations of extract electrode geometry and pole pieces: ISIS standard geometry with old pole pieces, ISIS standard geometry with new pole pieces, terminated standard geometry with new pole pieces and Pierce geometry with new pole pieces. In each case the source parameters were kept as constant as possible, with an extract voltage of 17 kV, a beam energy of 35 keV, a pulse width of 250 μs and a beam current of between 45 and 55 mA. Previous experiments on the ISDR[1] have shown that there is very little space charge compensation in the diagnostics chamber when the pressure (of 1.7×10^{-5} mbar) is determined solely by the transmission of H2 from the ion source chamber. The introduction of Kr as a buffer gas has been shown to provide space charge compensation and improve the emittance measurements. To investigate this effect further in the present measurements Kr was introduced into the diagnostics chamber to raise the combined pressure of H2 and Kr to levels of 2.2×10^{-5} mbar, 4.0×10^{-5} mbar and 1.0×10^{-4} mbar. The results are shown in Table 2, where all values are for the normalized rms emittance in πmm mrad.

TABLE 2. Emittance values for the four combinations of extract electrode and pole pieces with varying pressures of Kr in the diagnostics chamber. All values are for normalized rms emittance in πmm mrad during a 10 μs interval 150 μs into the 250 μs pulse, with an extract voltage of 17 kV, beam energy of 35 keV and a beam current of between 45 and 55 mA.

	H2 (1.7×10^{-5} mbar)	H2+Kr (2.2×10^{-5} mbar)	H2+Kr (4.0×10^{-5} mbar)	H2+Kr (1.0×10^{-4} mbar)
Standard Geometry Old Pole Pieces	$\varepsilon H = 0.97$ $\varepsilon V = 0.94$	$\varepsilon H = 0.99$ $\varepsilon V = 0.84$	εH no measurement $\varepsilon V = 0.91$	no measurements
Standard Geometry New Pole Pieces	$\varepsilon H = 0.97$ $\varepsilon V = 0.90$	$\varepsilon H = 0.91$ $\varepsilon V = 0.92$	$\varepsilon H = 0.92$ $\varepsilon V = 0.90$	$\varepsilon H = 0.83$ $\varepsilon V = 0.72$
Terminated Standard New Pole Pieces	$\varepsilon H = 0.91$ $\varepsilon V = 0.98$	$\varepsilon H = 0.85$ $\varepsilon V = 0.97$	$\varepsilon H = 0.82$ $\varepsilon V = 1.06$	$\varepsilon H = 0.77$ $\varepsilon V = 0.97$
Pierce Geometry New Pole Pieces	$\varepsilon H = 0.73$ $\varepsilon V = 0.80$	$\varepsilon H = 0.72$ $\varepsilon V = 0.83$	$\varepsilon H = 0.71$ $\varepsilon V = 0.79$	$\varepsilon H = 0.62$ $\varepsilon V = 0.73$

The emittances in both planes display the general trend that the values get smaller with each successive geometry refinement and the introduction of more Kr, but this is not always the case. For instance replacing the ISIS standard geometry extraction electrode with the terminated standard leads to an increase in the vertical emittance values. However, it can be seen that overall the worst values are those for the ISIS standard geometry and old pole pieces with no introduction of Kr (worst case, shaded orange in Table 2 worst case. It is immediately obvious that the spatial extents of $\approx\pm50$ mm in b) and the best values are those for the Pierce geometry and new pole pieces with Kr introduced to 1.0×10^{-4} mbar (best case, shaded green in Table 2). Figure 9 shows the worst case. It is immediately obvious that the spatial extents of $\approx\pm50$ mm in both the horizontal and vertical planes justify the modification of the emittance scanners to scan over a wider range. Indeed there is evidence that previous emittance measurements quoted for the ISDR[1] may have been too small because the edges of the beam were being missed.

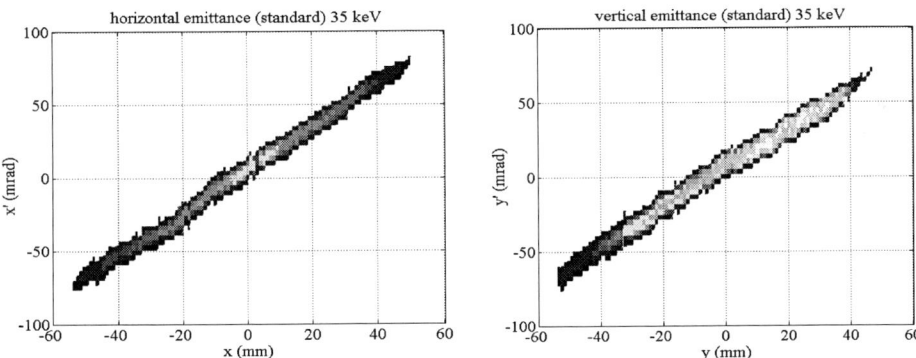

FIGURE 9. ISDR horizontal and vertical emittance plots for the ISIS standard geometry and old pole pieces with no introduction of Kr.

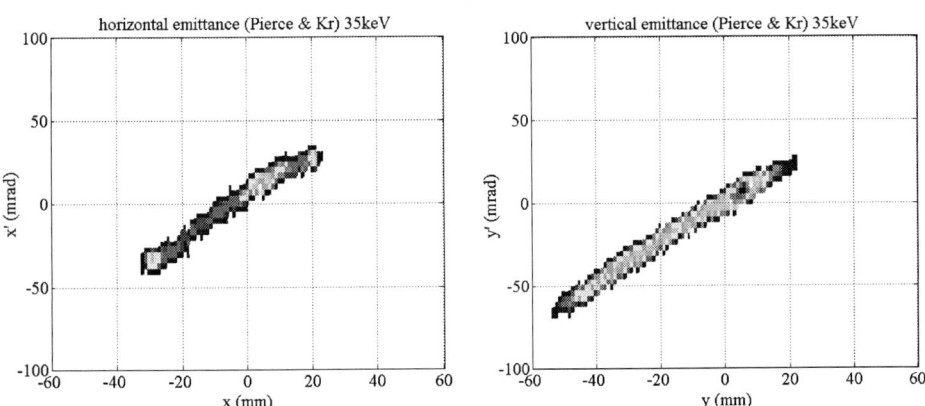

FIGURE 10. ISDR horizontal and vertical emittance plots for the Pierce geometry and new pole pieces with Kr introduced to raise the diagnostics chamber pressure to 1×10^{-4} mbar.

Figure 10 shows the best case. The spatial extent of the beam in the horizontal plane has been reduced to ≈±30 mm, and in this plane the beam displays three distinct peaks. This structure is evident in all of the scans taken with the Pierce geometry, irrespective of the amount of Kr, and is taken to be indicative of a slight overfocusing in the horizontal plane of the extract electrode. This will be investigated at a later date by reducing the angle of the recess in the Pierce geometry aperture plate (see Figure 8). In the vertical plane the beam is asymmetric, and has been positioned to maximize the charge on axis, but again covers a range of ≈±60 mm. The asymmetry of the vertical profiles in both the best and worst cases may be evidence that there is not an even distribution of charge across the slit in the aperture plate when the ions are extracted. Examination of many used ion source cathodes has shown erosion concentrated towards the area where Cs is fed through the anode into the source. If this is a consequence of the plasma being localized near the Cs feed this could well result in an asymmetry in the vertical plane. A new Cs delivery system, which should give a more even plasma distribution has been designed and manufactured, and awaits testing on the ISDR. Although the changes in normalized rms emittance values (from $\varepsilon H = 0.97$, $\varepsilon V = 0.94$ πmm mrad in the worst case to $\varepsilon H = 0.62$, $\varepsilon V = 0.73$ πmm mrad in the best case) are not as dramatic as those predicted by MAFIA modeling[6], there is still a marked improvement in both the horizontal and vertical planes as a result of the Pierce geometry, new pole pieces and the introduction of Kr. It is hoped that future refinements of the extraction geometry and Cs delivery system will improve the situation still further.

CONCLUSIONS

To increase the duty on the ISIS ion source the cathode must be cooled more directly by replacing the mica sheet with a better thermal conductor. This may introduce the requirement for a cathode heater during source startup.

If the electrode surface temperature rise is critical it will be necessary to move to larger electrodes. The simplest way of doing this is to scale the size of the entire source.

Improved extraction electrodes and beam transport have improved beam quality, though not to the extent predicted by the modeling; this indicates space charge plays a significant role in the beam optics.

ACKNOWLEDGMENTS

This work was supported by the European Union High Performance Negative Ion Source (HP-NIS) network, contract number HPRI-CT-2001-50021.

REFERENCES

1. J. W. G. Thomason, et al., "Performance Of The H⁻ Ion Source Development Rig at RAL", Paper THPRI012, Conf. Proc. EPAC, Paris, France, p.1738-1740, June 2002.
2. C. P. Bailey, et al., "The ISIS RFQ Emittance System", Paper THPLE048, Conf. Proc. EPAC, Paris, France, p.1846-1848, June 2002.
3. ALGOR Inc., 150 Beta Drive, Pittsburgh, PA 15238-2932 USA. www.algor.com
4. D. C. Faircloth, et al., "Thermal Modeling of the ISIS H⁻ Ion Source", Review Of Scientific Instruments, Volume 75, Number 5, p. 1738-1740, May 2004.
5. H. Vernon Smith Jr., "H⁻ and D⁻ Scaling Laws for Penning Surface-Plasma Sources", Review Of Scientific Instruments, Volume 65, Number 1, January 1994.
6. D. C. Faircloth, et al., "Electromagnetic Modeling of the Extraction Region of the ISIS H⁻ Ion Source", Review Of Scientific Instruments, Volume 75, Number 5, p.1735, May 2004.
7. MAFIA, CST Ltd., Bad Nauheimer Strasse 19, 64289 Darmstadt, Frankfurt, Germany www.cst.de
8. R. Sidlow, et al., "Operational Experience of Penning H⁻ Ion Sources At ISIS", Paper THP084L, Conf. Proc. EPAC, Sitges, Spain, p.1525-1527, June 1996.
9. J. W. G. Thomason and R. Sidlow, "ISIS Ion Source Operational Experience", Paper THP4A07, Conf. Proc. EPAC, Vienna, Austria, p.1625-1627, June 2000.
10. J. R. Pierce, Theory and Design of Electron Beams, 2nd ed. (Van Nostrand, Princeton, N.J., 1954).
11. A. P. Letchford, et al., "Measured Performance of the ISIS RFQ", Paper THPLE045, EPAC 2002, Conf. Proc. EPAC, Paris, France, p.927-929, June 2002.
12. J. D. Sherman, et al., "Review of Scaled Penning H⁻ Surface Plasma Source with Slit Emitters for High Duty Factor Linacs", Paper THALA002, Conf. Proc. EPAC, Paris, France, p.284-286 June 2002.

Physical Insights and Test Stand Results for the LANSCE H⁻ Surface Converter Source

Joseph Sherman, Edwin Chacon-Golcher, Ernest G. Geros, Edward Jacobson, Patrick Lara, Bruce J. Meyer, Peter Naffziger, Gary Rouleau, Stuart C. Schaller, Ralph R. Stevens, Jr., and Thomas Zaugg

LANSCE Division, Los Alamos National Laboratory, Los Alamos, NM87545

Abstract. The Los Alamos Neutron Science Center (LANSCE) H⁻ surface converter source upgrade project has been ongoing for several years to reach 25-40 mA current with 7(πcm-mrad) lab emittance (95% beam fraction). The duty factor is 12% (120 Hz, 1ms pulse length). Summary test stand results and interpretations for a six filament axial extraction H⁻ source are presented. This source did produce 40-mA H⁻ current, but with unacceptable emittance growth. More recently a fourth modified LANSCE H⁻ production source with radial H⁻ extraction system has been constructed, and is presently undergoing tests. Currents up to 25mA H⁻ have been observed with 20% emittance growth. This emittance growth may be acceptable for 800 MeV linac operations. A summary of physical principles of emittance growth mechanisms and converter physics are given.

INTRODUCTION

Significant upgrades to the LANSCE 800-MeV linac and Proton Storage Ring (PSR) operations can be realized by development of an H⁻ ion source with laboratory emittance of 7(πcm-mrad) at 95% beam fraction with 20-40mA H⁻ current. The source beam energy is 80keV and the duty factor (df) is 12%. A six-filament version of the surface converter source with axial H⁻ beam extraction was developed in a collaborative effort between Lawrence Berkeley National (LBNL) and Los Alamos National Laboratories (LANL) [1]. Although this source did produce 40-mA H⁻ current, an unexpected emittance growth factor of 2.5 made these higher current beams unacceptable for LANSCE operations [2]. A subsequent decision was made to fabricate and develop a fourth production source to produce 25-mA H⁻ current with an emittance growth of no greater than 20% [3]. The source upgrade technology would be used in the 750-keV H⁻ injector B at LANSCE.

Experiments and analysis on the six-filament axial extraction source will be reviewed in the next section. Evidence is given that at the higher discharge power characteristic of this source, two beams are formed at extraction. The two beams originate from surface (converter electrode) and volume processes. The third section

CP763, *Production and Neutralization of Negative Ions and Beams*, edited by J. D. Sherman and Y. I. Belchenko
© 2005 American Institute of Physics 0-7354-0248-5/05/$22.50

summarizes recent work completed on the fourth production source with radial extraction. Effort is directed at increasing the H⁻ current without significantly increasing the discharge power, thus avoiding the two-beam emittance growth mechanism. All beam measurements were made on the LANSCE Ion Source Test Stand (ISTS), which is now computer controlled and may be operated on a 24 hour, 7-day/week basis.

SIX FILAMENT AXIAL EXTRACTION SOURCE

Figure 1 shows a photo of the copper axial extraction H⁻ source. Most of the six-filament development was done with this source. A higher current column with focus electrode was developed at LBNL for 80keV, 40-mA operations [4], although

FIGURE 1. Photo of the copper prototype axial source with the electron repeller assembly and emission aperture (Pierce electrode) mount plate removed. The molybdenum converter with its quartz insulator and six tungsten filaments are visible.

many measurements were also done with the LANSCE production column [5]. (A scaled drawing for the LBNL accel column is shown below in Fig. 5.) H⁻ is produced at the cesiated converter surface [6], and subsequently accelerated towards the emission aperture by the –250 to –300V bias of the converter electrode. The axial geometry permits a full-line, cusp-magnet configuration for plasma confinement, and the copper housing provides good temperature control even at the high discharge powers used in the six-filament source. No temperature related problems were observed in the copper source operation. The converter surface has a curvature radius $\rho_{cnv} = 12.5$cm to focus H⁻ ions toward the emission aperture. Emission aperture radii were either $r_p = 0.8$ or 0.5cm. The filaments operate in an emission-limit mode with up to 175A discharge current in hydrogen-cesium gases where the H_2 gas pressure is 3-7mTorr. All measurements were made at 80keV on the ISTS, shown in Fig. 2. The ISTS is a reproduction of the LANSCE injector B 80-keV beam system [7]. Slit and collector emittance stations are located immediately after 80-kV column, between the two focus solenoids, and near the end of the LEBT. These are labelled emittance

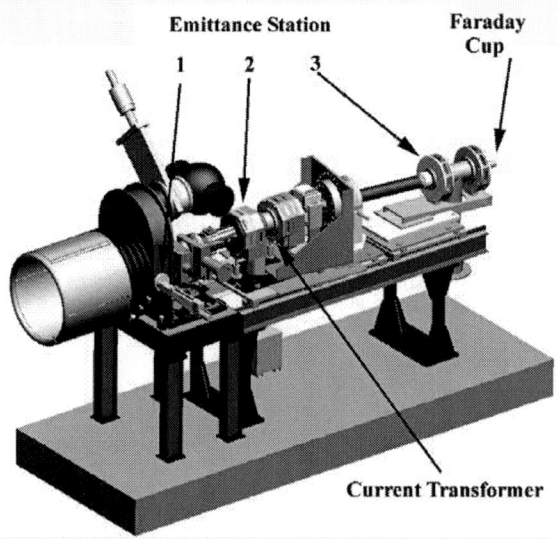

FIGURE 2. Shows the ISTS and low-energy beam transport system (LEBT) diagnostics used in these measurements.

stations 1, 2, and 3 in Fig. 2. The end of the LEBT corresponds to the injector B 670-kV column injection point. Beam current measurements are made between the solenoids (beam current transformer), and then at the end of the LEBT in a Faraday cup.

Figures 3(A) and 3(B), respectively, show H⁻ beam current transformer measurements for r_P = 0.8 and 0.5cm. Four different repeller magnet assemblies were used in this source development [1,2]: line cusp magnets, two ring solenoid magnets with on-axis fields of 500 and 250G, and an undulator magnet comprised of two

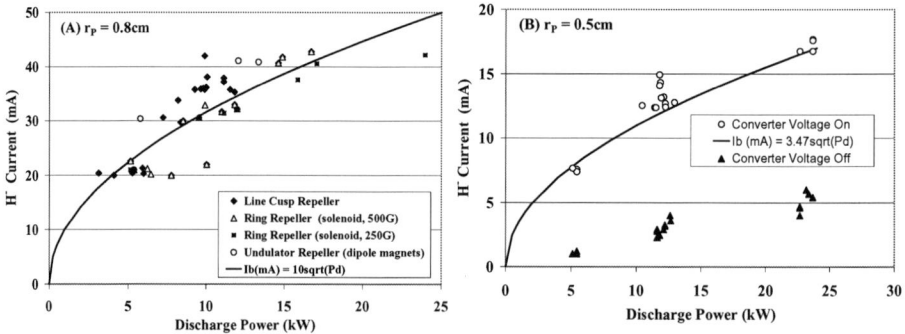

FIGURE 3. (A) Shows H⁻ current measurements made with four different repeller magnets in place with the r_P = 0.8cm emission aperture. (B) H⁻ current measurements made with r_P = 0.5cm, and with the converter voltage turned on and off. All measurements made at 80keV beam energy.

dipole magnets. The e/H⁻ ratios for the line cusp, ring solenoid, and undulator repeller magnets gave e/H⁻ = 4, 2.4, and 1, respectively. Fig. 3(A) shows that all four repeller magnet configurations gave 40mA H⁻ current. The solid curve shows the H⁻ beam current follows an approximate square root dependence on the discharge power. The discharge voltage was typically –140V. Data in Fig. 3(B) were taken with the converter voltage on (open symbol), and then with the converter voltage off (filled symbols). The undulator repeller configuration was used, and the data again shows H⁻ currents with converter on following a square root dependence on discharge power. Fig. 3(B) also shows a linear increasing H⁻ current with converter voltage off, thus suggesting H⁻ production mechanisms other than H⁻ surface conversion are increasing with discharge power. The ratio of the normalizing coefficients for the H⁻ current dependence on discharge power is 10/3.47 = 2.9, which nearly equals the ratio of Pierce aperture radii, squared, $(8/5)^2 = 2.6$. Thus the emission aperture appears to be uniformly illuminated by H⁻ from the plasma with maximum H⁻ current density j_{H^-} = 21mA/cm².

Fig. 4(A) shows a summary of the total lab emittance (ε_l) extracted from the slit and collector phase-space measurement gear. The area is calculated at the 2% threshold level, which usually corresponds to beam fractions greater than 98%. The solid symbols show data from r_P = 0.5cm, while the open symbols show the emittance

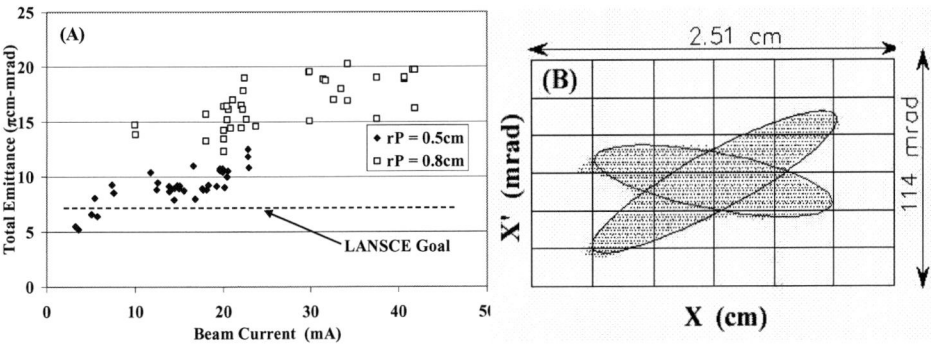

FIGURE 4. (A) Summary of emittance measurements made with the axial copper source using r_P = 0.5 and 0.8cm emission apertures. (B) Slit and collector phase space scan at ISTS emittance station 3. The full scales on the phase space scan are shown. The ellipses drawn on the data are discussed in the text.

data from the r_P = 0.8cm aperture. Above approximately 16mA H⁻ current, there is an emittance increase, which makes tuning of the LANSCE injector B 750keV beam line increasingly difficult [8]. The emittance data in Fig. 4(A) are selected from the three emittance stations in the 80keV LEBT while the ion source was operating with the 80kV LBNL column. The emittance limits based on the converter source admittances (see refs. [1,2] and below) for r_P = 0.5 and 0.8cm sources are respectively 5.4 and 7.5(πcm-mrad). Thus most of the measured emittances are greater than the source admittance limits. Fig. 4(B) shows a 30-mA emittance scan acquired at station three with the LANSCE 80keV column. The total emittance derived from this phase-space scan is 18(πcm-mrad). The dots represent current intensity at the 2% threshold analysis level. A two-beam structure is particularly evident in Fig. 4(B). Superposition of two beams formed at extraction is thought to be the dominant emittance growth mechanism when the surface converter source is operated at higher discharge power.

A theoretical estimation of the two-beam emittance growth mechanism may be made using the H⁻ plasma option model of the PBGUNS code [9,10]. The PBGUNS plasma model for the high current accel column is shown to scale in Fig. 5. Relative to ground potential, this simulation has the Pierce electrode ($r_P = 0.8$cm) at –80kV, the

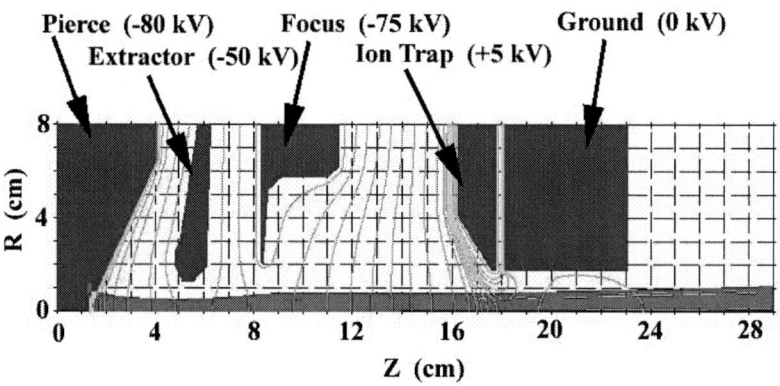

FIGURE 5. Shows a scaled drawing of the LBNL accel column with predicted trajectories for a 30-mA H⁻ beam. Electrode potentials relative to local ground are given.

extractor electrode at –50kV, the focus electrode at –75kV, the ion trap electrode at 5kV, and the ground electrode at 0 kV. The code calculates a self-consistent sheath (cf ref. [10] for an example of the sheath location and shape). The accelerated H⁻ beam current is 30mA, and is composed of injected H⁻ ions with 50% 12eV (energy characteristic of volume H⁻ in the pre-sheath), and 50% 300eV H⁻ ions (energy characteristic of surface converter H⁻).

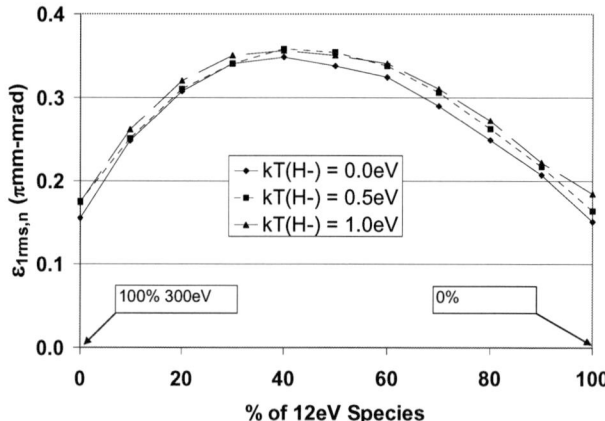

FIGURE 6. Predicted $\varepsilon_{1rms,n}$ vs. % of 12eV species in 30 mA H⁻ current. The remaining current % is in the 300eV injected species.

258

The predicted one rms normalized emittances($1\varepsilon_{rms,n}$) as a function of the % input 12eV (volume) H⁻ species are shown in Fig. 6. The relation between normalized emittances and total lab emittances [1] for the Fig. 4(A) data is $\varepsilon_l = 7(\varepsilon_{1rms,n})/\beta$, where $\beta=.013$ is the relativistic velocity factor of 80keV H⁻ ion. A maximum emittance growth by the PBGUNS two-beam model is factor 2.2, which agrees with the emittance growth shown in Fig. 4(A). The maximum predicted $\varepsilon_{1rms,n} = 0.35$ (πmm-mrad), which corresponds to 19(πcm-mrad) lab total emittance, is also in good agreement with maximum emittances of Fig. 4(A). The three curves in Fig. 6 are parametric in plasma H⁻ temperature. Increasing the H⁻ temperature from 0 to 1eV does not describe the observed emittance growths within this model. An emittance growth calculation derived from mismatch formalism [11] may be attempted. The ellipses drawn in Fig. 4(B) have the following Courant-Snyder parameters: for the convergent beam; $\alpha_c = .795$, $\beta_c = .0779$ (cm/mrad), and for the divergent beam; $\alpha_d = -2.45$, $\beta_d = .0775$(cm/mrad). An emittance growth factor of 3.5 is derived, which on the $1\varepsilon_{rms,n}$ scale of Fig. 6 corresponds to 0.45 (πmm-mrad). This approach appears to overestimate emittance growth.

TWO FILAMENT RADIAL EXTRACTION SOURCE

A fourth radial extraction (LANSCE production) source has been assembled for development purposes. Goals are to produce a 25mA H⁻ source, < 85mA total pulsed current, < 20% emittance growth as compared to operations source, and 12% df (120Hz, 1ms). The specification of 85mA total pulsed current implies e/H⁻ ratio < 2.4. Reasons for this approach are: (1) the present LANSCE 80-kV accel column is thought to be sufficient for 25-mA H⁻ beam production [5], (2) there is tremendous operational experience with the radial extraction source, (3) there is clear upgrade path to 25-mA facility operation, and (4) higher power operation available in the 6-filament source does not appear to be desirable short-term solution for LANSCE injector B. A

FIGURE 7. LANSCE production source. Present development is using a simple modification of this source.

side view photo of the radial production source is shown in Fig. 7. The converter electrode is seen near the center of the source with the cesium oven feed tube just below. Both of these components are made of molybdenum. The electron repeller assembly is located opposite the converter electrode on the stainless steel wall. The H⁻ converter beam (250 – 300eV) is extracted radially through a broken line cusp magnet row. This source was developed 20 years ago for the proton storage ring at LANSCE [12,13].

Table 1 contains summary data relating to the development program. The first column is a source parameter number, the second column contains the source

TABLE 1. Comparison of LANSCE production source, status of the development source, and the development goal.

	Ion Source Parameter	Production Source	Development Status	Development Goal
1	r_P (cm)	0.50	0.60	TBD**
2	r_{rep} (cm)	0.64	0.86	0.86
3	r_{cnv} (cm)	1.9	1.9	TBD
4	ρ_{cnv} (cm)	12.5	12.5	TBD
5	Admittance (cm-mrad)	304	379	TBD
6	B_c (kG)	2.0	3.4	TBD
7	Electron repel $(I_{H-})_{max}$	Line cusp	Line cusp	TBD
8	Discharge power (kW)	8	7.6	8
9	$(I_{H-})_{max}$ (mA)	18	25	25
10	e/H- ratio (line cusp)	3.0	5.9	2.4
11	Electron repel (for ε_l)	Line cusp	PM solenoid	TBD
12	e/H⁻ ratio (ring PM)		4	2.4
13	ε_l (πcm-mrad), measured	7	8-9*	8.4

*(I_{H-} = 20mA), **TBD = to be determined

parameter, the third column the production source parameter value, the fourth column the development source status, and the last column the development source goal. The development approach is to increase the emission aperture to r_P=0.6cm, thus increasing the emission area by a factor 1.44. The repeller aperture, r_{rep}, is also increased in the development source to prevent converter beam interception on this electrode. The production source typically produces 16-18mA H⁻, thus the new source should produce 23-26mA H⁻. This procedure maintains constant discharge power in the new source, thus filament lifetime between the two sources should be the same, and the 28-day accelerator run time between source recycles should be preserved. Additional development effort is being made to understand H⁻ production efficiency, especially as regards converter processes, with the goal of producing more H⁻ current at a fixed discharge power.

All results reported in this section have been obtained with the LANSCE production accel column. Fig. 8(A) shows the PBGUNS simulation for the LANSCE column with $r_P = 0.6$cm, limiting repeller aperture $r_{rep} = 0.86$cm, and the converter radius $r_{cnv} = 1.9$cm. The repeller housing can contain a variety of magnet geometries, and can also be biased to tens of volts to suppress electrons. The distance from the converter to the Pierce aperture is 12cm in this case – slightly less than the converter's machined curvature radius $\rho_{cnv} = 12.5$cm. In a ballistic model of the source, no spreading of the converter H⁻ beam occurs, and the converter beam comes to a point focus near the Pierce aperture. This PBGUNS model uses the H⁻ sputter option where beam leaves the converter with 260eV and arrives at the emission aperture with uniform longitudinal energy. At 12eV sputter energy the repeller and Pierce apertures are fully illuminated with H⁻ beam, whereas at 6eV sputter energy the converter beam does not spread sufficiently to intercept the Pierce aperture. A discussion of the influence of sputter energy on simulations for this source is found in ref. [3]. The

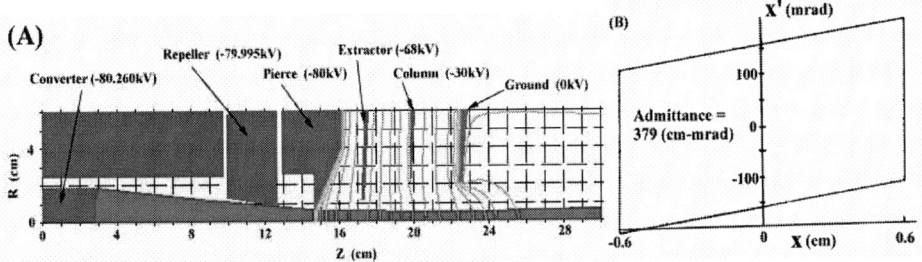

FIGURE 8. (A) PBGUNS simulation of the development source converter geometry and the LANSCE 80-kV accel column. Electrode potentials relative to local ground are given. (B) Converter source admittance diagram with $r_P = 0.6$cm.

admittance diagram based on the development source is shown in Fig. 8(B), and the area is $A_{ad} = 379$ (cm-mrad) (cf Table 1, entry 5). For 300eV H⁻, $\beta = 8.0 \times 10^{-4}$, thus the limiting normalized emittance $= \beta A_{ad}/\pi = 0.96$ (πmm-mrad). Converting this normalized emittance to 80keV laboratory emittance, one finds a lower emittance limit of 7.4(πcm-mrad).

First operation of the development source led to unstable discharge voltages. This

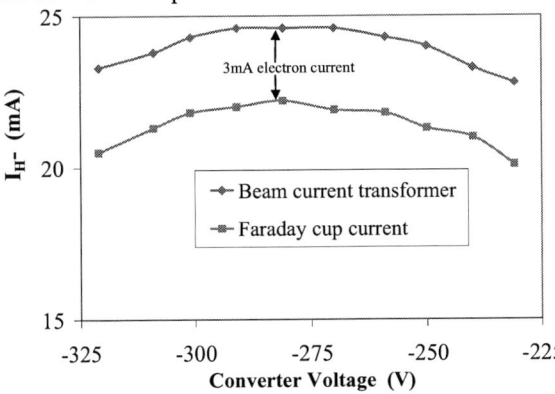

FIGURE 9. H⁻ currents measured in the 80-keV LEBT. The beam current transformer located after solenoid 1, and the Faraday cup current measurement is made at the end of the LEBT.

261

phenomenon has been previously observed in surface converter sources [14]. Stable arc discharge voltages at −190V were obtained by reducing the magnetic cusp confinement field at the source vacuum wall [3]. Using a line cusp magnet in the repeller assembly (cf entry 7 in Table 1), up to 25-mA H⁻ current has been obtained. The H⁻ currents measured in the beam current transformer after solenoid 1 and then in a Faraday cup at the end of the LEBT are shown in Fig. 9 as a function of the converter voltage. The H⁻ current has reached the 25-mA design at the beam current transformer, but decreases to 22mA at the Faraday cup (cf Fig. 2). The difference in current is thought to be residual electron current at the beam current transformer. The e/H⁻ ratio was 5.9, see Table 1, parameter number 10. For the LANSCE injector B operations, this e/H⁻ ratio is too large by more than a factor two.

A permanent magnet (PM) ring solenoid (500G maximum on axis field) was installed in the development source repeller assembly. For the 6 filament source discussed above, this magnet configuration reduced the e/H⁻ ratio by about 50% [1,2]. On the two-filament development source, this magnet produced e/H⁻ = 4/1, a 32%

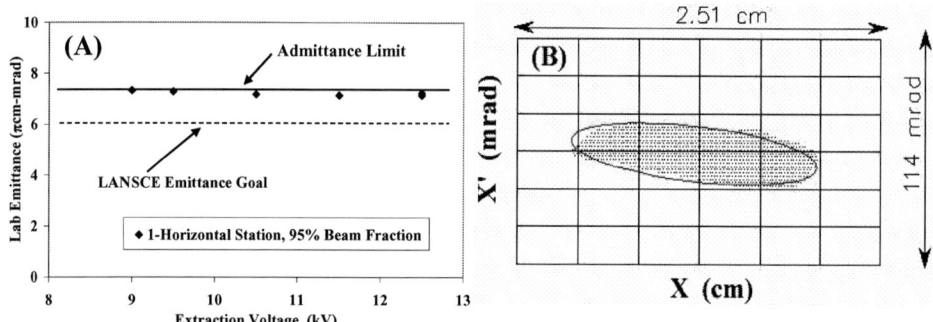

FIGURE 10. (A). Measured emittances at station 1 on the ISTS. Taken on the development source for r_P = 0.6cm and 80keV beam energy. (B) Station 3 emittance scan for the development source corresponding to 17mA H⁻ current.

reduction (cf line 12, Table 1). However the maximum H⁻ current at the ISTS current transformer was reduced to 20mA with 17mA being transported to the Faraday cup. Although this current is below the H⁻ specification (cf entry 9, Table 1), an emittance scan study was carried out at the three emittance stations as a function of the extractor electrode voltage. Emittance station one lab emittances shown in Fig. 10(A) are taken at 95% beam fraction. In a Gaussian beam emittance model, 95% beam fraction corresponds to six times the rms value. The development source results are near the predicted emittance minimum based on the admittance calculations (line 5, Table 1). The ratio of admittances in line 5, Table 1, for the development and production sources gives an emittance growth prediction = 379/304 = 1.25, while the average measurements in Fig. 10(A) to the LANSCE emittance goal (dashed line in Fig. 10(A), production source emittance) gives the emittance growth = 1.2. The development source emittance limit of 7.4(πcm-mrad) is also shown in Fig. 10(A) as solid line. This prediction and emittance station 1 measurements are in agreement.

Thus by source admittance limit arguments, there appears to be no unexpected emittance growth at station 1. At emittance station 2, the measured emittances are 8-9 (πcm-mrad). This ISTS station corresponds to the injector B measurement which is typically 7(πcm-mrad). Thus the emittance station 2 results are quoted in Table 1, line 13. At emittance station 3, measured emittances are 8-10 (πcm-mrad). The station 3 scan taken in the horizontal plane shown in Fig. 10(B) has laboratory emittance of 8(πcm-mrad). Comparing Fig. 10(B) with Fig. 4(B) shows that the multi-beam component at this location is greatly reduced for the development source case. The ISTS focus solenoids had similar current settings for the Fig. 4(B) and 10(B) measurements. A systematic difference between the ISTS horizontal and vertical emittance results was noted earlier [3] in the development source work. Since the last development source data acquisition, a misalignment was found in the LEBT beam line, and this may be the cause of the asymmetric emittances [3] observed at stations 2 and 3.

Summarizing, the development source has met the design current (25mA) and design emittance, although the latter needs to be confirmed at the higher design current. The present situation is that the e/H⁻ ratio is too great for injector B operations. Further efforts to reduce the e/H⁻ ratio are; first, reduce the magnetic field strength at the ion source wall; second, the wiggler (opposed dipole fields) repeller magnet may be tested in the development source; and third, careful measurements of the repeller voltage effect on e/H⁻ ratios can be made [15]. Modelling of cusp field confinement schemes has shown that reducing the confinement field will increase the source anode area [16,17], thus reducing electron current extraction at the emission aperture. The wiggler repeller magnet was used in the six-filament source, and it demonstrated a 75% reduction of extracted electron currents as compared to the line cusp repeller magnet [1,2].

A second approach for a more comprehensive solution to enhanced surface converter source performance is the improvement of the H⁻ production efficiency. This H⁻ surface converter source falls into the general category of cathodic surface

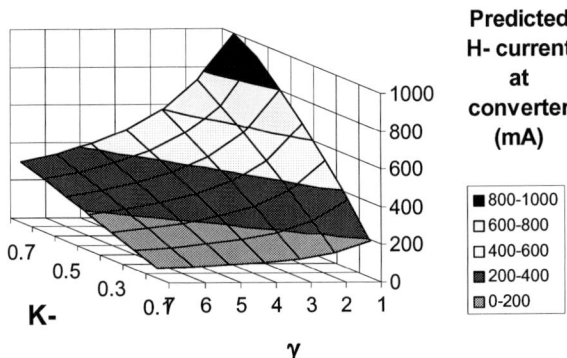

FIGURE 11. Prediction for H⁻ converter current, I_{H^-cnv}, using the measured converter current, I_{cnv} of 4A, and published γ and K⁻ factors for SPS.

plasma source (SPS) H⁻ production (cf. Fig. 1f in ref [18]). Total converter currents, I_{cnv}, in both the production and development sources are measured to be 4A. Secondary electron production coefficient $\gamma = I_e/I^+$ at the cathode may vary from 1 to 7 while the secondary H⁻ production coefficient $K^- = I_{H-cnv}/I^+$ may vary from .1 to .7 in cesiated SPS [6]. I_{H-cnv} is the H⁻ current produced at the converter. Since $I_{cnv} = I^+ + I^-$ = 4A, a prediction for possible I_{H-cnv} may be made over the limits of the γ and K^- parameters. Here $I^- = (\gamma + K^-)I^+$. The prediction is shown in Fig. 11, and $(I_{H-cnv})_{min}$ = 40mA is found at $\gamma = 7$, $K^- = 0.1$ while $(I_{H-cnv})_{max}$ = 1000mA is found for $\gamma = 1$, $K^- = 0.7$. For a well-cesiated molybdenum surface the parameters γ and K^- may be 7 and 0.7 [6] which yields I_{H-cnv} = 300mA. This H⁻ converter current is order factor ten greater than the PBGUNS sputter model currents used in Fig 8(A).

In addition to H⁻ sputter energy at the converter, another cause of converter efficiency reduction is H⁻ converter beam expansion by incomplete neutralization of the H⁻ beam space charge. Such expansion may occur in a localized area around the converter sheath, and/or in beam transport from the converter to the emission aperture. An approximate plasma density of 10^{11} (cm)⁻³ has been derived in the six-filament source by using the converter as a floating probe. Using this plasma density and an assumed electron temperature of 1eV, a converter plasma sheath thickness of 1.7 mm is derived [19]. A 2-D particle in cell (PIC) code is being developed at Los Alamos for application to ion source plasma problems [20]. A preliminary result from the PIC code simulation as applied to this H⁻ surface conversion source is shown in Fig. 12.

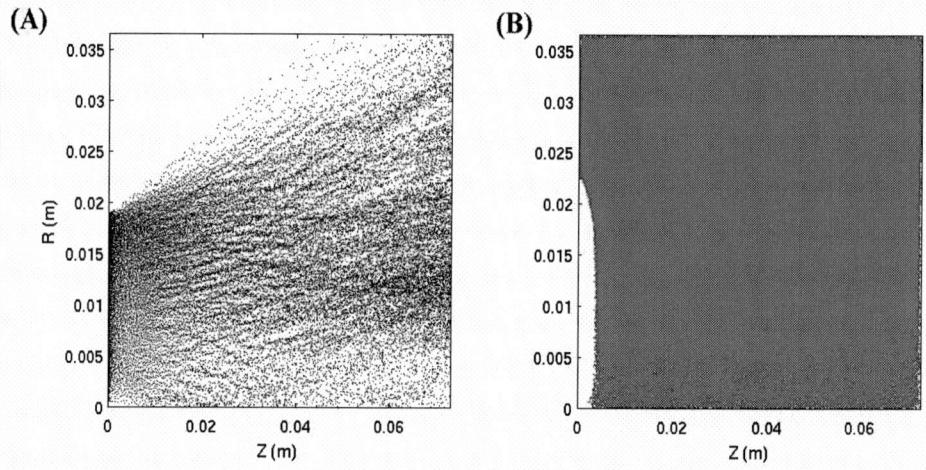

FIGURE 12. Preliminary 2-D PIC code simulation for H⁻ beam being accelerated off the surface converter located on the left in Fig. 12(A). Fig. 12(B) shows the plasma electrons, and formation of the sheath at about 3 mm from the converter.

The plasma density in this simulation is 3×10^{10} (cm)⁻³. The H⁻ beam is born on the plasma converter on the left of Fig. 12(A). The sheath region shown in Fig. 12(B) has formed approximately 3mm downstream from the converter. The 300eV H⁻ beam is indeed predicted to have a strong divergence at the converter from residual negative space-charge, and from a defocusing electric field at the converter edge. These sheath predictions are suggestive of further experimental work with shaped converters. Not until experiments on electron repeller options and increased H⁻ production efficiency are completed, will the parameters labeled TBD in Table 1 be established.

ACKNOWLEDGEMENTS

We acknowledge the U.S. National Nuclear Security Administration (NNSA) and LANSCE Division operations for continued support in the field of H⁻ source development.

REFERENCES

1. Benjamin A. Prichard, Jr., and Ralph R. Stevens, Jr., "Status of the SPSS H⁻ Ion Source Development Program", Los Alamos National Lab Report LA-UR-02-547 (2002).
2. G. Rouleau, E. Chacon-Golcher, E. G. Jacobson, B. J. Meyer, and E. G. Geros, B. A. Prichard, Jr., J. D. Sherman, J. E. Stelzer, R. R. Stevens, Jr., "H⁻ Surface Converter Source Development at Los Alamos", Proceedings of the 2003 Particle Accelerator Conference (Portland, Oregon), IEEE Catalog Number 03CH37423C, 73 (2003).
3. J. Sherman, A. Arvin, E. Chacon-Golcher, E. Geros, E. Jacobson, B. Meyer, P. Naffziger, G. Rouleau, S. Schaller, J. Stelzer, and T. Zaugg, "Development of a 25-mA, 12% Duty Factor H⁻ Source for LANSCE", to be published in Proc. of the 2004 European Particle Accelerator Conference (Lucerne, Switzerland).
4. R. Keller, J. M. Verbeke, P. Scott, M. Wilcox, L. Wu, and N. Zahir, "A Versatile Column Layout for the LANSCE Upgrade", Proceedings of the 1999 Particle Accelerator Conference (New York), IEEE Catalog Number 99CH36366, 1926 (1999).
5. R. R. Stevens, Jr., W. Ingalls, O. Sander, B. Prichard, and J. Sherman, "Beam Simulations for the H⁻ Injector Upgrade at LANSCE", Proc. of the XIX International Linac Conference (Chicago, IL), Argonne National Lab Report ANL-98/28, 514 (1998).
6. Yu. I. Belchenko, G. I. Dimov, and V. G. Dudnikov, "Physical Principles of the Surface Plasma Method for Producing Beams of Negative Ions", Proc. of the Symposium on the Production and Neutralization of Negative Hydrogen Ions and Beams (Brookhaven, New York), Brookhaven National Lab Report BNL 50727, 79(1977).
7. W. B. Ingalls, M. W. Hardy, B. A. Prichard, O. R. Sander, J. E. Stelzer, R. R. Stevens, K. N. Leung, and M. D. Williams, "Enhanced H⁻ Ion Source Testing Capabilities at LANSCE", Proc. of the XIX International Linac Conference (Chicago, IL), Argonne National Lab Report ANL-98/28, 887 (1998).
8. R. C. McCrady, M. S. Gulley, and A. A. Browman, "Evaluation of the LANSCE H⁻ Low-Energy Beam Transport System for High Emittance Beams", LANSCE Activity Report LA-13943-PR, 58(2001).
9. Jack E. Boers, "PBGUNS: An Interactive IBM PC Computer Program for the Simulation of Electron and Ion Beams and Guns", Thunderbird Simulations, (August, 2001).
10. R. F. Welton, M. P. Stockli, J. E. Boers, R. Rauniyar, R. Keller, J. W. Staples, and R. W. Thomae, "Simulation of the Ion Source Extraction and Low Energy Beam Transport Systems for the Spallation Neutron Source", Review of Sci. Instrum. 73 (2), 1013(2002).
11. J. Guyard and M. Weiss, Proc. of the 1976 Linear Accelerator Conf., (Chalk River, Ontario, Canada), AECL-5677, 254(1976).
12. Ralph R. Stevens, Jr., Rob L. York, John R. McConnell, and Robert Kandarian, "Status of the High-Intensity H⁻ Injector at LAMPF", Proc of the 1984 Linear Accelerator Conf., (Seeheim, Germany), GSI-84-11, 226(1984).

13. R. L. York, Ralph R. Stevens, Jr., R. A. Dehaven, J. R. McConnell, E. P. Chamberlain, and R. Kandarian, Nucl. Instrum. And Methods in Phys. Res. B10/11, 891(1985).
14. J. W. Kwan, G. D. Ackerman, O. A. Anderson, C. F. Chan, W. S. Cooper, G, J. DeVries, A. F. Lietzke, L. Soroka, and W. F. Steele, Rev. Sci. Instrum 57(5), 831(1986).
15. K. W. Ehlers and K. N. Leung, "Electron Suppression in a Multicusp Negative Ion Source", Appl. Phys. Lett.38 (4), 287(1981).
16. Dan M. Goebel, "Ion Source Discharge Performance and Stability", Phys. Fluids 25(6), 1093(1982).
17. A. P. H. Goede and T. S. Green, "Operation Limits of Multipole Ion Sources", Phys. Fluids 25(10), 1797(1982).
18. Vadim Dudnikov, Rev. Sci. Instrum. 73(2), 992 (Feb. 2002).
19. Claude LeJeune, Advances in Electronics and Electron Physics, Academic Press, Part C, Very High Density Beams, 207 (1983).
20. Edwin Chacon-Golcher, K. J. Bowers, and J. D. Sherman, "Progress in the Computer Simulation of LANSCE's Production H⁻ Source", to be published in the Proc. of the 31st European Plasma Physics Physical Society (London, England), (June, 2004)

H⁻ Source Developments at CERN

C. Hill*, D. Küchler‡, R. Scrivens* and T. Steiner*

*CERN, Geneva, Switzerland

Abstract. Future CERN programmes for LHC and ISOLDE require increasing the beam intensity and brightness from the Proton Synchotron Booster (PSB). This could be achieved by injection from a higher energy H⁻ linac. A new injector will require a high performance, high reliability, negative hydrogen ion source. This paper will present the requirements for such a source together with the first results for a prototype microwave driven source.

INTRODUCTION

Present and future CERN programmes, like LHC[1], CNGS[2] or ISOLDE[3], require an intense and bright proton beam from the injector. The present combination of Linac2 and Proton Synchotron Booster (PSB) limits the performance of the whole complex (see Fig.1). Different upgrade options are under investigation (for the complete list see [5]):

Accelerator chain of CERN (operating or approved projects)

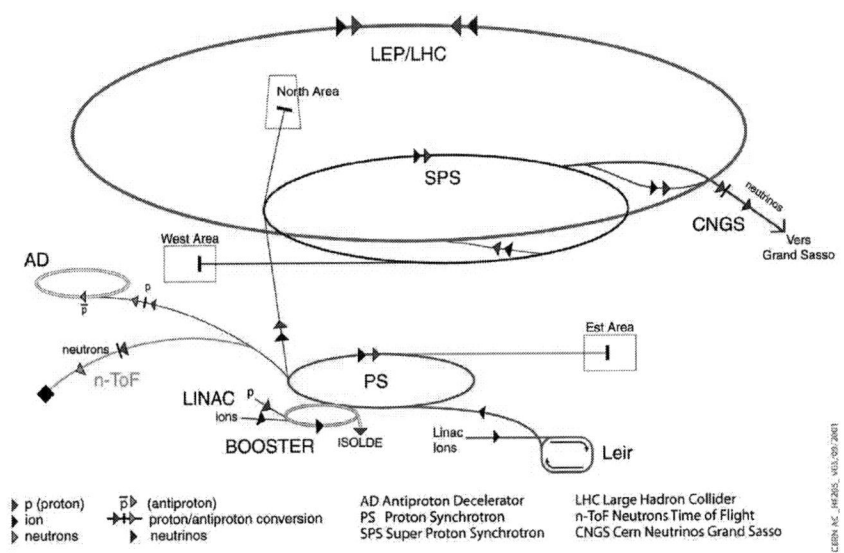

FIGURE 1. CERN accelerator complex[4].

---

1 detlef.kuchler@cern.ch

CP763, *Production and Neutralization of Negative Ions and Beams*, edited by J. D. Sherman and Y. I. Belchenko
© 2005 American Institute of Physics 0-7354-0248-5/05/$22.50

- modifications and optimization of existing machines (e.g. change of basic period)
- injection Linac4[6] into PSB (at an injection energy of 160 MeV)
- injection Superconducting Proton Linac (SPL)[7] into PS (at an injection energy of 2.2 GeV)

Common for several of the modifications is the need of a high performance, high reliability, negative hydrogen ion source. The source parameters for Linac4 and SPL are given in Table 1.

TABLE 1. Parameters of an H⁻ ion source for Linac4 and SPL.

	Linac4	SPL
Instantaneous current	50 mA	>40 mA
Pulse length	0.5 ms	2.8 ms*
Repetition rate	2 Hz	50 Hz
Extraction voltage	95 kV	95 kV
Emittance (rms normalized)†	0.25 π mm mrad	0.25 π mm mrad
Availability for tests	2007	2012
Assumed start of operational use	~2008	~2013
Further features (not essential but desired)	cesium free no antenna or filament in the plasma chamber high pulse-to-pulse stability mean time between failure: 100 days easy maintenance	

* 1.5 ms for the new SPL Layout (2005)
† Emittance at RFQ input

The developments are based on a microwave driven source due to the good and extensive experience with the ECR technology for the heavy ion physics programme. The first experiments are meant to investigate the behavior of the source and the extracted beam while changing different source parameters and certain other conditions.

EXPERIMENTAL SET-UP

The source body is a water-cooled plasma chamber of 10 cm inner diameter and 20 cm length. For the first experiments a configuration of 10 permanent magnets around the plasma chamber was used (see Fig. 2). This provides a field of 0.21 T at the magnet position inside the plasma chamber.

The microwave of a frequency of 2.45 GHz and a maximum power of 1.5 kW (pulsed) is injected with a movable antenna (range of the antenna tip into the plasma chamber is $\lambda/4$).

A magnetic filter of two permanent magnets separates the source into the plasma and the production region. The field in the middle of the filter is ~0.007 T. The filter is 7.8 cm away from the plasma electrode.

The extraction system consists of two electrodes. The experiments were done with 20 kV extraction voltage. The plasma electrode is insulated in respect to the plasma

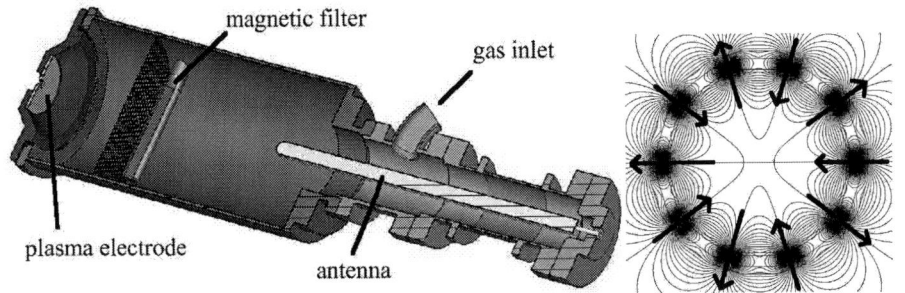

FIGURE 2. Schematic view of the source and the magnetic field.

chamber and can be biased (see MEASUREMENTS section).

MICROWAVE SIMULATION AND MEASUREMENTS

The response of the actual plasma chamber on microwave injection was studied with some simulations in MICROWAVE STUDIO[8]. Figure 3 shows the calculated S_{11} parameter of the plasma chamber and the measurements of the S_{11} parameter are shown in Fig. 4. Marker 1 is at the generator frequency of 2.45 GHz, which is not an eigen mode, leading to a high reflected power. Figure 5 shows the distribution of the electrical field in the chamber for the frequency of 2.45 GHz.

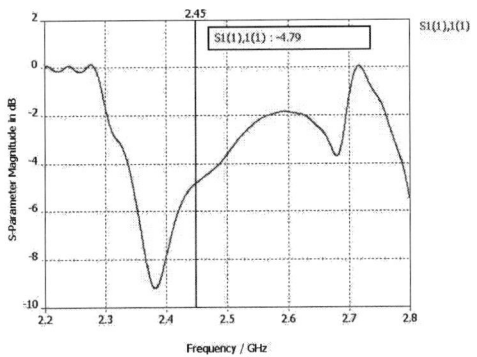

FIGURE 3. Simulation of the S_{11} parameter of the plasma chamber.

In Fig. 5 a high field at the foot of the antenna indicates the bad matching. To improve the situation in the future the transition from the cable to the antenna will be modified with a taper.

The simulations and the measurements were done with an empty plasma chamber, i.e. without a plasma. The behavior with plasma is expected to be different.

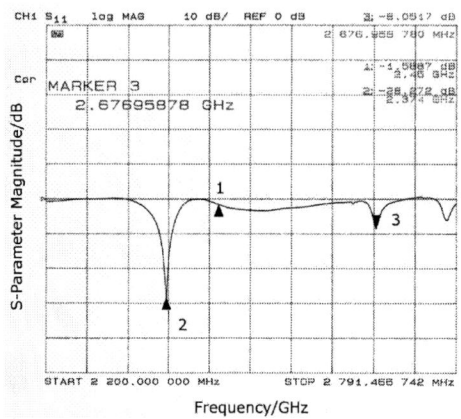

FIGURE 4. Measurement of the S_{11} parameter.

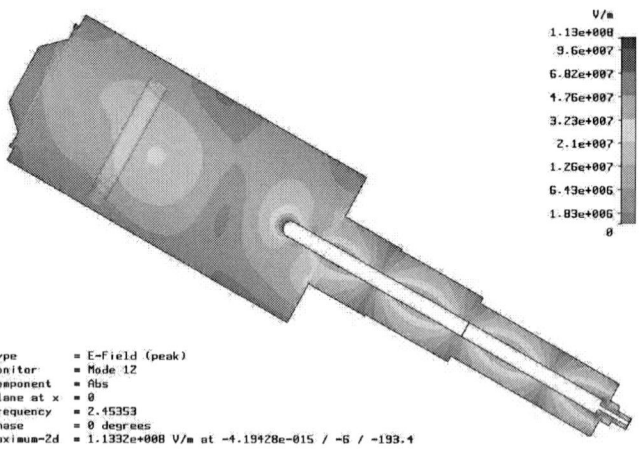

Type	= E-Field (peak)
Monitor	= Mode 12
Component	= Abs
Plane at x	= 0
Frequency	= 2.45353
Phase	= 0 degrees
Maximum-2d	= 1.1332e+008 V/m at −4.19428e−015 / −6 / −193.4

FIGURE 5. Simulation of the absolute value of the electric field at the frequency of 2.45 GHz.

MEASUREMENTS

Many parameters of the source were varied and for all measurements the H^- current, the electron current, the pressure in the extraction tank and the reflected power were recorded (experimental setup see Fig. 6, RF setup see Fig. 7).

First the influence of the RF power, the gas flow and the position of the antenna were studied (Figs. 8 and 9).

The influence of the gas was difficult to obtain because there is no differential pumping between the extraction tank and the source. An increase of the pressure in the source resulted in an increase of the pressure in the extraction tank as well. Due to the sensitivity of the H^- ion to higher pressures the results for high gas flow rates are most likely

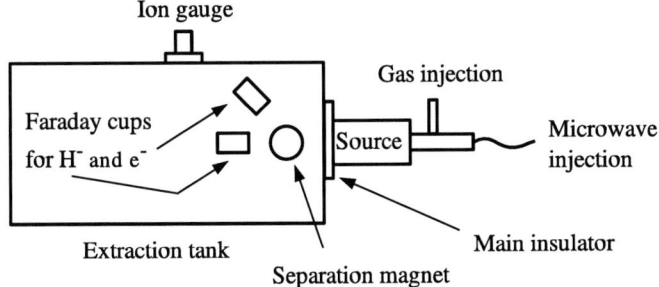

FIGURE 6. The experimental setup.

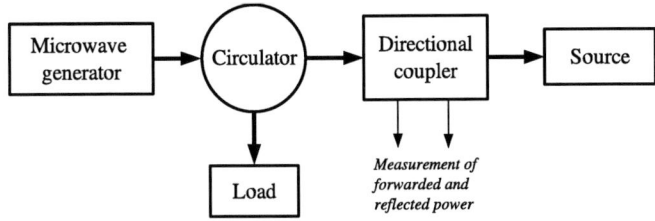

FIGURE 7. The setup of the RF components.

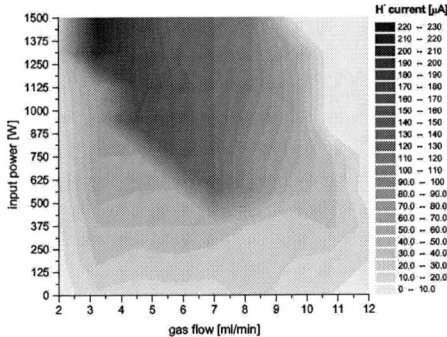

FIGURE 8. Dependency of the H⁻ current (in μA) from RF power and gas flow (bias voltage +30 V, antenna position $z = 25$ mm).

too low (see top right corner in Fig. 8).

Figure 10 shows the maximal H⁻ current measured up to now ($I_{H^-} = 0.29$ mA, $I_{e^-} = 4.6$ mA, $e^-/H^- = 15.9$). The rise time of the beam is 6 ms. The drop at the end of the pulse is due to the per pulse limited power available from the RF generator.

The plasma electrode is insulated and can be biased in respect to the plasma chamber. With the bias the beam current could be increased by about a factor 10 and there is also a big influence on the e^-/H^- ratio (see Fig. 12). This ratio was for the most of the measurements (positive bias of the plasma electrode) in the range 20–80 (see Fig. 11).

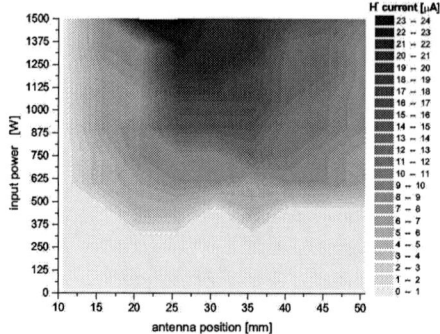

FIGURE 9. Dependency of the H⁻ current (in μA) from RF power and antenna position (bias voltage 0 V, gas flow 3 ml/min).

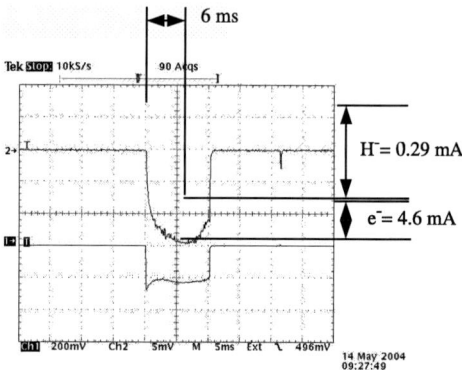

FIGURE 10. Oscillogram of the H⁻ (top trace) and the e⁻ (bottom trace) beam.

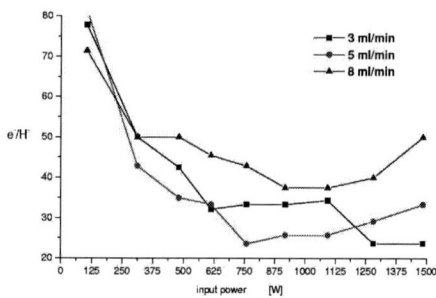

FIGURE 11. The ratio e⁻/H⁻ as dependence from the RF power (bias voltage +30 V, antenna position z = 25 mm).

Several recipies for ion improvement were copied from other sources. The group at CEA/Saclay had improved the ion currents after installing a grid between the plasma and the production chamber [9]. The idea is to prevent microwave heating in the production chamber. Introducing a grid in the CERN source, reduced the H⁻ current significantly.

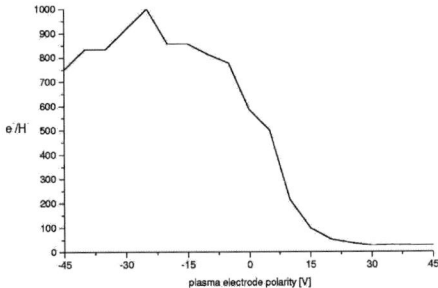

FIGURE 12. The ratio e^-/H^- as dependence from the plasma electrode voltage (RF power 1500 W, antenna position $z = 25$ mm, gas flow 3 ml/min).

This may be an effect of a too low electron density. In the present scenario the additional heating in the production region is useful. This may change for other magnetic configurations.

Several groups reported a positive effect of tantalum in the source [9]. This was tested in two ways. First a tantalum sheet was placed inside the plasma chamber. Second a collar made out of tantalum was used. The tantalum sheet did not improve the output on H^- significantly, but it reduced the reflected power and changed the optimal plasma electrode bias to lower voltages. The reduction of the reflected power could be a result of the change of the inner diameter of the plasma chamber due to the foil. This "tunes" the cavity and the resonances move. Three different lengths of the collar have been tried: 78 mm (up to the magnetic filter), 39 mm and 20 mm. They were electrically connected to the plasma electrode and could so be biased as well. No improvement was found. The H^- current was rather reduced to 50 % for the shortest collar and to less than 5 % for the longest collar.

Bacal et al. [10] reported some success improving the H^- density in the source by using a small argon addition. At the CERN source the use of argon destroyed the H^- current.

CONCLUSION AND FUTURE PLANS

In the next step the permanent magnets will be replaced by two solenoids (Fig. 13). The stronger magnetic field (Fig. 14) should result in a higher electron density and therefore in a higher H^- beam current. Optimization of all the above parameters will then be performed.

A small additional experiment is to place a disk of boron nitride at the end of the plasma chamber and to set the magnetic field onto the disk to a value corresponding to the electron cyclotron resonance (2.45 GHz $\rightarrow$ 0.0875 T). This method showed good results in the ECR H^- source at CEA/Saclay.

On the basis of all the experiences made with the present experimental setup it is planed to design a new source. This work will start at the end of this year and will go in parallel with the design of the Low Energy Beam Transport (LEBT) of Linac4.

FIGURE 13.   The source body with the new solenoids.

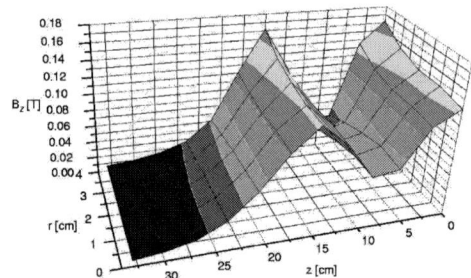

FIGURE 14.   The field of the solenoids.

ACKNOWLEDGMENTS

The authors would like to thank for the support by C. Mastrostefano, M. O'Neil, K. Fritioff. F. Caspers, R. Gobin.

This work is supported by the European Commission under contract n°: HPRI-CT-2001-50021.

REFERENCES

1.   LHC Design report, CERN-2004-003, Geneva, 2004.
2.   K. Elsener (editor), The CERN Neutrino Beam to Gran Sasso (NGS), CERN 98-02 and CERN-SL/99-034(DI), Geneva, 1998 and 1999.
3.   http://isolde.web.cern.ch/ISOLDE/
4.   http://user.web.cern.ch/user/Index/Accelerators.html
5.   M. Benedikt, K. Cornelis, R. Garoby, E. Métral, F. Ruggiero, M. Vretenar, Report of the High Intensity Protons Working Group, CERN-AB-2004-022 OP/RF, Geneva, 2004.
6.   R. Garoby, K. Hanke, A. Lombardi, C. Rossi, M. Vretenar, F. Gerigk, Design of the Linac4, a new Injector for the CERN Booster, to be published at Linac2004, Lübeck (Germany), 2004.
7.   The SPL Study Group, Conceptual Design of the SPL, CERN 2000-012, Geneva, 2000.
8.   http://www.cst.com
9.   HP-NIS collaboration, annual meeting, private communication, Abington, 2004.
10. M. Bacal, A. Ivanov, C. Rouille, J.M. Buzzi, Camembert III: Argon additive effect on hydrogen plasma and project of ECR H⁻ production, http://www.hpnis.dcu.ie/Bacal.pdf, HP-NIS collaboration, annual meeting, Dublin, 2003.

Source of Negative Hydrogen Ions with Hot Cathode

G.I.Kuznetsov and M.A.Batazova

Budker Institute of Nuclear Physics, Novosibirsk 630090, Russia,
E-mail: G.I.Kuznetsov@inp.nsk.su

Abstract. In the report an H⁻ ion source with following parameters is described. Ion energy is up to 30 keV, current is up to 5 mA, pulse duration varies from 1 to tens microseconds. H⁻ ions are produced in magnetron discharge near a hot cathode surface made of LaB_6 or IrCe alloy, which is 6 mm in diameter. H⁻ ions are extracted through 0.8x7 mm slot. An electromagnetic valve was used for pulsed gas feed. Ion current depends on discharge current that in turn is defined by cathode temperature, discharge voltage, and hydrogen pressure. Two permanent magnets fixed on iron yoke generate magnetic field for magnetron discharge. There are a magnetic and electrostatic corrections of trajectories of extracted H⁻ ions. The H⁻ ion source has been used as an injector for tandem accelerator with the following proton beam parameters: energy is 1.4 MeV, current is 3mA, and pulse duration is 2 microseconds. Simulation of ion beam trajectories in the source, transport channel, and tandem accelerator shows good correlation with experiments.

INTRODUCTION

In order to raise the synchrotron output current for therapy of cancer (TRAPP) developed at the Institute of Nuclear Physics, Novosibirsk [1], a reliable H⁻ ion source was required. It was decided to mount the source on a tandem accelerator [2] for production of proton beam with energy 1.4 MeV. Thus, to have 10^9 protons per pulse at TRAPP output, 10^{10} particles during 1 µs was demanded from the source, which means an ion current of about 3 mA. The 25 kV extraction energy was chosen for source matching with tandem accelerator input.

The initial idea was to create H⁻ ions in the vicinity of a hot cathode surface where electrons have high energy enough to dissociate the hydrogen molecule, and the cloud of low energy electrons exists to be captured by excited molecules for negative ions production. To create the thin sheath, a flat magnetron discharge in transverse electrical and magnetic fields is used. In this case we have no electrons in the extraction area because they are collected to the discharge chamber wall.

CP763, *Production and Neutralization of Negative Ions and Beams*, edited by J. D. Sherman and Y. I. Belchenko
© 2005 American Institute of Physics 0-7354-0248-5/05/$22.50

DESIGN

The total scheme of negative ion source for TRAPP device is presented in Fig.1. The source consists of the H⁻ extraction system, of transport system with two pulsed lenses, of electrostatic and magnet correctors to direct a beam into the accelerating tube(not shown in Fig.1), of a Faraday cap to measure H⁻ current, and of a pumping system. The front and side views of the extraction system are correspondingly shown in Fig.2 and Fig.3. The cathode inserted into discharge chamber is the standard 6 mm in diameter cathode [3], developed at BINP for commercial electron accelerator of ELV (electrostatic linear diode) type. Transverse magnetic field is induced by two permanent magnets fixed on movable yoke that allows to vary magnetic field in the chamber. Hydrogen gas was injected into the chamber by the electromagnetic valve [4] designed and produced in our institute.

Negative discharge voltage is applied to the cathode. The high extraction voltage was applied to the "focus" electrode with emission slot. Extraction of H⁻ is performed from the cathode area through the slot with 0.8-0.9 mm width and 6-8 mm length. Two electrostatic lenses were used in the transport system between ion source and tandem accelerator at first. Considering their aberrations and difficulties with their power supply, they have been changed to magnetic lenses.

Magnetic correction is used to compensate ion beam mis-steering caused by the transverse magnetic field in the extraction area. An electrostatic system of parallel shift and deviation of ion trajectories is mounted between the magnetic lenses of transport channel to direct the beam to the axis of charge-exchange tube in the tandem accelerator. A movable sectioned Faraday cup is situated at the source output. Vacuum tanks of the source and of tandem accelerator are connected with the pipe of a small diameter to separate their volumes. NMD-0.1 ion pumps with 100l/s pumping speed were used.

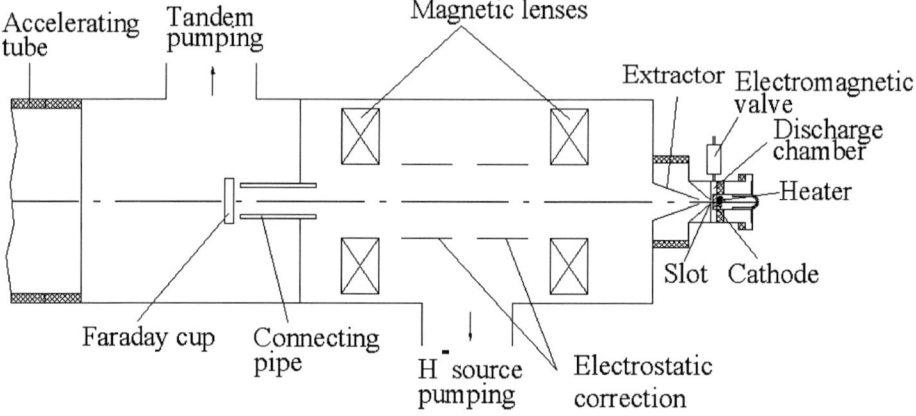

FIGURE 1. General scheme of experimental setup.

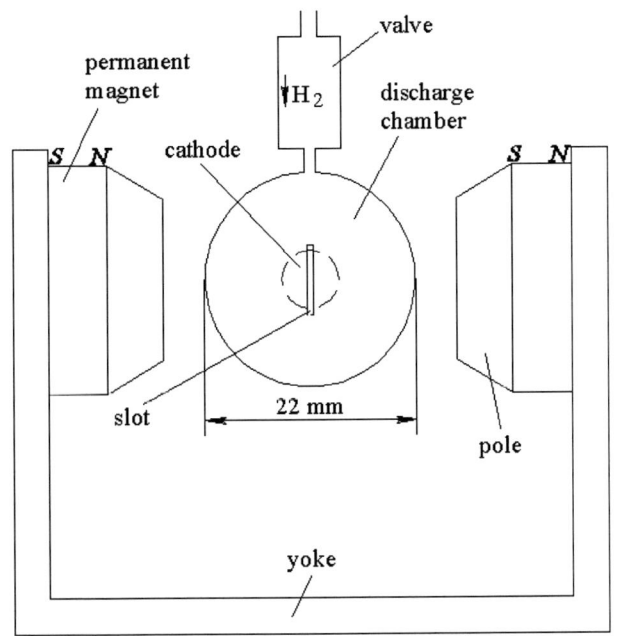

FIGURE 2. Front view of H⁻ extraction system.

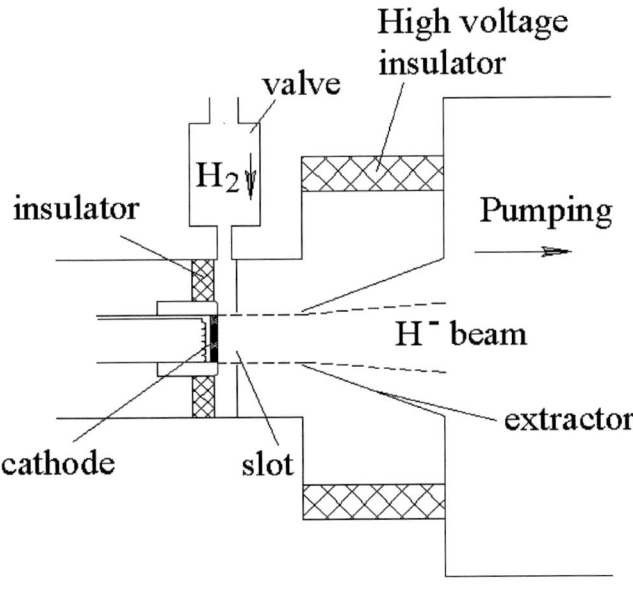

FIGURE 3. Side view of H⁻ extraction system.

SIMULATION

Simulation of extractor ion optics has been made in cylindrical approximation with the electrode geometry, presented in Fig.4. The 1 mm gap between cathode surface and the electrode with emission slot was used in simulations. The H⁻ ion emission supposed to be non-limited and was estimated by Child-Langmuir law. The space charge of ions, accelerated by discharge voltage, is negligible in the extraction area, and the optics between slot and the extractor can be regarded as a pure geometrical one.

The results of simulation show, that 6 mm cathode can deliver 5 mA H⁻ beam with divergence less than 100 mrad at the extractor outlet (Fig.4). With the increase of the slot–extractor distance, one could reduce the divergence, but it will increase the beam shift due to turn in the transverse magnetic field of discharge chamber. It should be mentioned, that this field was not taken into account in simulations.

The rectangular pulse of H⁻ current was transported through the system with two magnetic lenses. One of the variants of transport channel simulation with lenses and tandem accelerator in axial approximation is shown in Fig.5. Adjusting the lens currents we can change input angle of the beam quite smoothly, thus tuning beam coordinates for match with tandem channel and discharge tube.

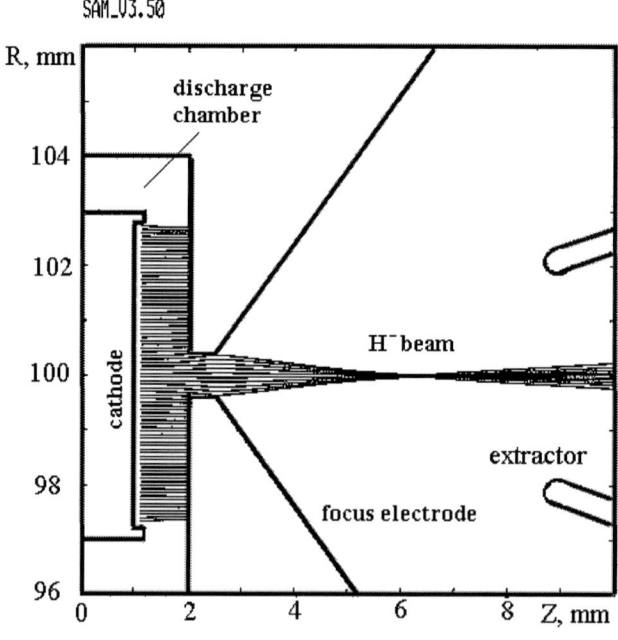

FIGURE 4. The extractor optics simulation.

278

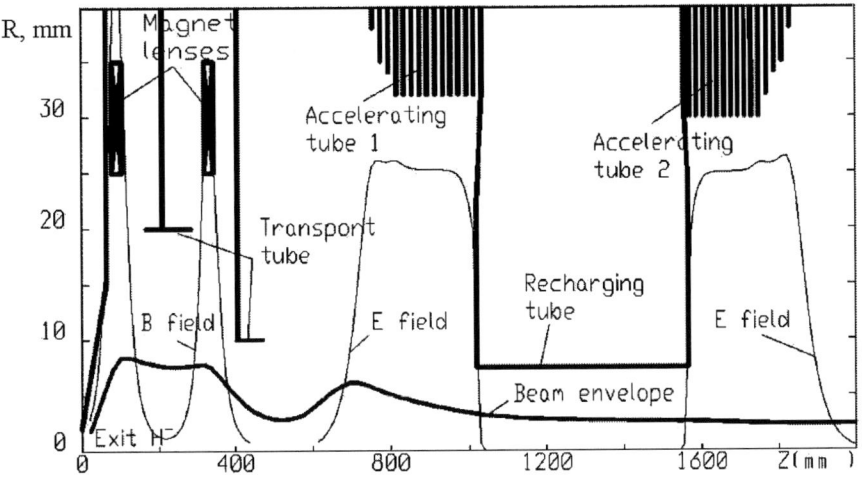

FIGURE 5. H⁻ source and tandem optics.

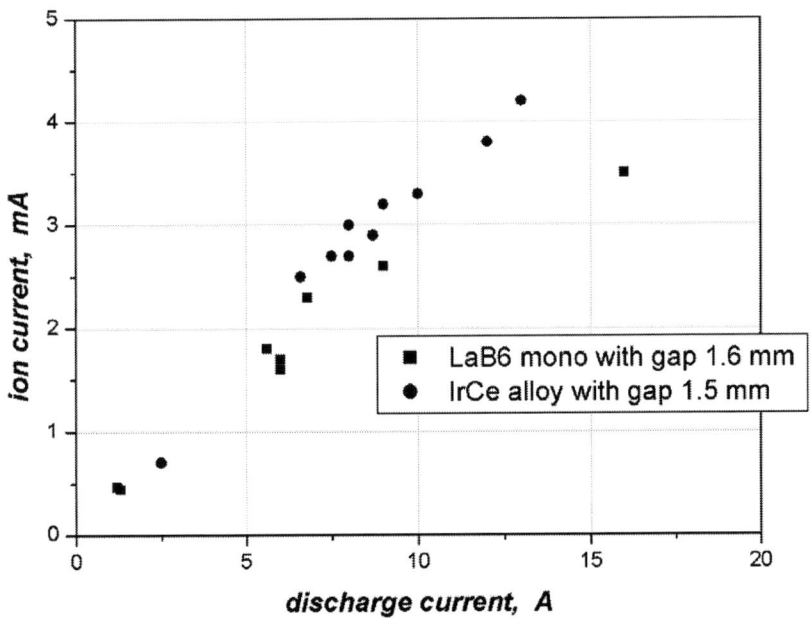

FIGURE 6. H⁻ current vs discharge current.

RESULTS

In Fig.6 the dependence of H⁻ current versus the discharge one is presented. Two kinds of cathode materials, the monocrystal of lanthanum hexaboride (LaB_6, face [100]), and the Iridium-Cerium (IrCe) alloy have been tested. The cold cathode-slot gaps were 1.6 mm and 1.5 mm accordingly. The H⁻ current was measured by Faraday cup at 48 cm distance from the cathode surface at the output of the tube that connects the ion source and tandem accelerator. The beam cross-section was elliptical one with the axis ratio 1:1.5.

The H⁻ beam was extracted and measured at the moment, when pressure in the discharge chamber attained the value high enough to start the discharge, but before the gas filled the transportation channel. The exit ion current reduced two times if the transportation channel was filled with gas. The valve operation time was 250-450 μs. The discharge could be supported during more then 2 ms after beginning of its ignition, but the discharge current was decreased to 0.6-0.7 of its initial value at the end of the 2 ms pulse. The range of the source parameters, used in the source study, is shown in Table 1.

TABLE 1. Studied range of source parameters

Discharge pulse duration	1÷15 μs
Discharge voltage	0.1÷1 kV
Magnetic field	0.02÷0.15 T
Duration of extraction voltage pulse	1÷6 μs
Extraction voltage	5÷30 kV

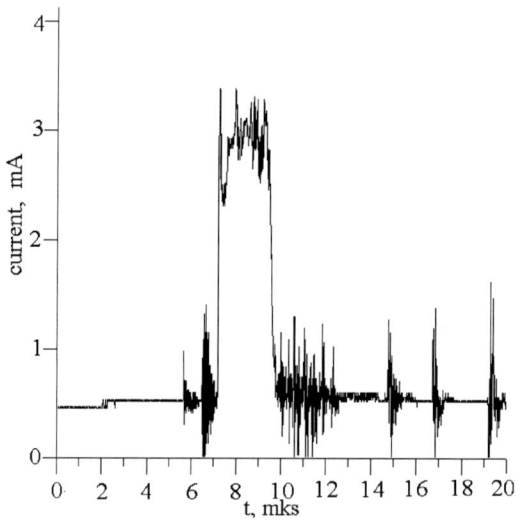

FIGURE 7. Proton current.

280

The maximal extracted H⁻ current corresponds to the value computed by 3/2 law for the plane diode. The gas feed to the discharge chamber is varied in the range $1\div3\cdot10^{-3}$ cm³ per the shot at atmosphere pressure. It is better to have the sharp front of injected hydrogen in the discharge chamber, produced by the high gas pressure (more than 3 atmospheres) in the gas valve box.

The pulsed proton beam with about 2.5 mA current, measured at the output of tandem accelerator is shown in Fig.5. These protons were produced by 3 mA H⁻ beam, extracted from the source. The beam diameter is about 8 mm, close to 6 mm simulated. It should be mentioned, that the 2 mA tandem proton output current was obtained in the first shot with no special conditioning, just after the tandem and the source parameters have been set by computer control.

CONCLUSION

The source of negative hydrogen ions with hot cathode has been designed, constructed, tested and mounted into the tandem accelerator of TRAPP synchrotron. The source allows one to have 3-5 mA of H⁻ ions with energy 25 kV and pulse duration of 0.5-10 µs. Cathode life-time is above 50 hours and is limited by interaction of gas hydrogen with lanthanides and hydrides formation. By reducing the source elements we can decrease the amount of gas hydrogen injected to the discharge volume by order, and to extend the source as well as the pumps life-time.

REFERENCES

1. Balakin, V.E., Skrinskiy, A.N., et al.. " TRAPP-Facility for Proton Therapy of Cancer". EPAC, Rome, **v.2**, 1988 , p.1505.
2. Kuznetsov, G.I., Balakin, V.E., Batazova, M.A., et al., "Tandem proton accelerator as injector for TRAPP", *Voprosi Atomnoj Nauki I Tehkniki*, **35**, 15 (2001).
3. Kuznetsov, G.I." High temperature cathode for high current density". *Nuclear Instruments and Methods in Physics Research*, **A 340**, 204 (1994).
4. Derevjankin, G.E., Dudnikov, V.G., Ghuravlev, P.A. "Electromagnetic valve for pulsed gas injection," *Pribori i Tehnika Eksperimenta,*, **35**, 168 (1975), in Russian

Polarized Negative Light Ions
at the Cooler Synchrotron COSY/Jülich

R. Gebel, O. Felden, P. von Rossen

Institute for Nuclear Physics - COSY/ Jülich, Germany

Abstract. The polarized ion source at the cooler synchrotron facility COSY of the research centre Jülich in Germany delivers negative polarized protons or deuterons for medium energy experiments. The polarized ion source, originally built by the universities of Bonn, Erlangen and Cologne, is based on the colliding beams principle, using after an upgrade procedure an intense pulsed neutralized caesium beam for charge exchange with a pulsed highly polarized hydrogen beam. The source is operated at 0.5 Hz repetition rate with 20 ms pulse length, which is the maximum useful length for the injection into the synchrotron. Routinely intensities of 20 μA are delivered for injection into the cyclotron of the COSY facility. For internal targets the intensity of 2 mA and a polarization up to 90% have been reached. Reliable long-term operation for experiments at COSY for up to 9 weeks has been achieved. Since 2003 polarized deuterons with different combinations of vector and tensor polarization were delivered to experiments.

INTRODUCTION

The accelerator facility COSY [1,2] consists of the injector cyclotron and the synchrotron and storage ring. Polarized and unpolarized protons and deuterons in the momentum range between 0.3 and 3.65 GeV/c are accelerated for the research of production and interaction of strange mesons. The operation close to the production threshold benefits from the effective use of beam cooling methods. Electron cooling increases the phase space density at injection momentum. At high beam momentum a stochastic cooling system is used for internal experiments to conserve the beam emittance. Polarized beams for fundamental research is provided by negative ion sources especially adapted for the use at the COSY facility. Figure 1 shows the floor plan of the facility with the internal and external experimental areas.

THE POLARIZED ION SOURCE

The polarized colliding beams source at COSY [3,4] comprises three major groups of components, the pulsed ground state atomic beam source [5-8], the caesium beam source and the charge exchange and extraction region. The set-up is shown schematically in Fig. 2.

The ground state atomic beam source produces an intense pulsed polarized atomic hydrogen or deuterium beam. The gas molecules are dissociated in an inductively coupled RF discharge. The atoms are cooled to about 30 K by passing an aluminium nozzle of 20 mm length and 3 mm diameter. A high degree of dissociation is kept by

CP763, *Production and Neutralization of Negative Ions and Beams*, edited by J. D. Sherman and Y. I. Belchenko
© 2005 American Institute of Physics 0-7354-0248-5/05/$22.50

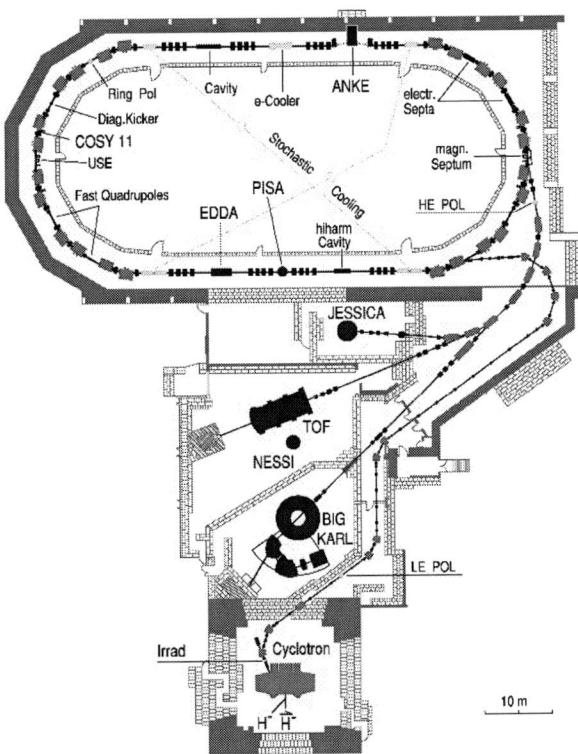

FIGURE 1. The COSY floor plan.

special admixture of small amounts of nitrogen and oxygen, reducing surface and volume recombination. The current output of the source depends sensitively on the relative fluxes of the gases and on their timing with respect to the dissociator radio frequency. The cooled beams are focussed by an optimized set of permanent hexapoles into the charge exchange region. By cooling down the supersonic atomic beam the acceptance of the hexapole system and the dwell time in the charge exchange region are increased in proportion to the decrease of the beam velocity. Gas scattering in the vicinity of the nozzle reduces partly these beneficial effects. A peak intensity of $7.5 \cdot 10^{16}$ atoms is measured in a diameter of 10 mm at the exit of the hexapole chamber [9].

The highly nuclear polarized atomic $\vec{H}^0$ beam meets inside the charge exchange region the fast neutral Cs^0 beam and charges are swapped according to the reaction $\vec{H}^0 + Cs^0 \rightarrow \vec{H}^- + Cs^+$. The negatively charged $\vec{H}^-$ ions are extracted from the charge exchange region by electric fields and are deflected magnetically by $90°$ into the beam line to the injector. The ions are transferred to the cyclotron, passing a Wien Filter to provide the proper spin alignment for injection into the cyclotron.

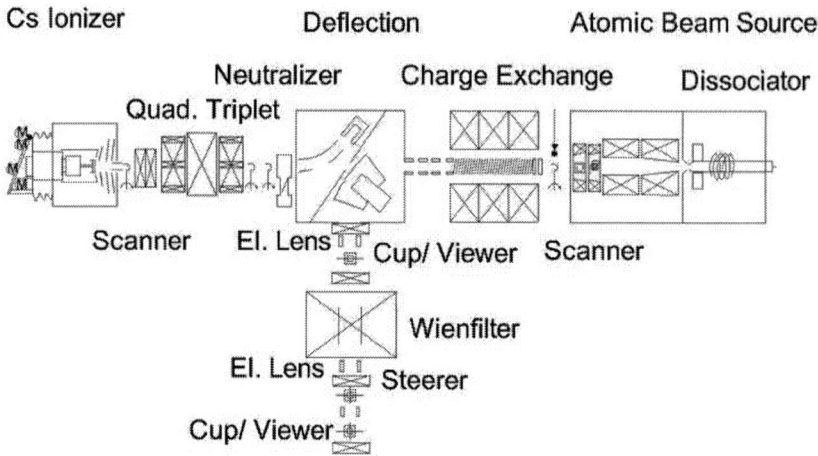

FIGURE 2. Scheme of the polarized ion source.

The fast neutral Cs^0 beam for the charge exchange reaction is produced in a two-step process. Caesium vapour is thermally ionized on a hot porous tungsten surface at a beam potential around 45 kV. The beam is focussed by a quadrupole triplet to the charge exchange region. Space charge compensation of the intense beam is improved by feeding 10^{-3} mbar l/s Argon to the beam tube following the extraction system. The neutralizer, a chamber filled with caesium vapour, is placed between the quadrupoles and the Cs deflector. The neutralizer comprises a caesium oven, a cell filled with caesium vapour and a magnetically driven flapper valve between the oven and the cell. The remaining Cs^+ beam is deflected in front of the solenoid to the Cs cup. Routinely a neutralizer efficiency of over 90 % was measured.

The highly selective charge exchange ionization produces only little unpolarized background that would reduce the nuclear beam polarization. In the charge exchange solenoid various beam properties can be adjusted. The transversal emittance can be traded for polarization by varying the solenoids magnetic field. The magnitude of the electrical drift field inside the solenoid can be tuned to optimize the energy spread of the beam. A monotonous gradient in combination with a double buncher system in the injection beam line to the cyclotron led to an improved bunching factor.

OPERATIONAL EXPERIENCES

Long-term operation revealed that several components limited the effectiveness for routine experiments. In order to provide reliable operation for experiments at COSY prototype parts, mainly of the caesium beam section [10] of the source were replaced

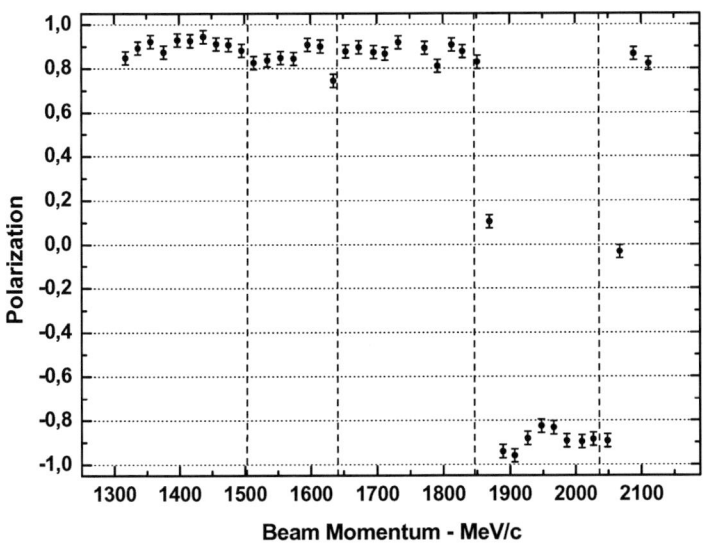

FIGURE 3. Polarization during the acceleration ramp with spin flips around 1.88 and 2.08 GeV/c.

through an improved design. Caesium sputtering and contamination is generally impeding long-term reliability. Therefore pulsed operation of the caesium ionizer has been included in the source [11]. The caesium pulses reached peak intensities of over 10 mA with reduced width around 10 ms. For routine operation caesium pulses with 5 mA flat shape of 20 ms width and a repetition rate of 0.5 Hz are used [8].

Because of the high facility use by the COSY user community only little experimentation on the sources is possible. Specific ion source development stands has been constructed to allow specific work for improvements. A dissociator stand is used for preparation of replacement parts, study of efficient cooling schemes and atomic beam production. An atomic beam part is in operation for the test of atomic beam focussing with PM-hexapoles and transition units for improving efficiency and polarization. On a common test bench a Cs - ionizer with beam line and diagnostic elements is in use for technical improvements.

A high degree of polarization for proton beams was achieved by working out a special tune of the machine so that jump quadrupoles were able to shrink losses to a negligible amount. The polarization inside COSY is depicted in Fig. 3. The polarization is measured during the acceleration ramp by the EDDA group [12,13]. Modifications of the colliding beams source contributed too to this high degree of polarization. A 90° bending magnet replaces the former electrostatic deflection system used to bend the ions into the beam line for injection into the cyclotron. The new alignment of the spin in this beam line reduced depolarizing effects and resulted in a polarization of over 90% after the cyclotron.

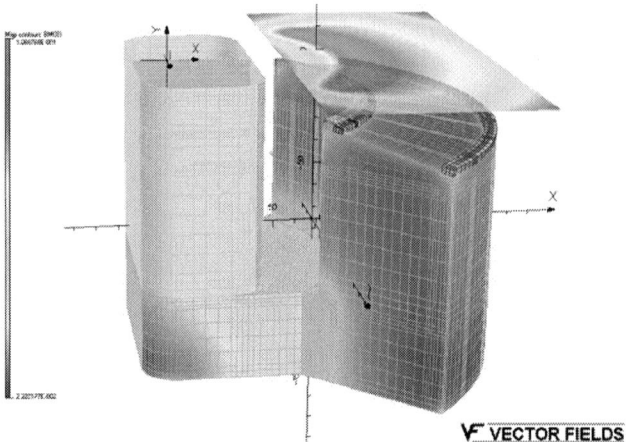

FIGURE 4. The shape of the lower iron yoke with a Fe-Nd-B permanent magnet layer with a height of 3 mm. The magnet has an indexed pole gap, slanted at 8.84° over the full radial width of the pole from R= 80 mm to R= 140 mm, without shims. The gap is 40 mm at the central radius of 110 mm. The magnet is excited by a 160 mm high coil with 29 turns of hollow copper wire around the back yoke. The radial field gradient was linearized using shims (5 mm wide and 1 mm high).

EXTRACTION MAGNET DESIGN AND CONSTRUCTION

Convinced by this improvement for the proton beam a hybrid magnet, employing Fe-ND-B permanent magnets, has been designed. Using permanent magnets with a high-energy product results in a superior field shape.

The colliding beams source provides either 4.5 keV polarized H⁻ or 7.6 keV polarized D⁻ in pulsed mode for axial injection into the injector cyclotron. For extracting the beam a C-shaped 90° bending magnet is located in the CBS vacuum chamber. The magnet coil is water-cooled. This electromagnet as well as a proposed hybrid design, which in addition makes use of permanent magnets, was analyzed numerically with the aim to evaluate and compare their excitation and bending characteristics [14]. The complexity of the different magnet configurations warranted the use of finite-element analysis and particle tracking [15]. Figure 4 shows the result of a simulation performed with the finite element program package TOSCA/ VECTORFIELDS. An indexed field has been chosen to get about the same spin rotation for all particles. The iron edges have been shaped to give a very good linear radial dependence of the field over the size of the beam.

The excitation of the magnet coil, and thus the heat transfer to the surrounding, can be considerably reduced by inserting permanent-magnet material at suitable symmetrical locations of the pole geometry above and below the median plane. These layers of permanent magnet material should be chosen such that the field without coil

TABLE 1. A subset of Polarization Modes for the polarized ion source for deuterons and protons. Modes P1 and P2 show the performance for polarized protons in COSY. All Data are taken from online measurements with the EDDA detector [16,17] at a beam momentum of 1.024 GeV/c.

Mode	P_z	P_{zz}	RFT1	RFT2	RFT3	Measured P_Z	Measured P_{ZZ}
				Transition		EDDA@1042MeV/c	
D1	0	0	Off	Off	Off	-0.002±0.003	
D2	-2/3	0	Off	Off	On	-0.533±0.004	0.057±0.051
D3	+2/3	0	Off	On	Off	0.438±0.014	
D4	+1/3	+1	Off	On	Off	0.285±0.032	0.594±0.050
D5	-1/3	-1	Off	On	On	-0.294±0.003	-0.634±0.051
D6	+1/3	-1	Off	On	Off	0.285±0.039	
D10	+1	+1	On	On	Off	0.764±0.022	0.545±0.050
D11	-1	+1	On	Off	On	-0.701±0.032	-0.537±0.052
D16	-1/2	-1/2	On	On	On	-0.349±0.022	-0.499±0.053
D17	1/2	-1/2	On	Off	Off	0.378±0.036	-0.282±0.052
P 1	+1	-	Off	On	Off	0.914 ± 0.022 ± 0.03	
P 2	-1	-	Off	Off	On	average in COSY	

is about halfway between the field values required for H⁻ and D⁻. The coil is then only used to add or subtract field.

The results of the analysis showed that NdFeB with a remanence of 1.12 T and a layer thickness of about 3.1 mm would fulfill these requirements. In fact, such permanent magnets are commercially available in the form of blocks of 31.75 mm × 14.224 mm × 3.175 mm (- easy axis). The analysis showed that a minimum pole-tip thickness of 5 mm is adequate for this purpose, and that without coil current a field value of almost 108 mT can be expected at the center of the pole gap. Even at a short distance from the pole faces, the field ripple is still insignificant.

The calculated fields were used for the final assessment of the bending properties of such a hybrid magnet configuration. The results show that the effective field boundaries are here much closer to the physical edge of the pole gap than for the electromagnet, but they differ by almost 20 mm for the two coil excitations. The final position of the magnet in the extraction chamber is fixed close to the calculated position.

CONCLUSION

The colliding beams type negative ion source can provide negative polarized hydrogen and deuteron beams without modification in comparable intensities. To prepare polarized deuterons with the desired combinations of vector and tensor polarization the atomic beam part of the source needed to be equipped with new high frequency transitions. These transition units are operated at the magnetic fields and radio frequencies to allow exchange of occupation of the different hyperfine states in deuterium. A set of three installed devices, RFT1 to RFT3, allows a large number of combinations to be delivered to experiments. Table 1 summarizes the deuteron polarization states and compares the measured polarizations to the theoretical limits for these states. Parts of the EDDA detector were used to monitor the polarization in COSY [16,17]. The unpolarized mode is used for normalization and no polarization is given in the table.

For experiments at the cooler facility COSY the polarized ion source delivers reliable polarized protons with increased intensities and a high degree of polarization close to the technical limit. The experimental options are successfully extended by a diversity of vector and tensor polarized deuterons.

ACKNOWLEDGMENTS

This work is partly funded by the European community under contract number HPRI-CT-2001-50021.

REFERENCES

1. R. Maier, Cooler Synchrotron COSY – performance and perspectives, NIM A 390 (1997) 1-8
2. H. Stockhorst et al., „Progress and Developments at the Cooler Synchrotron COSY", Proc. of the 8th European Particle Accelerator Conference (EPAC 03)
3. W. Haeberli, Nucl. Instrum. Methods, 62, 355 (1968)
4. J. Alessi et al., Proc. of the Int. Workshop on Pol. Sources and Targets, Montana 1986, Helv. Phys. Acta 59 (1986) 563
5. P.D. Eversheim et al, Proc. of Polarized Beams and Polarized Gas Targets, Köln; 224,(1995).
6. P.D. Eversheim et al., The Polarized Ion-Source for COSY, Proc. SPIN 1996 Symposium, Amsterdam.
7. R. Weidmann et al., Rev. Sci. Instrum.,67 p.II,(1996), 1357.
8. O. Felden et al., in Proc. of the 9th International Workshop on Polarized Sources and Targets, Nashville, IN,   2001, edited by V.P. Derenchuk and B. von Przewoski (World Scientific, Singapore, 2002), p. 200.
9. A.S. Belov, IKP Annual Report 2003
10. S. Lemaitre et al., Nuclear Instruments & Methods in Physics Research, Section A, vol. 408, (1998), 345
11. M. Eggert et al., Nuclear Instruments & Methods in Physics Research, Section A, vol. 453; (2000), 514
12. V. Schwarz et al., in Proc. SPIN 98, eds. N.E. Tyurin et al., World Sci., (1998), 560.
13. H. Rohdjess et al., "EDDA@COSY Results", AIP Conf. Proc., (2003), 665.
14. OPERA-3d with TOSCA, Vector Fields Ltd, U.K.
15. G. de Villiers, Collaboration Report (2002), IKP / iThemba LABS., Faure, South Africa
16. K. Yonehara et al., in Proceedings of the 8th Conference on the Intersections of Particle and Nuclear Physics, New York, NY, 2003, edited by Z. Parsa, AIP Conf. Proc. No. 698 (AIP, Melville, NY, 2003), p. 763.
17. V.S. Morozov et al., Phys. Rev. ST Accel. Beams 7, 024002 (2004)

Status of the Negative Hydrogen Ion Test Stand at CEA Saclay

R. Gobin[1], K. Benmeziane[1], O. Delferrière[1], R. Ferdinand[1], A. Girard[2], and F. Harrault[1],

[1]Commissariat à l'Energie Atomique, DSM / DAPNIA / SACM - 91 191 Gif/Yvette – France
[2]Commissariat à l'Energie Atomique, DSM / DRFC / SBT – 38 000 Grenoble – France

Abstract. At CEA-Saclay, in 2003, the new 2.45 GHz ECR source, based on pure volume H⁻ ion production, showed a dramatic increase of the H⁻ extracted ion beam. In fact, since the rectangular plasma chamber is separated in two different parts by a stainless steel grid, the extracted H⁻ current rose from few µA to 1.5 mA. Of course the grid position and its potential with respect to the plasma chamber were optimised. Ceramic plates allow increasing the electron density and lead to an improvement of the negative ion production. Plasma characterization by using Langmuir probes and positive charge analysis are also presented. The last results are reported and discussed. This work is part of the High Performance Negative Ion Sources (HP-NIS) program and supported by the European Union under contract HPRI-CT-2001-50021.

INTRODUCTION

Some of the future high power proton accelerators (HPPA) like SNS, ESS or neutrino factories require reliable and efficient H⁻ ion production to inject in compressor rings. These high intensity facilities plan to work in pulsed mode with high duty cycle requiring source improvements. Moreover, existing machines are also interested in new source developments.

In this framework, the CEA decided an important R&D program to develop high intensity light ion sources. At CEA/Saclay, the SILHI team develops a 2.45 GHz source [1] delivering a high intensity proton beam (more than 100 mA in CW mode) with a good efficiency for several years. The proton source performance encouraged the development of a negative ion source program. So a test stand has been built to study a new 2.45 GHz ECR negative ion source. This source has been initially designed with SILHI spare parts, to operate in pulse mode (mostly 1 or 2 ms – 10 Hz). Up to the beginning of 2003, only a small amount of H- ions was observed. Since then, measurements performed with the rectangular plasma chamber separated in 2 zones showed an important improvement. This upgrade is discussed in the next section. The third section will summarize the last results obtained by testing different materials and by varying the production zone volume. To improve the source behaviour understanding, plasma characterisation measurements were performed. So, recent Langmuir probe measurements and positive ion extraction analysis will be presented. Finally the source now regularly produces a 1.5 mA pulsed negative

CP763, *Production and Neutralization of Negative Ions and Beams*, edited by J. D. Sherman and Y. I. Belchenko
© 2005 American Institute of Physics 0-7354-0248-5/05/$22.50

hydrogen ion beam through a 5 mm diameter aperture.

PLASMA CHAMBER SEPARATION

To produce negative hydrogen ions, the sources are generally based on the double stage principle [2]. First, in the plasma creation zone, energetic electrons of about 20 to 40 eV interact with the gas to excite the molecules and then in a second zone, slow electrons (about 1-2 eV) react with these excited molecules to give an H atom and an H⁻ ion. This process is called the dissociative attachment. Generally, both zones are separated by a magnetic filter to avoid the high energy electron flux enters into the negative ion production zone.

At Saclay, the preliminary design of the source based on plasma generated by electron cyclotron resonance, followed this principle. The 2.45 GHz RF power was injected in the plasma production zone where the axial magnetic field provided by 2 coils reached close to 1000 Gauss. Then a tunable C-shape magnetic filter separated this zone and the negative ion production area. With this magnetic configuration, only few μA of H⁻ ions have been observed [3] while the plasma chamber was biased to a - 10 kV power supply.

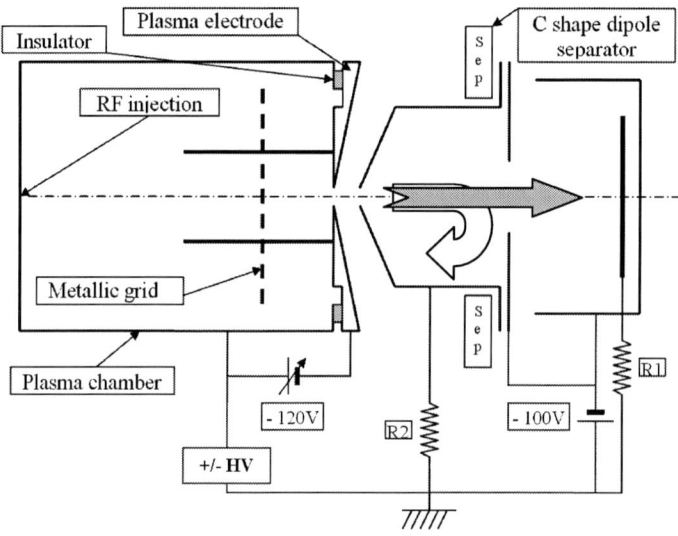

FIGURE 1: Scheme of the H⁻ ECR source

Previous spectroscopic plasma measurements showed for example the $\lambda = 674.0$ nm line indicating the presence of excited molecules able to produce H⁻ ions by dissociative attachment. So this small H⁻ ion production may be attributed to negative ion destruction close to the plasma electrode. The microwave power, not completely absorbed by the plasma, could contribute to H⁻ loss. Simulations showed that a metallic grid can stop the microwave penetration. So the magnetic filter was replaced by a 5 mm mesh stainless steel grid. As a result, an important improvement has been observed when the plasma chamber has been effectively separated in 2 zones. Then the grid and the plasma electrode have been simultaneously negatively polarised and the H- ion current increased. Figure 1 presents the grid in the plasma chamber and summarizes the wiring of the test stand.

After optimisation of the grid position, the maximum H⁻ current occurred while the grid was located at 25-30 mm from the plasma electrode. By tuning simultaneously the potential of the grid and plasma electrode from 0 to – 120 V compared to the plasma chamber, the 10 kV H⁻ extracted current rose from the precedent maximum value (84 µA) to 950 µA. Then both the plasma electrode and the grid were biased by independent negative power supplies in order to allow a slight voltage difference. But no improvement was observed.

MATERIAL DEPENDENCE

The above-mentioned results were obtained with a stainless steel grid and a molybdenum plasma electrode. The rectangular plasma chamber is made of water-cooled copper and a 2 mm thick boron nitride disc is inserted between the RF ridged transition and the plasma chamber [4]. The source was typically working in pulsed mode (1-2 ms – 10 Hz) 5 days a week for several months. And no degradation has been observed. The first stainless steel grid, installed in June 2003, has never been changed excepted during the Tantalum grid test reported hereinafter.

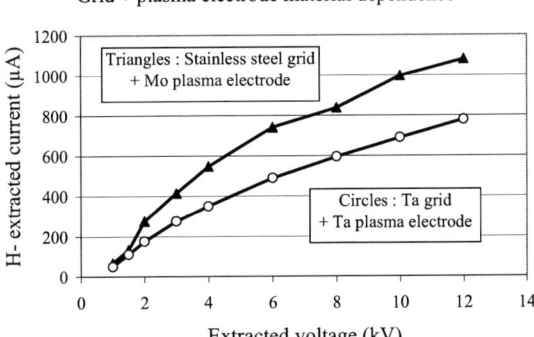

Grid + plasma electrode material dependence

FIGURE 2: Extracted H⁻ current versus extraction voltage with: (i) Ta grid and Ta plasma electrode (circles); (ii) Stainless steel grid and Mo plasma electrode (triangles)

Several authors already reported hydrogen negative ion production improvements while using Tantalum material inside the plasma chamber [2, 5]. So a Tantalum grid has been tested while the Saclay source operated at 10 kV. The H- extracted current was plotted as a function of the plasma chamber pressure (5 points between 2 and 3.5 mTorr) with both Stainless steel and Tantalum grid. The performances did not change dramatically but the results with the Tantalum grid always reach few percents lower than those obtained with the stainless steel grid. Moreover, if both the grid and the plasma electrode are made of Tantalum, the extracted H⁻ current decreases by about 25 % (Fig. 2) compare to the typical design.

The extracted current continuously increases from 140 to 850 µA while the RF power rises from 380 to 950 W [3]. This continuous increase of the RF power can be associated to an increase of the electron production in the plasma generator zone. And simulations [6] predict the increase of the H⁻ ion production as a function of the electron density. Ceramic materials like quartz, alumina or boron nitride produce an important amount of secondary electrons under plasma particle bombardment. So to confirm the dependence of the H⁻ ion production with respect to the primary electron density, 4 boron nitride plates have been installed in the plasma creation zone. And the H⁻ extracted current increased from 950 µA to 1.32 mA with the same source running conditions.

VOLUME AND SURFACE EFFECTS

The extracted negative ion current strongly depends on the grid position [3]. By reducing the distance between the grid and the plasma electrode, the volumes of both the production zone and the plasma generation zone change. So to verify the production zone volume effect, a 30 mm diameter stainless steel tube was installed around the 2 rods which maintain the grid. The volume was thus limited at 21 cm3 compare to 73 cm3 for the typical design. The tube was electrically linked to the grid and the plasma electrode. As a result, the smaller volume led to reduced performance.

Then the tube was replaced by a 5 mm mesh stainless steel grid cylinder. Once more the performance of the source was largely reduced. These results confirm the important influence of the plasma chamber shape. So, new designs with circular plasma chamber are envisaged.

PLASMA CHARACTERISATION

To facilitate the negative hydrogen ion production, several experiments have already been performed in order to increase the knowledge of the plasma. First the spectroscopic measurements showed the presence of excited hydrogen molecules [4]. Then a Langmuir probe has been successively installed in both parts of the plasma chamber [7]. And finally, by polarizing the source at positive high voltage, the extracted positive charge analysis gives important information on the plasma characteristics.

For the Langmuir probe measurements, the axial magnetic field led to difficult probe characteristic interpretation. Comparative curves indicate the tendency. The 2 mm diameter molybdenum probe has been placed on the axis of the source but perpendicularly to this axis. When the probe is installed at 58 mm from the plasma electrode (28 mm upstream to the grid), the axial magnetic field reaches 500 Gauss and the electron temperature increases when the grid potential varies from − 0 to − 90 V (fig. 3).

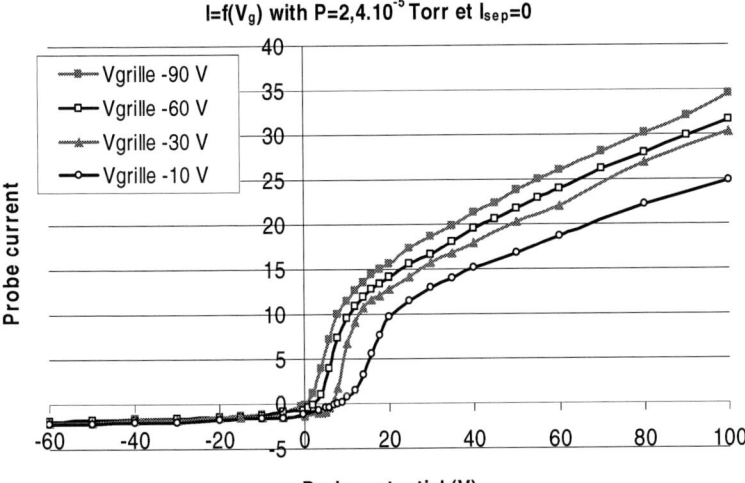

FIGURE 3: Probe characteristics in the plasma generation zone

The probe was then installed in the negative ion production zone at 6 mm from the plasma electrode. Here the axial magnetic field is much lower (around 200 Gauss). By comparing the probe curves on both sides of the grid, for the same voltage, the electron temperature is largely lower in the H⁻ production zone. This electron energy reduction could explain the increase of the H⁻ ion current.

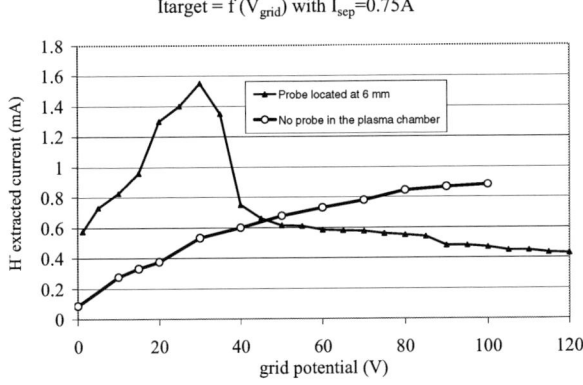

FIGURE 4: Extracted H- current vs grid potential (with or without grid).

The probe made of a 2 mm Molybdenum wire covered with a 0.5 mm Al2O3 layer largely influences the extracted ion current. Figure 4 presents the H- current versus the grid polarization. The empty circle curve represents the ion current with no probe in the plasma chamber. For this measurement, the Boron Nitride plates were not installed in the chamber but the curve looks similar with a lower level. The triangle curve represents the ion current with the probe located at 6 mm from the extraction aperture. The plasma perturbation is probably mainly due to the secondary electron emitted by the probe alumina. As a result, the maximum extracted H- ion current (1.5 mA) was then finally obtained in these conditions.

The electron energy reduction is confirmed by the positive charge analysis. A small dipole magnet (10 mm between poles) located just after the extraction system, in the vacuum chamber, allows us to check the different species (H^+, H_2^+, H_3^+) extracted from the source. A 5 mm diaphragm limits the current entering in the dipole and then the particles are collected on a Faraday cup (triangle curve). The measurements indicate a very low amount of H_2^+ while the grid is installed in the chamber (Fig. 5). Like in the cold plasma where the following reaction (1) takes place, this analysis shows the H_3^+ peak becomes the highest one.

$$H_2 + H_2^+ \rightarrow H_3^+ + H + 1.71eV \qquad (1)$$

Otherwise, the distribution of positive charges depends on the plasma chamber pressure. The above triangle curve represents the different species when the pressure equals 3 mTorr. The species fractions are: $H^+ = 22\%$, $H_2^+ = 5\%$ and $H_3^+ = 72\%$. Whereas, with a 2.5 mTorr pressure, the fractions are: $H^+ = 34\%$, $H_2^+ = 13\%$ and $H_3^+ = 52\%$.

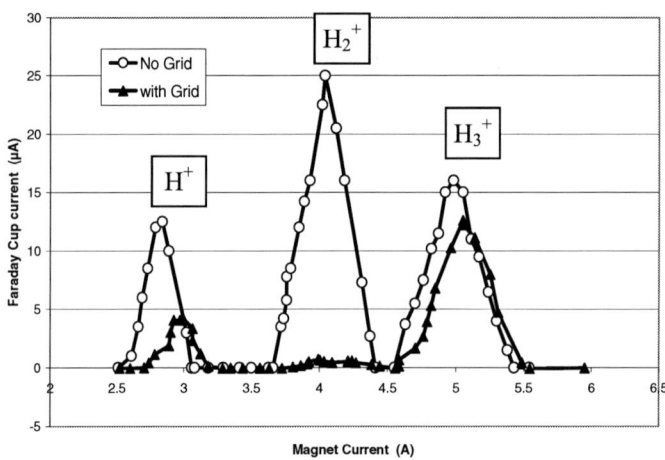

FIGURE 5: Positive charge analysis (extracted current vs analyzer magnet) with or without the grid.

ACKNOWLEDGMENTS

The authors would like to express their acknowledgements to the European Union which support this development (Contract HPRI-CT-2001-50021). The very fruitful discussions engaged with the HP-NIS network collaborators allow us to improve our knowledge in term of source and beam optimization and plasma physics. Moreover, G. Charruau, M. Desmons, A. France and Y. Gauthier are also thanked for their active contribution. The participation of G. Gousset (Orsay University), J. Sherman (Los Alamos National Laboratory) and T. Steiner (CERN) was also greatly appreciated for this development.

REFERENCES

1. R. Gobin et al, RSI, Vol 75, N° 5, 1414 (May 2004)
2. J. Peters, RSI, Vol 71, N° 2, Feb. 2000.
3. R. Gobin et al, RSI, Vol 75, N° 5, 1741 (May 2004).
4. R. Gobin et al, Proceedings of the 9th Intern. Symposium on Negative Ion Sources and Beams, Saclay, France, May 2002, AIP Conference Proceedings, 639, pg: 177.
5. D. Spence et al, Proceedings of LINAC Conference August 1996.
6. K. Benmeziane et al, "2D PIC-MCC code for electron-hydrogen gas interaction study in H- ion sources", this Symposium
7. R. Gobin et al, "Recent results of the 2.45 GHz ECR source producing H⁻ ions at CEA/Saclay", Proceedings of the Linac conference, Lubeck, Germany (August 2004)

Recent Advances in the Performance and Understanding of the SNS* Ion Source

R. F. Welton, M. P. Stockli, S. N. Murray and R. Keller#

SNS, Oak Ridge National Laboratory, P.O. Box 2008, Oak Ridge, TN 37831, USA
*SNS, Lawrence Berkeley National Laboratory, 1 Cyclotron Rd.., Berkeley, CA, 94720, USA

Abstract. The ion source developed for the Spallation Neutron Source* (SNS) by Lawrence Berkeley National Laboratory (LBNL), is a radio frequency, multi-cusp source designed to produce ~ 40 mA of H$^-$ with a normalized rms emittance of less than 0.2 pi mm mrad. To date, the source has been utilized in the commissioning of the SNS accelerator and has already demonstrated stable, satisfactory operation at beam currents of ~30 mA with duty-factors of ~0.1% for operational periods of several weeks. Once the SNS is fully operational in 2008, a beam current duty-factor of 6% (1 ms pulse length, 60 Hz repetition rate) will be required in order to inject the accelerator. To ascertain the capability of the source to deliver beams at this high duty-factor over sustained time periods, several experimental runs have been conducted, each ~1 week in length, in which the ion source was continuously operated on a dedicated test stand. The results of these tests are reported as well as a theory of the Cs release and transport processes which were derived from these data. The theory was then employed to develop a more effective source conditioning procedure as well as an improved Cs collar design. Initial results of tests employing a Cs collar with enhanced surface ionization geometry are also discussed.

INTRODUCTION

The Spallation Neutron Source (SNS) will be a large multinational user facility dedicated to the study of the dynamics and structure of materials by neutron scattering and is currently under construction at Oak Ridge National Laboratory (ORNL) [1]. Neutrons will be produced by directing 1 GeV pulses of protons from a chain of linear accelerators and a compressor ring onto a liquid Hg target. In order to meet the baseline requirement of 1.4 MW of proton beam power on target, the ion source must produce ~40 mA of H$^-$ within a 1-ms pulse at a repetition rate of 60 Hz (6% duty-factor).

To date, the ion source has been extensively utilized in commissioning the SNS Front-End (FE) both at LBNL [2] and ORNL [3] as well as the Drift Tube Linac (DTL) [4]. In this role, the source has performed very well, with availability of the source and Low Energy Beam Transport (LEBT) reaching 98% during the last commissioning period which spanned several weeks. During this time, the ion source

---

* SNS is a collaboration of six US National Laboratories: Argonne National Laboratory (ANL), Brookhaven National Laboratory (BNL), Thomas Jefferson National Accelerator Facility (TJNAF), Los Alamos National Laboratory (LANL), Lawrence Berkeley National Laboratory (LBNL), and Oak Ridge National Laboratory (ORNL). SNS is managed by UT-Battelle, LLC, under contract DE-AC05-00OR22725 for the US Department of Energy.

CP763, *Production and Neutralization of Negative Ions and Beams*, edited by J. D. Sherman and Y. I. Belchenko
© 2005 American Institute of Physics 0-7354-0248-5/05/$22.50

was mainly operated at beam currents of ~30 mA with a very low duty-factor of ~0.1%, although the design goal of 38 mA (6% duty-factor) was briefly demonstrated both at LBNL and ORNL. On one occasion an endurance test was performed at LBNL in which the ion source was operated for ~5 days with a duty-factor of 2-3% delivering ~25 mA of beam current.

Given the lack of performance data [5] taken over long, sustained run-periods at beam duty-factors of ~6%, we are currently performing such tests on a dedicated test stand capable of unattended continuous operation. Two identical ion sources have been tested over the course of 8 experimental runs, each lasting approximately 1 week, while having performance and operational parameters electronically logged [6]. These data have then been used to develop a formulation of the Cs release and transfer process, which was then used to construct a more effective source conditioning procedure as well as guide the next iterations of SNS ion source development. Given the recent interest shown at this symposium in scaling of various types of Cs enhanced H⁻ ion sources to long-pulse, high-duty-factor operation, the Cs transport model developed for the SNS ion source is described in sufficient detail to allow applicability to other H⁻ ion sources. Information gleaned from an initial test of the integrated Cs collar / outlet aperture, which employs an optimized ionization surface, is also discussed.

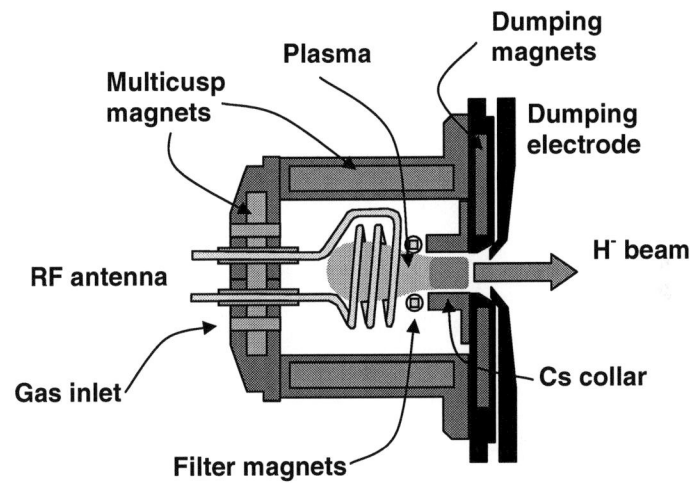

FIGURE 1. Schematic diagram of the SNS ion source.

THE MULTICUSP H⁻ SOURCE AND LEBT

A schematic diagram of the H⁻ ion source is shown in Fig. 1. The source plasma is confined by a multicusp magnetic field created by a total of 20 samarium-cobalt magnets lining the cylindrical chamber wall and 4 magnets lining the back plate. Pulsed RF power (2 MHz, 20-60 kW) is applied to the antenna shown in the figure through a transformer-based impedance-matching network. The plasma is sustained between high-power RF pulses by continuous application of ~200 W of 13.56 MHz power to the same antenna. A magnetic dipole (150-300 Gauss) filter separates the main plasma from a smaller H⁻ generation region where low-energy electrons facilitate the production of large amounts of negative ions. An air heated/cooled collar, equipped with eight cesium dispensers, each containing a mixture of Al, Zr and Cs_2CrO_4, surrounds this H⁻ generation volume. The RF antenna is made from copper tubing that is water cooled and coiled to 2 ½ turns and centered in the plasma chamber. A porcelain enamel layer insulates the plasma from the oscillating antenna potentials [7]. More details of this source design can be found in reference [8].

Ions are extracted from the source by applying a continuous DC bias of -65kV to the ion source and accelerating ions through a circular extractor electrode held at or near ground potential. The pulse structure of the ion beam is achieved by pulsing the 2 MHz high-power RF creating the plasma. Closely coupled to the source is the Low Energy Beam Transport (LEBT) section of the SNS accelerator which matches the Twiss parameters of the extracted beam to that of the Radio Frequency Quadrupole, the first accelerator in the SNS chain. The LEBT, ~10 cm in length, consists of two Einzel-type electrostatic lenses which can be independently voltage-controlled to provide optimal matching and is described in detail in Reference 2 and references therein. Identical LEBTs exist on the SNS front end as well as the ion source test stand which was employed in these investigations.

HIGH-DUTY FACTOR ION SOURCE TESTS – RUNS 1-7

Three nearly identical ion sources were shipped from LBNL to ORNL along with the front end system [2, 3]. One source has been in nearly continuous service on the SNS accelerator serving to commission the machine. The other two sources have been employed in this study, each being run four times on the ion source test stand for a period of ~1 week (experimental Runs 1-8). The beam current was measured using a toroidal Beam Current Monitor (BCM) located a few cm downstream of the second Einzel lens of the LEBT and a Faraday cup located ~20 cm downstream of the BCM. Prior to each run, the ion source was thoroughly cleaned using 15 μm diamond-grit sandpaper, fresh Cs dispensers were installed and a new antenna was added. Each time a new source was mounted, the system was thoroughly leak-checked with a He leak detector.

Following the guidance of LBNL, the following ion source conditioning procedure was used for experimental Runs 1-7. The source was started with a low-duty-factor plasma of ~0.1% achieved by pulsing 20-30 kW of the 2 MHz RF power with a pulse width of ~100 μs and a repetition rate of ~10 Hz for several hours. During this time the Cs collar was gradually heated to a nominal operating temperature of ~300C by increasing the collar air temperature using an external air heater. Next, Cs was released into the source by raising the plasma duty-factor to ~3% (pulse width: ~1 ms, repetition rate: ~30 Hz) and restricting the Cs collar air flow, allowing the collar to be heated to temperatures of 500-550 C for ~½ hour. Finally, after 1 cesiation, the source was ramped to the nominal duty-factor of 7.3% used in these studies by increasing the

RF pulse length to 1.2 ms and the repetition rate to 60 Hz. It was necessary to exceed the beam duty-factor requirement of 6% since the SNS will be unable to accelerate the initial peak due to excessive emittance in this region of the pulse. Once high-duty factor operation was established, the RF power was adjusted to give maximum beam current, typically 50-60 kW. The Cesiation process described above was repeated as needed to keep the beam intensity as high as possible. The beam current was then sustained for ~1 week, long enough to resolve the beam current trend, unless the beam could not be increased above 20 mA, in which case the run was terminated. After the source was run it was disassembled and inspected and, in each case, the source was found in generally the same condition as when installed with the exception of the Cs collar and RF antenna. The Cs collar was typically found discolored, noticeably black in appearance, with some evidence of sputtering. The condition of the antenna after each experimental run is given in the last column of Table I.

Figure 2 shows the electronically logged beam current for experimental Runs 5-7 measured with the BCM corrected for droop and averaged over the entire pulse. Similar beam current plots of experimental Runs 1-4 can be found in an earlier report [9]. The ion source downtime observed in Fig. 2 resulted from system trips or source instabilities which occurred during unattended periods. Had an operator been present on a 24-hour basis very little downtime would have been observed. The time of each source cesiation as well as changes in RF power and duty-factor are noted in the plot area of Fig. 2.

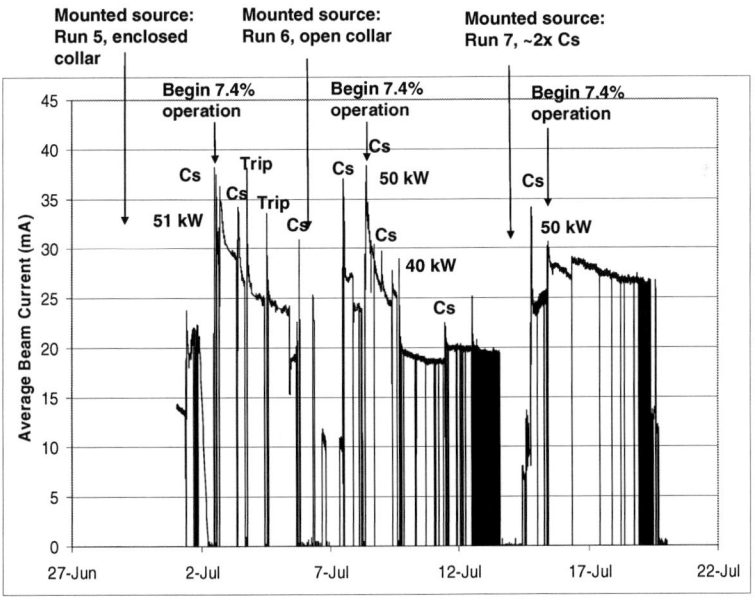

FIGURE 2. Measured beam current (averaged over the 1.2 ms pulse) produced from the source using the conditioning and cesiation procedures described above for experimental runs 5-7.

299

Experimental Run-5 is typical of each of the earlier tests (Runs 1-4) employing the original LBNL source configuration. The plot shows a maximum measured just after the source's initial cesiation followed by a steady beam attenuation rate (~4 mA/day) which could only temporarily be abated with subsequent cesiations. In experimental Run-6, the downstream aperture of the Cs collar, shown later in Fig. 6, was removed to allow H⁻ ions produced to the collar surface to directly enter the ion beam.

This configuration required a higher H_2 gas flow rate and resulted in a comparable initial beam intensity of 38 mA but a much greater beam attenuation rate of 9 mA/day. In experimental Run-7 additional Cs dispensers were added to the collar to approximately double the Cs supply. During this run the maximum beam current was somewhat less then earlier runs, but the beam attenuation rate of 2 mA/day was comparable to the best earlier runs.

Table I contains a statistical summary of each experimental run in which the source was operated with a 6% or higher duty-factor. Six percent duty-factor could not be reached in experimental Run-3 due to insufficient cooling of the LEBT extractor electrode and is therefore omitted from the table. The maximum beam current shown in the table is the largest current measured during the experimental run. The average beam current also shown in the table is computed by averaging all measured beam current values from the beginning of the run to termination excluding the down-periods which resulted from system trips. The power efficiency is computed from the average beam current divided by the 2 MHz RF power used to generate the plasma during the run period. The run length, also shown, is defined as the total time the source ran with a duty-factor of 6% or greater which accounted for the majority of most run-periods. The average beam attenuation rate is determined by subtracting the maximum beam current achieved during the experimental run from the end-of-run beam current value and dividing by the time interval between these two points, shown as Δt in the table.

TABLE I. Statistical summary of the high-duty factor experimental runs.

Run #	Config.	RF power (kW)	Max current (mA)	Average current (mA)	Power Efficiency (mA/kW)	Run length (days)	# of Cesia-tions	Beam attenuation (mA/day)	Δt days	Antenna condition
1	Normal	40	28	18	0.45	1.6	4	10	1.8	Punctures outside plasma region / light conductive coating
2	Normal	40	30	25	0.63	8.0	6	1.5	4	Punctures in plasma region / light conductive coating
4	Normal (fwd ant)	60	30	28	0.47	4.1	4	2.3	3	No punctures / light conductive coating
5	Normal (fwd ant)	51	38	28	0.55	3.5	2	4.6	3	No punctures / light conductive coating
6	Open collar	50	38	27	0.54	1.7	3	9	1.5	No punctures / light conductive coating
7	~2x Cs	50	34	27	0.56	4.5	1	2	5	No punctures / heavy conductive coating

Experimental Runs 1-5 show that using the original LBNL source configuration and conditioning procedure the source will, on average, initially produce 32 mA which immediately begins to decay at an average rate of 4 mA/day. We also note that 1 out of 7 antennas were found punctured in the plasma region. This clearly falls short of the SNS requirement of ~40 mA sustained for ~21-day operational periods. In the following section we carefully examine these data and develop a qualitative model of Cs transport in order to guide design and procedural modifications to the source necessary to realize this goal.

DATA ANALYSIS

Since we have already shown that beam currents close to the SNS requirement can be produced only for short periods of time, this analysis will focus on determining the cause of the large observed beam attenuation rates. It is widely held that Cs-enhanced, multi-cusp ion sources produce H^- ions both on plasma facing surfaces and in the plasma volume. It is also believed that the addition of Cs lowers the work function of the surface thereby enhancing the H^- yield [10]. In our source, during the nominal cesiation procedure a maximum of $~2.7 \times 10^{17}$ Cs atoms are delivered to the source (see below) with the H^- enhancement persisting for several hours to several days. If Cs were to significantly partake in volume processes it would have to be present in such a high volume density it would be pumped from the source on much shorter time scales. In this analysis we therefore consider the Cs-enhancement to be purely a surface effect. Without the addition of Cs the SNS source will produce about 10-15 mA of H^- current. This can only be increased to ~40 mA when Cs is added. We therefore consider the 25-30 mA of additional beam current to be entirely surface-produced.

We monitor the hydrogen spectral lines emitted from the plasma during each experimental run using a fiber-coupled optical spectrometer (300-1400 nm). Since the fiber is affixed in a stationary position to a plasma-viewing window we can observe changes in plasma density over time by monitoring changes in spectral intensities. We can also detect changes to the plasma temperature by monitoring the ratio of the hydrogen lines. Plasma impurities can be tracked by observing the intensity of the background peaks. Cs-lines cannot be observed since Cs resides mostly on the surface rather than in the plasma volume, as discussed above. During the experimental runs we see essentially no changes in these plasma properties (at constant RF power) while observing considerable beam attenuation rates of ~4 mA/day (see Fig. 2). Since the plasma conditions apparently remain the same, we can attribute most of the observed beam attenuation to a reduction in surface rather than volume production.

Theoretically, efficient surface ionization requires (i) an intense flux of ions and/or fast neutrals arriving at the surface from the plasma core (Kishinevskii probability [10] increases rapidly with bombardment energy), (ii) a coating of Cs, on the order of a monolayer, creating a low work-function ionization surface and, (iii) efficient extraction of the H^- ions once they are produced. Since optical spectroscopy suggests plasma conditions are essentially unchanged during an experimental run, conditions i and iii likely remain unchanged. This leaves the most likely cause of the observed beam attenuation as failure to maintain a low work function surface. The apparent increase in work function cannot be attributed to impurity contamination since, as we will show later, the flux and energy of hydrogen ions from the plasma is sufficient to rapidly kinetically eject adsorbed Cs and other reactive impurities from the surface. Thus we conclude that the increase in work function most likely results from limitations in the Cs supply. We will now look closely at the two processes required to deliver Cs to the ionization surface: Cs release from the dispensers and subsequent transport to the ionization surface.

The Kinetics and Thermochemistry of Cs Release

Cs is introduced into the ion source using the commercially available alkali metal dispensers from SAES Getter Corporation [11]. Each source contains eight Cs cartridges (dimensions: 12x1.4x1 mm) containing a compressed powder mixture of 17% Cs_2CrO_4, 70% Al and 13% Zr with the later two species configured as intermetallic compounds forming the low-temperature getter ST-101 [11]. Each cartridge contains 5.2 mg of elemental Cs. In order to determine the Cs production rate, several experiments were previously conducted in which the dispensers were heated and the quantity of released Cs was monitored using atomic absorption spectroscopy (852.1 nm) [12]. The Cs release rate determined from these measurements, scaled to 8 dispensers (the nominal Cs load in the ion source) is given by Eqn. 1.

$$\phi(atoms/s) = 1.125 \times 10^{24} \exp(-18500/(273 + T(C))) \qquad (1)$$

The minimum flow-rate observed from a single dispenser cartridge was 7.5×10^{13} atoms/s measured in the 480-560 C range, depending on the particular dispenser sample.

Analysis of this mixture using the chemical thermodynamic equilibrium computer code HSC [13] shows the following dominant reactions:

$$4\ Cs_2CrO_4 + 5\ Zr \rightarrow 8\ Cs\ (g) + 5\ ZrO_2 + 2\ Cr_2O_3 \qquad (2)$$

$$6\ Cs_2CrO_4 + 10\ Al \rightarrow 12\ Cs\ (g) + 5\ Al_2O_3 + 3\ Cr_2O_3 \qquad (3)$$

The thermodynamic equilibrium concentration of each species in the Cs dispenser subjected to an H_2 atmosphere (held at the ion source pressure of 10 mTorr) was also computed by the code as a function of temperature and is shown in Fig. 3. We see that the formation of elemental Cs is thermodynamically favored at all temperatures shown, and above 200 C Cs vapor dominates over the condensed state. Running the same simulation without the ST-101 getter material (Al and Zr) reveals elemental Cs does not form at temperatures below ~950 C (as shown in Fig. 4). Thus, in the absence of the getter material ST-101 the Cs_2CrO_4 will not release Cs until nearly 1000 C. This suggests that we should be concerned about chemically reacting the ST-101 material with reactive gases present in the source - precisely what the gettering material is intended to do!

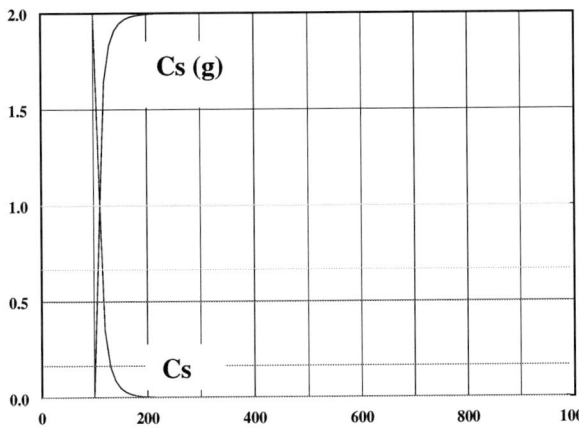

FIGURE 3. HSC calculation of the equilibrium concentrations of the components in the SAES Cs dispensers in H_2 gas (10 mTorr). Note for simplicity hydrogen compounds are not plotted.

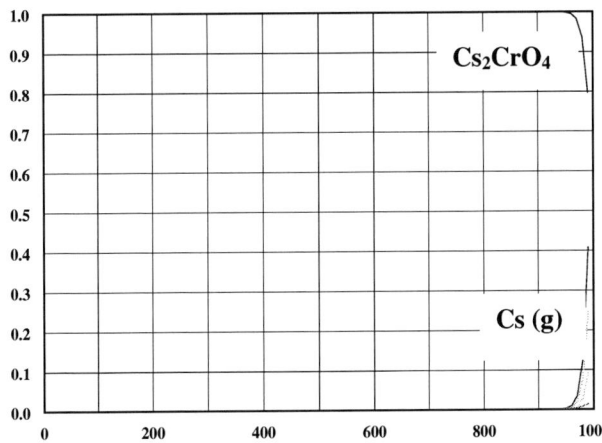

FIGURE 4. HSC calculation of the equilibrium concentrations of the components in the SAES Cs dispensers assuming the ST-101 getter is completely inactive. Note: for simplicity compounds containing hydrogen are not plotted.

The ST-101 product literature [11] states that reactive gases such as O_2, CO, CO_2, H_2O and H_2 begin to be sorbed by the getter material at temperatures about 250 C and are readily sorbed at 300 C. The literature also states that opposed to the other reactive gases listed, H_2 does not chemically react with the getter material, but instead forms a solid solution which has the characteristic of reversibility. Other reactive gasses are particularly harmful to the Cs dispenser since they are sorbed irreversibly by the ST-101 getter material and react with Al and Zr forming refractory compounds which inhibit Cs release. Unfortunately, analysis of the residual gas evolved by the source

303

reveals that O_2, CO, CO_2 and H_2O are abundantly released during source outgassing each time the plasma chamber is heated to a higher temperature by increasing the average RF power (pulse RF power and duty-factor). Thus, care should be taken to outgas the source thoroughly at or above the desired operational duty-factor and power level while maintaining collar temperatures below 250 C.

Experience has shown that exposing the source to very small vacuum leaks while the Cs collar is at operating temperature not only immediately reduces the H⁻ beam intensity to pre-cesiation levels but also renders any subsequent cesiation attempts completely ineffective. The characteristic beam attenuation curve which occurs after the initial cesiation is shown in Fig. 2 for each source. Here an approximately exponential decay is observed. Subsequent cesiations only temporarily increase the beam current which rapidly decays, also in an exponential fashion, to the trend line set by the initial cesiation. Eventually, subsequent cesiations are completely ineffective at enhancing beam current. This behavior is consistent with the above picture of Cs release being limited by the chemical activity of the ST-101 getter material. As the ion source operates the evolved reactive gases are chemically gettered by the ST-101 material. This forms refractory compounds which decrease the amount of Zr and Al which are available for reducing the Cs_2CrO_4. Since most of the ST-101 getter material is available at the beginning of the experimental run, the cesiations undertaken then are most effective. As the run proceeds, cesiations become less effective due to the decreased quantity of the ST-101 material. Theoretically, each newly mounted-source contains enough Cs for ~500 standard cesiations, but in practice only ~5 cesiations are effective at boosting the H⁻ current.

High-Duty Factor Ion Source Tests – Run-8

In order to test the above picture of Cs release and determine if the average beam attenuation rate of 4 mA/day could be improved we implemented a new ion source conditioning technique during experimental Run-8. Unlike the operating procedure employed for experimental Runs 1-7, in the new technique the source was outgassed at ~7% duty-factor and 50 kW of RF power for ~1 day while maintaining the collar temperature below 150 C. Source high voltage was off during this time to prevent excessive electron heating of the LEBT, which would have occurred if an uncesiated source was run at high beam duty-factor. Fig. 5 shows the beam current over the course of a 10 day run.

During the run the Cs collar was operated between 70-300 C and we observed the lowest beam attenuation rate of 1.1 mA/day. This led to our highest average beam current of any experimental run: 30 mA, suggesting much more Cs was, in fact, delivered to the source. Unfortunately, an H⁻ current of 30 mA is still less beam than required for SNS operation, so we must now consider in detail the Cs transport process.

Cs Transport to the Ionization Surface

Once elemental gaseous Cs has been released by the dispensers it must migrate to the active ionization surfaces. Fig. 6 shows an enlargement of the Cs collar shown in Fig. 1.

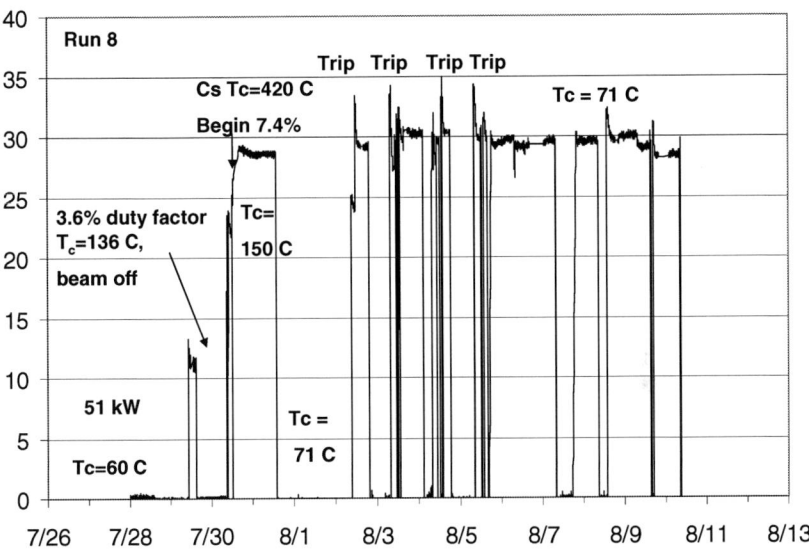

FIGURE 5. Measured beam current (averaged over the 1.2 ms pulse) during Run-8 which employed the new source conditioning technique.

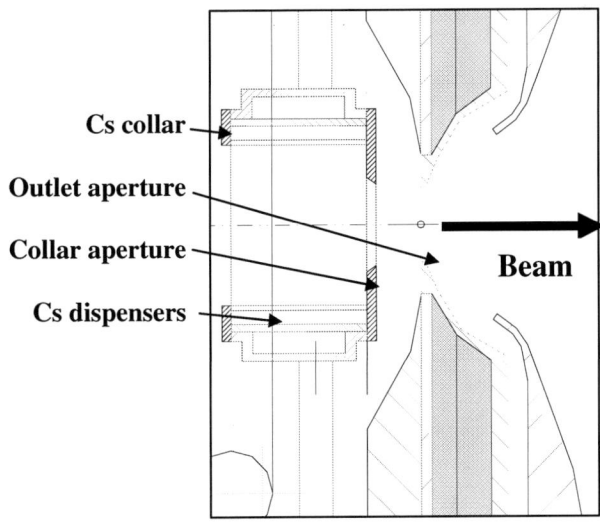

FIGURE 6. Close-up view of the standard Cs collar.

Before discussing the Cs transport process we must first clarify the location of the principal ionization surface(s). The NIETZSCHE Monte-Carlo code has been used to estimate the distance an H⁻ ion can travel without being destroyed by collisions with plasma particles. For discharges having parameters similar to the SNS ion source, calculation reveals ~90% of H⁻ ions having energies in the 1-10 eV range will be destroyed after transiting the plasma a distance of ~1 cm [14]. We therefore consider only surfaces within this short distance from the outlet of the source, which includes the interior surfaces of the Cs collar and outlet aperture.

During a typical cesiation the collar temperature is raised to 550 C. The extracted H⁻ beam typically shows no evidence of Cs enhancement while being held at this temperature. If the source has previously been cesiated and the collar is again raised to cesiation temperature, the beam current falls to levels of a completely uncesiated system until the temperature is lowered. It is unlikely that raising the temperature of the collar significantly affects the temperature of the outlet aperture since the latter is effectively water-cooled and radiant heat transfer is negligible in this temperature range. Thus we would expect the outlet aperture to be fully coated with Cs. Yet we see no enhancement strongly suggesting the outlet aperture plays only a minor role in surface production. In addition, Fig. 6 shows that most of the outlet aperture is shielded from direct plasma bombardment by the collar aperture. During Run-6 the Cs collar aperture was removed, and only then did we actually see beam enhancement while the collar was at cesiation temperatures. Therefore, we consider the plasma-facing surfaces of the Cs collar structure to be the primary ionization surface.

In order to elucidate Cs transport phenomena we must now estimate the rate at which Cs is lost from the collar in order to determine the rate at which the system needs to supply Cs. The bond strength U_o between a metallic surface and a Cs ad-atom can be estimated by calculating the potential energy stored by the Cs⁺ ion resting on the surface, separated from its image charge by 2 ionic radii r_o summed with the energy lost by electron transfer to the surface (surface work function ϕ – Cs ionization potential I) [15]. This is expressed in Eqn. 4 and was evaluated to be U_0~3.3 eV under conditions of near-zero coverage for Cs on a metallic surface.

$$U_0 = \frac{e^2}{4r_0} - (I - \phi) \tag{4}$$

Under conditions of increasing Cs surface coverage θ (in units of mono-layers) the Cs-Me bond weakens and has been described empirically for the case of Cs ad-atoms on polycrystalline W [15].

$$U_0 = \frac{2.78}{1+0.714\theta}(eV) \ (0.06 < \theta < 0.6) \tag{5}$$

Alternatively, others have measured the dependence of U_0 for Cs coverage of monocrystalline W(110) [16] and found

$$U_0 = 3.37 - 2.78\theta \ (eV) \ (0 < \theta < 1).$$ (6)

Under operational conditions the Cs coverage is adjusted for optimal H⁻ production, which occurs when the work-function of the Cs-Me surface is minimal [10]. It is generally accepted that this condition occurs at $\theta \sim 0.5$, yielding $U_0 = 2$ eV evaluated using either Eqn. 5 or 6 [17]. The Frenkel equation can now be used to estimate the mean thermal lifetime of Cs chemisorbed on the surface at the operating temperature of ~300 C, taking the enthalpy of adsorption to be $\Delta H \sim U_0 \sim 2$eV [18].

$$\tau \sim 6 \times 10^{-13} \exp\left(\frac{\Delta H}{kT}\right) \sim 63 \, hours$$ (7)

Thus for an optimally filled monolayer Cs will thermally desorb from the surface as a result of evaporation at the rate of ~10^9 Cs/s/cm². We will shortly see that this loss rate is negligible compared with the Cs losses which result from kinetic ejection of ad-atoms from the surface as a result of ionic impact from plasma particles.

In the SNS ion source the H⁺ ions primarily bombard the plasma facing surfaces of the Cs collar ejecting adsorped Cs. Fortunately, data exist characterizing positive ion beams extracted from RF-multicusp ion sources of virtually identical geometry to that of the SNS ion source and operated under essentially the same plasma conditions [19]. Beam current densities of ~1.5 A/cm² or 10^{19} ions/s/cm² of H⁺ ions were observed from these sources at ~50 kW of RF power. Other species such as H_2^+, H_3^+ and Cs⁺ are only observed as very small fractions of the primary H⁺ ion beam. Beam energy distributions of the extracted protons have also been measured and were found to have a FWHM = 3.2 eV for 0.5 kW of applied RF power and 3.6 eV at 1 kW [20]. Assuming a linear increase in the ionic energy spread with applied RF power, as suggested in reference 20, we would expect a FWHM ~ 40 eV at 50 kW of applied RF power. This corresponds to a Maxwellian distribution characterized by kT=17 eV of bombarding H⁺ ions striking the plasma-facing surfaces of the Cs collar.

The rate at which Cs is ejected from the metallic collar surface can be estimated using the approach of van Amersfoort[18] which is based on the Bohdansky formula developed to calculate light ion, near-threshold sputtering yields of a variety of materials used in fusion research [21]. In this approach, the Cs ejection rate was calculated by inputting the mass properties of the bulk metal target material along with the surface bond energy of the ad-atom, in our case, the Cs-Me bond energy of Uo = 2 eV. According to this formulation the sputtering threshold energy E_t is calculated using

$$E_t = \frac{U_0}{\gamma(1-\gamma)} \ (m_1 < 0.3m_2) \quad E_t = 8U_o\left(\frac{m_1}{m_2}\right)^{2/5} \ (m_1 > 0.3m_2) \quad \gamma = \frac{4m_1 m_2}{(m_1 + m_2)^2}.$$ (8)

Here m_1 and m_2 are the mass of the projectile and target, respectively. In our case the target material is 304 stainless steel which contains mainly Fe (~72%), Ni (~9%) and Cr (~19%) which has an weighted average mass of m_2=55. This yields a Cs-ejection threshold of E_t=31 eV. The Bohdansky formula, which is valid for up to ~20 times threshold, is given as a function of projectile energy E [21],

$$Y(E)=3.2\times10^{-3}\,m_2\left(\frac{4m_1m_2}{(m_1+m_2)^2}\right)^{5/3} E^{1/4}\left(1-\frac{E_t}{E}\right)^{7/2} atoms/ion. \qquad (9)$$

The numerical coefficient in Eqn. 9 has been modified to fit experimental sputtering yields measured for the specific case of H^+ bombardment of 304 stainless steel [22]. Folding the Maxwell-Boltsmann H^+ distribution into Eqn. 9 yields an average Cs ejection rate of ~2 x 10^{-4} atoms/ion for a Cs-filled monolayer.

We can now estimate the Cs ejection rate from a unit area (1 cm^2) in the collar region of the source under conditions of optimal coverage: 10^{19} H^+/s x $2x10^{-4}$ Cs/H^+ x 0.07 (plasma duty-factor) x ½ monlayer = ~$7x10^{13}$ Cs / s. Comparing this loss rate to that of thermal desorption, calculated above, we see losses are dominated by surface sputter-ejection. Also we observe that at this ejection rate an optimally-filled monolayer would be depleted in ½ x $3.56x10^{14}$ Cs/cm^2 / $7x10^{13}$ Cs/cm^2 s ~2.5 s if Cs were not continuously supplied to this surface! An unknown fraction of the sputter-ejected Cs must certainly return to the collar surface.

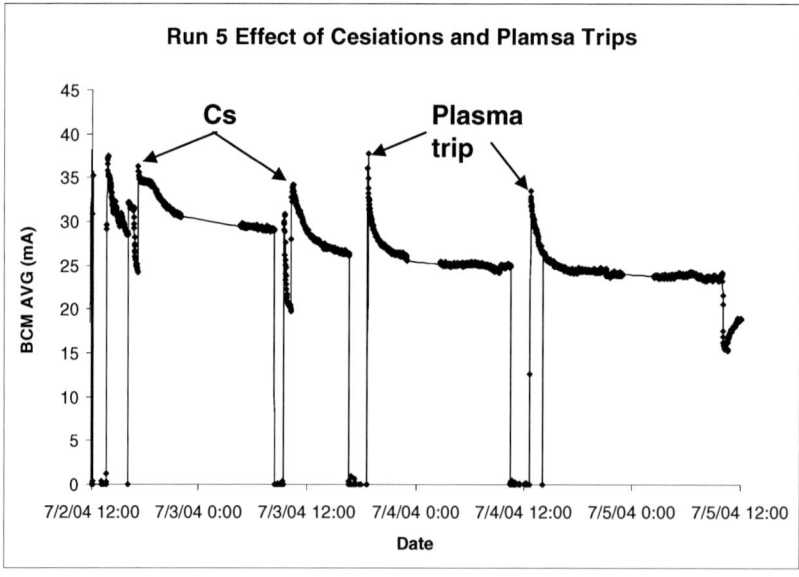

FIGURE 7. Detailed view of the H$^-$ beam current during Run-5 showing a characteristic response to cesiations and plasma outages.

Thus, we have shown that in order to maintain optimum coverage of the Cs collar, the system needs to continuously provide a flux of neutral Cs onto the surface $\sim 10^{14}$ Cs/s/cm^2.

Having described the location of the principal ionization surface in the source as well as the rate in which Cs needs to be delivered to that surface, we can now discuss the process of Cs transport between the dispenser and ionization surface. During a normal cesiation a maximum of $\sim 2.7 \times 10^{17}$ Cs atoms leave the dispensers. Approximately 25% of the released Cs is immediately lost from the source through the outlet aperture as a result of the close proximity of the dispensers to the outlet aperture. The balance of the released Cs enters the source and accumulates in cool areas. The accumulation rate depends on the difference between condensation rate, determined from the density of Cs atoms in the volume, and the evaporation rate, which can be determined from known Cs vapor pressure versus temperature curves. During this process Cs condensing on plasma-facing surfaces will likely be rapidly sputter-ejected and will accumulate in "shady regions" of the source. Most evidence of accumulated Cs was found behind the heat shield of the outlet aperture. After cesiation is complete and the collar temperature is lowered to the nominal operating temperature of ~ 300 C, Cs flux from the dispensers is greatly reduced and Cs is now mainly supplied to the collar from the quantity which has accumulated within the source interior. As this 'dose' of Cs which was supplied to the source during cesiation diminishes the rate at which the ionization surface is fed Cs also diminishes. As the Cs feed-rate is reduced it becomes insufficient to maintain an optimal coating on the ionization surface and the H$^-$ beam current begins to attenuate.

In order to determine the relative importance of the direct Cs transport pathway (dispensers $\rightarrow$ ionization surface) versus the indirect Cs transfer pathway (dispensers $\rightarrow$ source body $\rightarrow$ ionization surface) a specific experiment was performed. During Run-8, after an initial cesiation, the Cs collar was maintained at a constant temperature of ~ 70 C for the ~ 10 day run. Fig. 5 shows that our best Cs-enhanced H$^-$ yield was maintained over that entire period with very small beam attenuation. Since the maximum Cs feed-rate from the dispensers at the temperature of 70 C could have only been ~ 2 atoms/s, as calculated by Eqn. 1, we conclude that the role of direct Cs transfer is small compared with the indirect process in the present source configuration. At the end of the run, the collar temperature was raised to the nominal 300 C and little change in the beam current was observed.

Fig. 7 shows a detailed view of the beam response to source cesiations and plasma outages (trips) during Run-5. The approximately exponential character of these beam attenuation curves is consistent with this physical picture of indirect Cs transport: during a cesiation a finite amount of Cs atoms are introduced into the source and the rate at which these atoms re-evaporate Γ is proportional (k=proportionality constant) to the number of remaining condensed Cs atoms n.

$$\frac{dn(surface)}{dt} = -k \times n(surface) \rightarrow n(t) = \exp(-k \cdot t) \rightarrow \Gamma = k \exp(-k \cdot t) \quad (10)$$

The approximately exponential beam decay also observed immediately after plasma outages (trips) is shown in Fig. 7 and can also be understood within the context of this physical picture. During the trip the entire ion source body cools with the exception of the Cs collar which is heated by an external air heating system. Cs continues to be fed to the source at a very small rate ($\sim 10^{10}$ Cs/s at 300 C), accumulating on the cool source interior. Once the plasma is restored, the chamber heats and Cs is fed to the ionization surface at elevated rates resulting in conditions of more favorable coverage which accounts for the observed beam enhancement. Since the accumulated Cs is limited, the beam decays in approximately the same exponential fashion as it would after a standard cesiation described by Eqn. 10. Figure 2 also shows approximately exponential beam attenuation curves over the course of entire experimental runs, which is also a consequence of indirect Cs transport. In Run-6 the attenuation is larger since Cs leaves the source faster due to removal of the Cs collar aperture. In Run-7 the attenuation was slower due to an increased supply of Cs.

Thus we have shown that in the current configuration of the SNS source indirect Cs transport plays a dominant role. It is also likely that many of the H⁻ ion sources used in fusion research that were discussed at this conference also operate in this fashion. There are a number of fundamental disadvantages to indirect Cs transfer: (i) During source operation the flow-rate of Cs evaporated from the source walls is essentially uncontrollable, dependent on the temperature distribution of the source and its construction details. (ii) Because the flow-rate is uncontrollable it cannot be well matched to the Cs loss rate from the ionization surface in order to insure optimal coverage. (iii) Cs transfer is also inefficient; most of the Cs will condense in regions which are too cold to re-evaporate at significant rates. A new collar for the SNS source is proposed in which Cs is directly transferred from the dispensers to the ionization surface.

A DIRECT-TRANSFER CS COLLAR

In the previous section it was shown that indirect Cs transfer is inherently inefficient and almost certainly will feed Cs to the ionization surface at rates that are less than optimal. One simple solution to this problem is to insert a cooled/heated ionization surface immediately adjacent to the outlet aperture and independently control both Cs dispenser and ionization surface temperatures. These two structures need to be closely coupled to allow efficient particle transfer while, to the highest degree possible, being thermally isolated from each other. The proposed design of the reentrant ionization surface is shown in Fig. 8 and was developed using heat transfer software [23].

The design features two air heating/cooling circuits which are independently controlled through two identical interfaces and allow operation of each component over an independent temperature range of 20-600 C. The existing Cs collar is now utilized purely as a Cs dispenser which is temperature-controlled using the existing system. The ionization surface has been re-located closer to the outlet aperture and has been inclined $10°$ toward the plasma to receive a greater flux of particles from the plasma core. Thermal calculations show that a temperature uniformity of $\Delta \sim 10$ C under maximum plasma heating conditions can be achieved over the ionization surface. The entire assembly is supported by a Ta holder to provide protection from sputter damage, which has been observed on the existing stainless steel collar, as well as thermal isolation. Vacuum calculations suggest that approximately 70% of the Cs released from the dispensers will enter the surface ionization region. The design also allows easily interchangeable ionization surfaces to facilitate testing of different materials such as Al and Ta which could have favorable ionization characteristics. Higher work function materials are of interest since they bond more strongly to the Cs as well as yielding a lower effective work function when cesiated [17]. The system has currently been designed and will be tested within the next year.

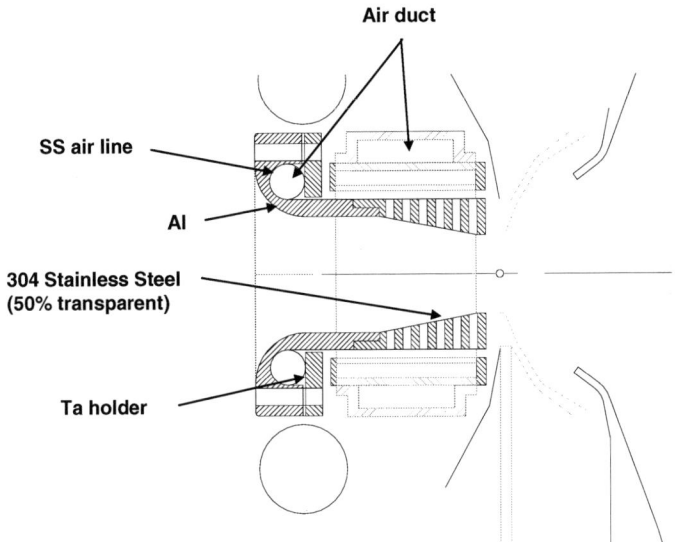

FIGURE 8. The design of the direct-transfer Cs collar.

Initial Tests of the Integrated Cs Collar / Outlet Aperture

The integrated Cs collar/outlet aperture has been conceived and developed as a result of collaborative ion source R&D work undertaken between LBNL [24] and ORNL [14] and will be described in more detail in subsequent publications. Figure 9 shows a schematic view of the assembly. In this work, initial tests of the system are reported in order to explore the benefits of inclining the ionization surface towards the plasma and decreasing the distance from the surface to the outlet aperture, both features of the design of the direct-transfer Cs collar. The integrated Cs collar / outlet aperture also features the capability of applying a potential bias to the collar in order explore the H⁻ enhancement which has been observed in the DESY multicusp source [25].

Fig. 10 shows a comparison between the H⁻ beam current extracted from the standard collar configuration shown in Fig. 6, and the integrated Cs collar / outlet aperture plotted as a function of applied RF power. The dramatically improved H⁻ yield is most likely a consequence of the improved location and orientation of the ionization surface. This increase in performance seems to have come at the price of higher emittance. Fig. 11 shows the first emittance scan from this source configuration. Note the broad shoulders in addition to the usual bright center of the scan. The entire distribution has an RMS normalized emittance value of 0.35 π mm mrad at 60 mA which is larger than emittances measured from the standard Cs collar at lower current levels of 40 mA. These shoulders disappear when the source is in an uncesiated state, suggesting that the surface-produced H⁻ is entering the beam with large energy and angular distributions as is typically seen with sputter ejected particles [10, 15]. We also biased the collar to -5V with respect to the plasma, drawing 35 A of steady state discharge current and found essentially no enhancement of H⁻ yield. This result contrasts with DESY data which show significant beam enhancement in this range of voltages and currents, emphasizing the differences between the Cs-enhanced SNS ion source and the Cs-free, presumably pure volume, DESY ion source [25].

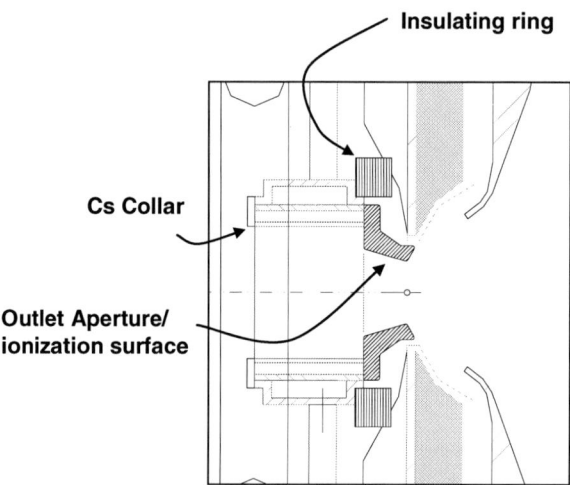

FIGURE 9. The Integrated Cs collar/outlet aperture.

Overall these results are very encouraging and show that prospects for improved H⁻ yield from the geometrical improvements to the direct-transfer collar are quite strong. As discussed above, the direct-transfer collar design also features an ionization surface which is inclined toward the plasma, as well as a reduced distance to the outlet aperture as compared with the original design. Compared to the integrated Cs collar / outlet aperture the ionization surface is located several mm further away from the outlet aperture in hopes that resonant charge exchange will effectively cool these ions before entering the beam as apparently occurs in the present design.

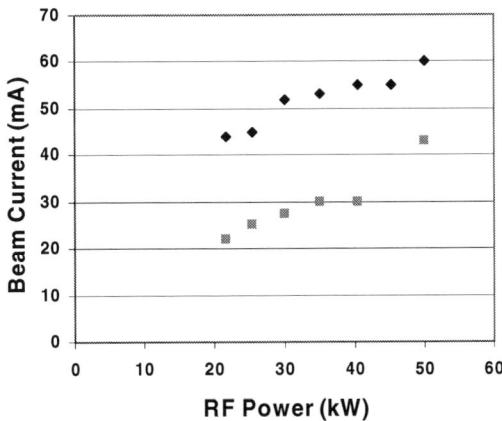

FIGURE 10. Beam current extracted from the integrated Cs collar/outlet aperture (diamonds) and from the standard source configuration (squares) versus applied RF power after 1 cesiation at full duty-factor.

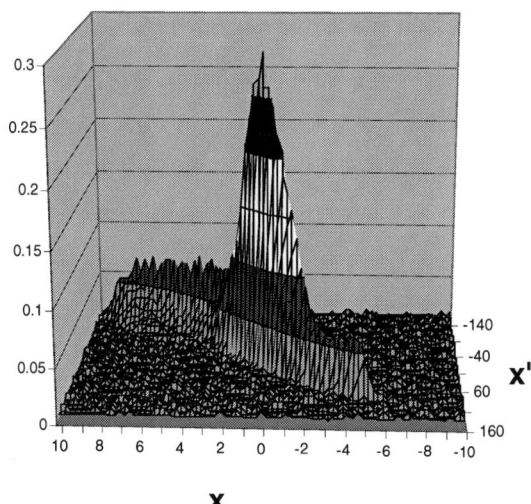

FIGURE 11. Emittance plot from the Integrated Cs collar/outlet aperture taken at 60 mA. Note the presence of 'shoulders' at the base of the bright core.

REFERENCES

1. N. Holtkamp, "Status of the SNS Project", PAC'03, Portland, USA, p. 11 ID: MOAL003.
2. R. Keller, et al., "Commissioning of the SNS Front End Systems at Berkeley Lab", EPAC'02, Paris, France, p. 1025, ID: THPLE012.
3. A. Aleksandrov, "Commissioning of SNS Front End Systems at ORNL," PAC'03, Portland, USA, p. 65, ID: MOPB002.
4. S. Henderson et al., "SNS Beam Commissioning Status", EPAC'04, p. 1255, ID: TUPLT168
5. R.F. Welton et al., "Development and Status of the SNS Ion Source", PAC'03 p. 3306 ID: FPAB009
6. R.F. Welton et al., Rev. Sci. Instrum. 75 1793 (2004).
7. R.F. Welton et al., Rev. Sci. Instrum. 73 1008 (2002).
8. R. Keller et al., Rev. Sci. Intrum. 73 914 (2002).
9. R.F. Welton, et al. EPAC'04, Lucerne, Switzerland, ID: TUPLT175
10. M.E. Kishinevskii, Sov. Phys. Tech. Phys. 20 799 (1976)
11. SAES Getters S.p.A. Via Gallarate, 215 20151 Milano, Italy
12. M. Succi, et al., Vaccum 35 579 (1985)
13. The HSC computer code, Outokumpu Research Oy, PO Box 60 FIN-28101 PORI, Finland
14. R.F. Welton, et al. EPAC'02, Paris, France, p.635 ID: THPLE019.
15. M. Kaminsky, "Atomic and Ionic Impact Phenomena on Metal Surfaces", Springer, New York, 1965
16. L.K. Hansen, "Thermionic Converters and Low Temperature Plasma", Techn. Inform. Center / US-DOE-tr-1 (1978)
17. G. Alton, Surface Sci. 175 226 (1986).
18. P.W. Van Amersfoot, et al., J. Appl. Phys. 58 2317 (1985)
19. KN Lueng, et al. Nucl. Instrum. And Meth. B74 291 (1993)
20. Y. Lee, et al., Rev. Sci. Instrum. 68 1398 (1997)
21. J. Bohdansky, et al., J. Appl. Phys. 51 2861 (1980)
22. J. Bohdansky, et al., Proc. 7th Int. Vacuum Congress and 3rd Int. Conf. on Solid Surfaces, ed R. Dobrozemsky, Wien, 1977.
23. Tera Analysis Ltd, Knasterhovvej 21, DK-5700 Svendborg, Denmark
24. R. Keller, et al, "Design, Operational Experiences and Beam Results Obtained with the SNS H- Ion Source and LEBT at Berkeley Lab," AIP Conf. Proc. 639, pp. 47-60 American Institute of Physics, Melville, NY (2002)
25. J. Peters, Rev. Sci. Instrum. 73, 900 (2002)

NEUTRALIZATION

Neutralization of H- Ions on the Plasma Target

Vladimir A. Baturin, Petr A. Litvinov

*Institute of Applied Physics, National Academy of Science of Ukraine, 58 Petropavlovskaya St.
Sumy, 40030 Ukraine, E-mail: baturin@ipflab.sumy.ua*

Abstract. This work describes a plasma neutralizer for ion beams and experimental results of conversion of H- ions into neutral atoms on the plasma target. The working medium of neutralizer is thermal plasma (potassium, cesium). The neutralizer is designed for an integrated plasma target density of a plasma target ~ 10^{15} cm^{-2}. A degree of plasma ionization may be adjusted. Dimensions of the neutralizer are 240 mm diameter and 350 mm length. Power consumption is 10 kW. The efficiency of H- ions conversion to neutral atoms and angular scattering of neutral atoms on plasma target in dependence on ion energy, target thickness and plasma ionization degree were measured.

INTRODUCTION

Accelerated beams of H^0 neutral atoms could be produced by neutralization of negative hydrogen ions in various targets. The plasma targets, providing a efficiency of H$^-$ ions conversion to atoms up to 80 - 85%, are rather promising for these purposes [1,2]. The important problem in formation of high energy neutral beams with high brightness is the angular scattering of neutrals in the target, as the emittance of produced H^0 beam includes a scattered particles component.

The present work is devoted to research of neutralization of hydrogen negative ions beam with energy 30 - 200 keV on an alkaline plasma target.

EXPERIMENTAL SETUP AND PROCEDURE

The plasma target was designed as a stand-alone component. A schematic of this target is shown in Figure1. The thermal alkaline plasma was formed by surface ionization on the hot tungsten ionizer 2. The ionizer was placed inside a heater 1. The heater was made from sheet tungsten. The heating power was supplied by the feedthrough electrodes 10, which tolerates the thermal expansion. The thermal screens 4 made of graphite were placed around the heater. The shell of a target 5 was water cooled. The working substance (potassium vapor) was supplied to the ionizer from the stainless steel crucible 7. Its design prevents the alkaline metal from dropping to the ionizer, and thus a stable operation of the target was provided. The crucible was connected with ionizer by the vapor line 8, which also represented the thermal bridge between ionizer and crucible. The temperature of crucible as well as vapor pressure in

CP763, *Production and Neutralization of Negative Ions and Beams*, edited by J. D. Sherman and Y. I. Belchenko

the ionizer, were adjusted with the help of the heater 9 and were measured by the thermocouple.

The alkaline metal was kept in a sealed glass ampoule 6. The ampoule was broken in vacuum after obtaining the necessary crucible and ionizer temperatures. The electrical insulation of plasma target elements was carried out with the help of high-temperature BeO ceramics. The water-cooled conic cups were placed at the target input and output to reduce the leakage of target working substance to the vacuum.

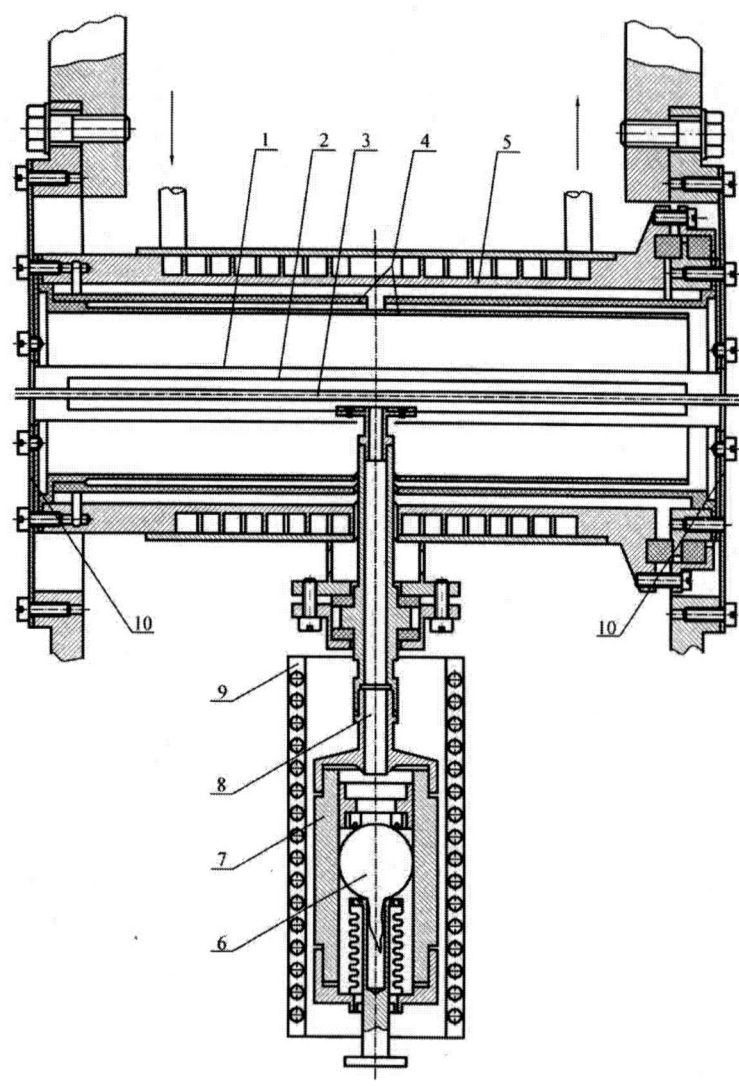

FIGURE 1. The schematic diagram of the plasma target.

A current to the heater 1 creates a magnetic field in the area of ion beam interaction with plasma. To exclude the influence of this field on ion beam, the heater power was pulsed with current up to 2500 A, pulse duration 10 ms and repetition rate 50 Hz. Such design of a plasma target and stabilized power supply system allowed heating the target up to 1000 - 2800 K temperature range, and to stabilize it for a long time with accuracy better than 5%. The ionizer temperature was measured by the W-Re thermocouple and also with the help of radiation pyrometer. The measurements of ion beam parameters were carried out in the intervals between power supply pulses.

The neutralizer design gives an integrated plasma target density of $t \sim 10^{15}$ cm^{-2}, while a degree of ionization of plasma may be adjusted. Dimensions of the neutralizer - diameter of 240 mm., length - 350 mm. Power consumption is 10 kW. It is difficult to study the plasma target parameters because of plasma channel high temperature and target design complexity. The thickness of plasma target was determined on H$^-$ beam attenuation after the interaction with a plasma target:

$$t_{pl} = -\frac{1}{\sigma_{-10}} \ln \frac{I_2^-}{I_1^-}, \qquad (1)$$

where I_1^- - H$^-$ ion current with the target off; I_2^- - H$^-$ ion current with the target on. The cross section of electron detachment from a negative ion by electron impact σ_{-10} was used [3,4]. We supposed that value of σ_{-10} for collisions with ions is the same, as for collisions with electrons. The parameters of a plasma target were also determined according to according alkaline metal consumption. The degree of plasma ionization was determined by efficiency of potassium ionization on tungsten [5] and it was also determined by the attenuation of H$^-$ and H$^+$ beams in a plasma target. The maximal thickness of a plasma target obtained in our experiments was about $t_{max} = 5 \cdot 10^{14}$ cm^{-2}. The degree of ionization is about $\leq$ 80 %, similar to that obtained in [6].

Schematic diagram of the experimental setup is shown in Fig.2. Flat beam of H$^-$ ions with energy 30 - 200 keV, angular divergence $2 \cdot 10^{-5}$ rad and ion current 10 μA was directed to the plasma target. After interaction with a plasma target, the components of a beam H$^-$, H^0, H$^+$ were separated in the magnet. A secondary-electron multiplier could register each of the beam component. The composition and spatial distribution of H$^-$ and H^0 beams density was carried out by a detector, consisting of a 0.01 by 5 mm diaphragm and a secondary-electron multiplier. The stepper motor shifted the detector discretely along the ion beam with a 5μm step. A distance of ion

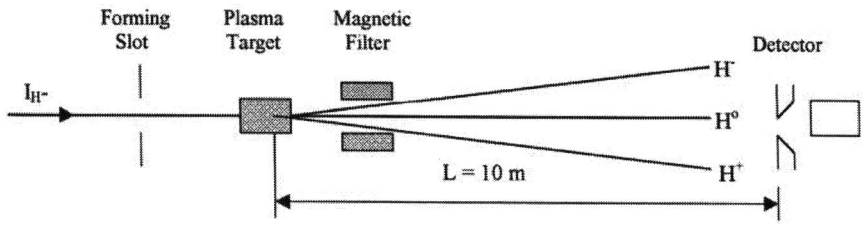

FIGURE 2. Schematic diagram of the experimental setup.

drift region between the interaction chamber and detector was 10 m, thus the angular resolution of experimental setup was $1 \cdot 10^{-6}$ rad. The vacuum in the ion beam drift region was less than 1×10^{-8} Torr in the presence of the beam. The integral thickness of the gas target in the beam drift region did not exceed $4 \cdot 10^{12}$ cm^2, which is about two orders of magnitude less than the thickness of change exchange target. Therefore the drift area did not lead to large errors in the measurements.

An integral H$^-$, H^o, H$^+$ ions currents were measured to determine the conversion efficiency η of H$^-$ ions into neutral atoms. The experimental value η was determined by formula

$$\eta = \frac{I^o}{I_1^-} \qquad (2)$$

where I^o is the equivalent current of neutral beam, I_1^- is the H$^-$ ion current with the target off. The scattering angle of neutral atoms $\theta_{1/2}$ at the fixed target thickness was determined by a widening of spatial distribution of H^o current density in comparison with the spatial distribution of primary H$^-$ current density, given by formula:

$$\theta_{1/2} = \frac{1}{2L}\sqrt{(\omega_\Sigma)^2 - (\omega_-)^2} \qquad (3)$$

where ω_Σ is the width of neutral beam distribution caused by scattering and by primary divergence of H$^-$ ion beam; ω_- is the width of distribution of H$^-$ ions current density; L is the distance from the plasma target center to detector.

RESULTS AND DISCUSSION

In our experiments the maximal thickness of a plasma target was $t_{max} = 5 \cdot 10^{14}$ cm^{-2}. H$^-$ ions with energy E_{H^-} = 200 keV have the conversion efficiency η = 40% for this thickness of plasma target. Therefore measurements of the scattering angle $\theta_{1/2}$ of the neutral atoms as a function of H$^-$ ion energy were carried out in two modes (Fig. 3):

- Conversion efficiency η = 10%
- Conversion efficiency η = 40%

The scattering angle of neutral atoms $\theta_{1/2}$ is also shown in Fig. 3 for the value η =80% (E$_{H^-}$= 30 keV). Scattering angle of neutral atoms $\theta_{1/2}$ measured for a hydrogen gas target is shown in Fig. 3 by solid line. It is possible to conclude from the submitted results, that the scattering angle of neutral atoms $\theta_{1/2}$ varied with energy as $\sim E^{-1/2}$ for H$^-$ ions neutralization on a thermal alkaline plasma target. Moreover, the scattering angle $\theta_{1/2}$ depends on thickness of a plasma target. The increase of the plasma target thickness with the corresponding growth of conversion efficiency from 10 up to 40 % results in increase of scattering angle by 30 - 40%. The scattering angle $\theta_{1/2}$ of H^o ions in the plasma target with maximal conversion efficiency had the value 2-3 times higher than that for the gas target.

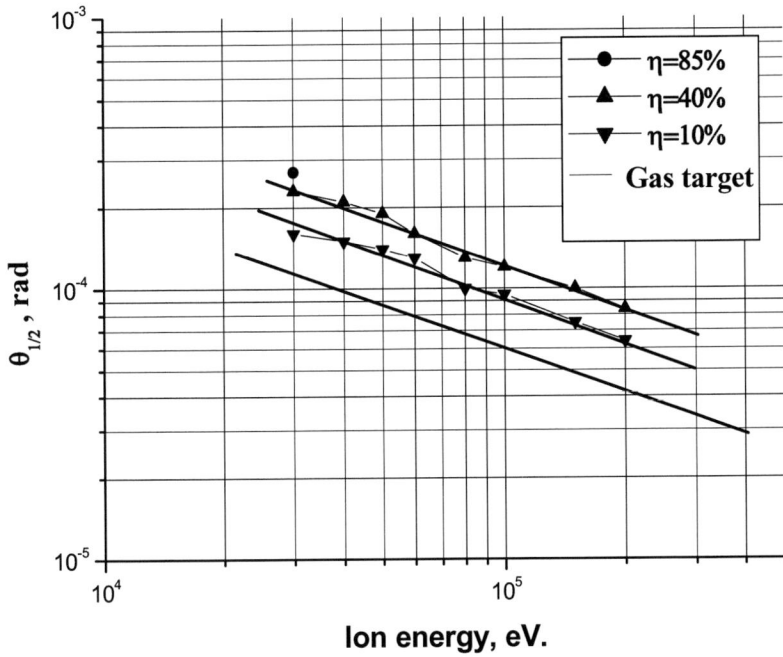

FIGURE 3. Scattering angle of the neutral atoms beam as a function of H⁻ ions energy.

For practical applications it is interesting to compare the quality of accelerated H^0 beams, produced in the gas and in plasma targets. The beam quality can be characterized by its brightness B. If the ion beam distribution is Gaussian, and B_{H^-} is a maximal brightness of H⁻ beam, then the maximal brightness of neutral beam B_o can be given in the form [7]:

$$B_0 = B_{H^-} \left[\frac{\eta_{max}}{1 + \left(\theta^0_{1/2} \Big/ \theta^-_{1/2} \right)^2} \right] \tag{4}$$

where $\theta^-_{1/2}$ is the H⁻ beam divergence.

The calculated B_O/B_{H^-} ratio as a function of H⁻ beam divergence for gas and plasma targets is shown in Fig. 4. The following experimental values were used in calculations: $E_{H^-} = 200$ keV; $\eta^{gas}_{max} = 52\%$; $\eta^{pl}_{max} = 80\%$; $\theta^{gas}_{1/2} = 5 \cdot 10^{-5}$ rad; $\theta^{pl}_{1/2} = 1 \cdot 10^{-4}$ rad.

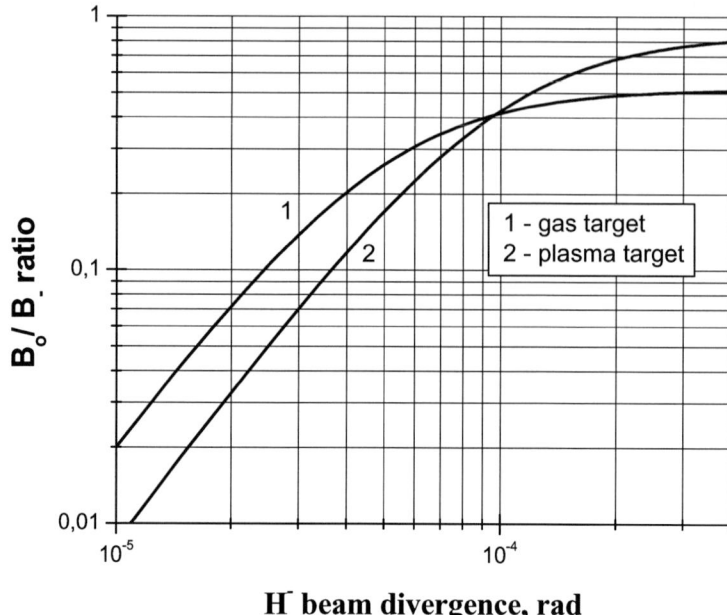

H⁻ beam divergence, rad

FIGURE 4. The B_0/B_{H^-} ratio as a function of H⁻ beam divergence for gas and plasma targets.

As it follows from the figure, the relative brightness of H° beam, produced in the plasma target, is higher than the relative brightness of H° beam, produced in the gas target, when the primary H⁻ beam divergence is $\theta_{1/2}^- \geq 10^{-4}$ rad.

REFERENCES

1. G.I. Dimov, A.A. Ivanov, R.G. Roslyakov, *Sov. Phys. Techn. Phys.* **22**, 1091 (1976).
2. A.I. Hershkovitch, B.M. Jonson, V.J. Kovarik, M. Meron, K.W. Jones and K. Prelec, *Rev. Sci. Instrum.* **55**, 1744 (1984).
3. D.F. Dance, M.F. Harrison, R.P. Rundel, *Prog. Roy. Soc.,* **A299**, 525 (1967)
4. W.L. Fite, R.T. Brakmann, *Phys. Rev.,* **112**, 1141 1958)
5. E. Y. Zandberg, N.I. Ionov, *Sov. Phys. UPN.,* **67**, 581(1959)
6. Sato M., *Phys. Fluids* **17**, 1903 (1974)
7. Fink I.H., *Production and Neutralization of Negative Ions and Beams*, ed. J.G. Alessi, Brookhaven NY, 1986, AIP Conference Proceedings No. **158** , 621 (1986)

APPLICATIONS

Study of Direct Current Negative Ion Source for Medicine Accelerator

Yu. Belchenko, I. Ivanov and I. Piunov

Budker Institute of Nuclear Physics, 63090, Novosibirsk, Russia

Abstract. Status of dc H⁻ ion source development for tandem accelerator of boron capture neutron therapy is described. Upgrade and study of the Penning surface-plasma source with hollow cathodes was continued. Results of source optimization, of ion optic computer simulation, and of emittance measurement are presented. The upgraded source delivers dc H⁻ beam with energy 25 kV, current 8 mA, 1rms emittance $\epsilon_X \sim 0.2 \; \pi$ mm·mrad, $\epsilon_Y \sim 0.3 \; \pi$ mm·mrad at discharge power ≤ 0.5 kW.

INTRODUCTION

A dc negative ion source delivering H- beam with intensity >10 mA is under development at Budker Institute for use at the tandem accelerator of boron neutron capture therapy (BNCT) device. It is important to have a long-term stability and an easy maintenance for the source. The surface-plasma source (SPS), based on the Penning discharge with hollow cathodes, can provide an efficient dc operation [1,2]. Experimental dc surface-plasma source, delivering 8 mA beam reliably, was developed recently [3]. The results of experimental source study and upgrade is presented below.

EXPERIMENTAL SOURCE

The scheme of experimental surface-plasma source for BNCT accelerator is shown in Fig.1. Negative Ions (NI) are extracted from plasma of the Penning discharge. The gas-discharge chamber consists of the cylindrical anode box with the massive cathode body enclosed. Penning discharge is supported by electron oscillations between the cathode protrusions in the horizontal magnetic field, produced by external electromagnet. The source uses no heated cathodes. Two hollow cathode bushings with small apertures (schematically shown in Fig.2) are inserted into the massive cathodes of Penning discharge. The hydrogen and small amount of cesium was fed to the discharge through the hollow cathode apertures.

The plasma injection from hollow cathodes supplies the high current Penning discharge operation at lower gas pressure and at smaller cesium consumption, as compared with the conventional cold cathode Penning SPS. Hollow cathode arc provides the Penning discharge stabilization: localization in the emission aperture vicinity, long term stability and the decreased level of fluctuations. In turn, electron

CP763, *Production and Neutralization of Negative Ions and Beams*, edited by J. D. Sherman and Y. I. Belchenko

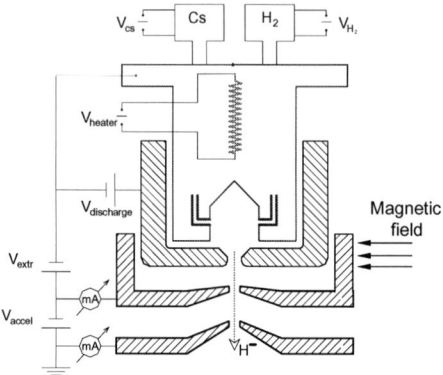

FIGURE 1. Experimental source scheme (cross section along magnetic field lines)

confinement in the Penning area favors the hollow cathode arc ignition.

The water cooling of discharge electrodes is applied for keeping an optimal temperature of the anode. Negative Ions (NI) are produced on the cesiated anode surface and by charge-exchange of surface-produced ions [1]. Both NI groups are extracted from the discharge through the emission aperture with diameter 3 mm, made in the anode bottom. The axisymmetrical ion-optical system is applied for beam extraction and acceleration. Solid bars at the extractor sides (Fig. 1) are used for the co-extracted electron flux interception at extractor potential. The experimental source is wholly located in a vacuum box. It improves the outgoing gas pumping from all the sides.

Experimental source operation was described in detail before [3]. Maximal H⁻ production was realized at the following source parameters: hydrogen feed 0.1 LTor/s, discharge voltage 80-90 V, anode temperature in the emission area 300-400 °C, cesium feed <1 mg/h. Beam current is increased proportionally to the emission aperture area growth and to the arc current increase while keeping the optimal anode cesium coverage. NI beam with current up to 8 mA was obtained with emission aperture diameter of 3 mm and discharge power 0.7 kW. The ohmic heater inside the cathode body provides the easy source discharge start. The full extraction voltage and the acceleration voltage of about 7-10 kV are applied at the start. The acceleration voltage is increased to values up to 25 kV within the next 3-5 minutes after the discharge on.

SOURCE OPTIMIZATION

Source experimental optimization and parameters enhancement were done recently. Both the triode and the tetrode ion optical system (shown in Fig.2) were studied and optimized. Independent cesium feed to each of hollow cathode units was applied for cesium consumption decrease. Enhancement of H⁻ production with the gas admixture to hydrogen was obtained and studied. Improved source parameters at 8 mA H⁻ beam production are listed in the Table 1. Data of Table 1 were obtained at lower values of

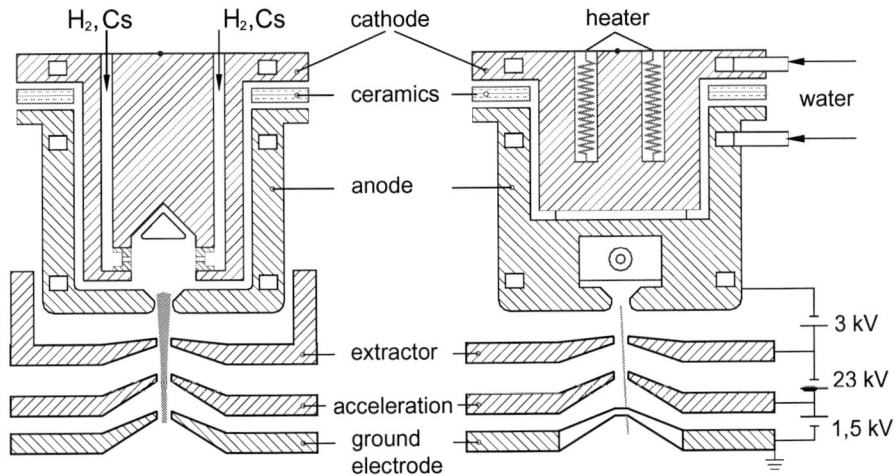

FIGURE 2. Layout of experimental source with tetrode ion-optical system and independent cesium feed *left- cross section along magnetic field lines, right- cross section across magnetic field).*

discharge current, magnetic field, cesium and gas feed, as compared with the previous source version [3], but with the same emission aperture diameter. The power efficiency of H⁻ production is increased to value of about 19 mA/kW, gas efficiency – to value of about 1 %.

Dependence of H- beam current vs acceleration voltage for the optimized triode ion-optical system is shown in Fig.3 by circles. Corresponding optimal values of the extraction voltage are shown in Fig. 3 by crosses. H⁻ beam current intensity is saturated at the acceleration voltage 18 kV (for 6.2 A discharge current and 0.7 kGs magnetic field). Optimal extraction voltage increases with the acceleration voltage growth.

The tetrode ion optical system produces similar H- beams, as that in the case of the triode system. Variation of its accelerating electrode potential (see Fig. 2) results in a small change of H- beam current and divergence.

H⁻ beam current (aperture 3 mm)	8 мА
Beam energy	25 (2 + 23) keV
Discharge voltage	60 - 80 V
Discharge current	6 A
Magnetic field	0.8 kGs
Cesium feed	< 0.5 mg/h
Gas feed	0.07 L Tor/s

TABLE 1. Improved source parameters.

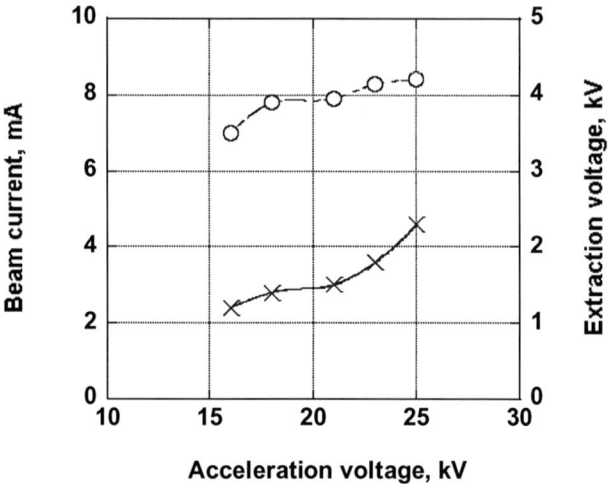

FIGURE 3. H- beam current (circles) and optimal extraction voltage (crosses) vs acceleration voltage (discharge current 6.2 A, voltage 80 V, magnetic field 0.7 kGs, gas feed 0.07 L Tor/s).

EMITTANCE

A movable emittance scanner was applied for the emittance measurement in X (along magnetic field) and Y (across magnetic field) directions across the beam. The scanner had the entrance collimator with an aperture 0.4 x 0.4 mm. The divergence of small elementary beam jet, formed by collimator, was analyzed at the 23 cm long flight base by the slit collectors with the help of electrostatic deflectors [3]. The scanner was shifted across the beam in order to determine the emittance diagram.

The elementary beam jets have an elliptical cross section while moving in the 90% beam area, and its current density distribution have the Gaussian waveforms in X and Y directions. The typical beam cross section and XX' diagram (along magnetic field) for 8 mA, 25 kV beam are shown in Fig. 4. Measured widths of Gaussian distribution in cross section of elementary beam jets are shown in Fig. 4 by triangles and by the corresponding oblique lines to guide the eye (solid lines – for 90% of elementary beam jet, dashed lines – at FWHM of elementary beam distribution). Corresponding normalized 1rms XX' emittance has the value $\epsilon_X \sim 0.2\ \pi$ mm·mrad.

The YY' diagram (across magnetic field) for 8 mA, 21 kV beam are shown in Fig. 5. The measured widths of cross section of elementary beam jets are shown in Fig. 5 by triangles and by the corresponding oblique lines to guide the eye (solid lines – for 90% of elementary beam jet Gaussian distribution, dashed lines – at FWHM of elementary beam jet Gaussian distribution). Corresponding normalized 1rms YY' emittance has the value $\epsilon_Y \sim 0.3\ \pi$ mm·mrad.

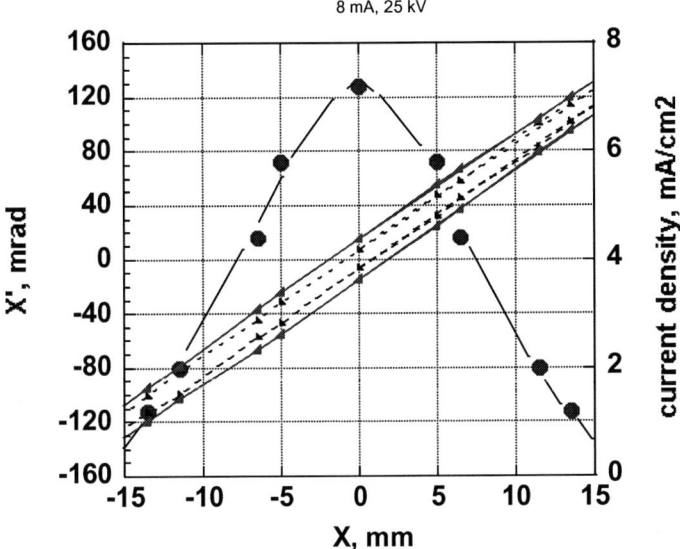

FIGURE 4. Beam current density distribution (circles) and XX' diagram for 8 mA, 25 kV beam. XX'-triangles and solid lines – 90% of elementary beam jet, triangles and dashed lines - FWHM of elementary beam jet.

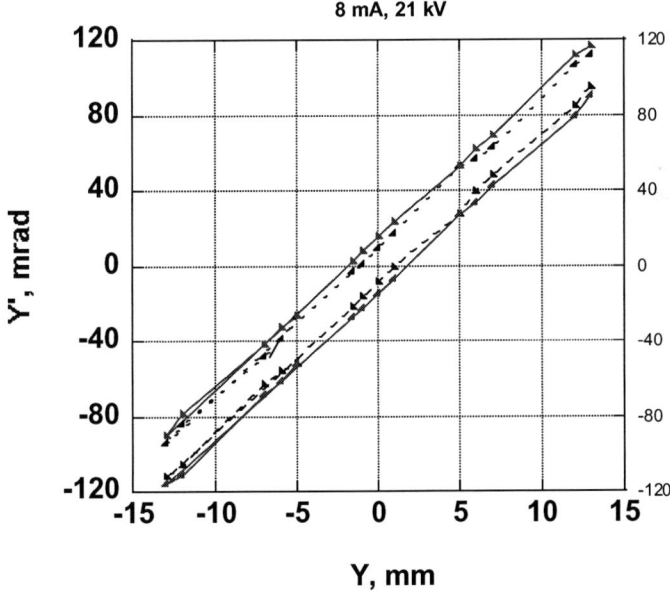

FIGURE 5. YY' diagram for 8 mA, 21 kV beam. Triangles and solid lines – 90% of elementary beam jet, triangles and dashed lines - FWHM of elementary beam jet.

RELIABILITY

The source dc operation was tested for 4-6 hours per day during about two months study. Two recognized problems in a long-term operation are shown in the Fig. 6.

FIGURE 6. a) Molybdenum flakes deposition on the anode and b) metal deposition on cathode-anode insulator (after one month of experiments)

Molybdenum, sputtered at low rate from the cathode was deposited and flaked at the anode plate (Fig. 6a). The metal deposition to the cathode-anode insulator is shown in Fig. 6b. Molybdenum sputtering could be depressed by prevention of poor-cesium discharge modes (with voltage > 100-120 V) at the source start.

BEAM FORMATION SIMULATION

PBGUNS computer program was used for computing of axisymmetrical beam extraction, formation and transport. It uses a fine matrix in the emission region, and computes the plasma boundary position with the self-consistent model of transient plasma sheath, modified for NI extraction. More than 100 parameters, including emitted NI drift energy and temperature, electron and positive ion density and temperature in the emission region can be interactively controlled.

Two types of plasma emission (uniform plasma emitter and emitter with the Maxwellian angular distribution) were tested and applied for simulation of H⁻ beam extraction from the source. Beam formation with various emission, plasma and ion-optical system parameters were simulated. The properties of transient plasma sheath, governing the plasma boundary position in the case of NI extraction were carefully studied as well. An example of PBGUNS output data for 10 mA H⁻ beam formation by the triode ion-optical system is shown in Fig. 7. The simulated data were used for the purpose of guiding the experimental optimization of the source ion-optical system.

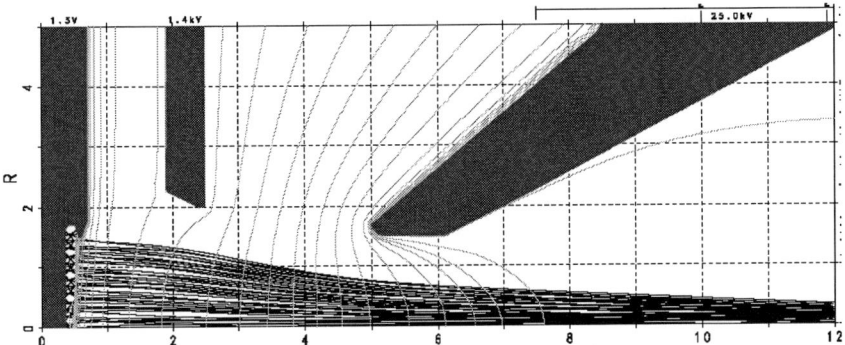

FIGURE 7. PBGUNS output data for 10 mA H⁻ beam formation. An optimal shape and position of the plasma boundary is obtained by variation of the extractor electrode form and potential.

FLANGE SOURCE DESIGN

The flange source to be installed at the test bed, and then at the low-energy beam transport line of BNCT accelerator is under manufacturing at BINP. The flange source has several novel features to be tested: most outside parts of the source (gas-discharge electrode flanges, discharge insulator, cesium system) will contact with air; source differential pumping will be produced from the source bottom only.

REFERENCES

1. Yu. Belchenko and A. Kupriyanov. *Rev. Sci. Instrum.* **65**, p.417 (1994)
2. Yu. Belchenko and M. Bacal. *Rev. Sci. Instrum.* **65**, p.1204 (1994)
3. Yu.Belchenko and V.Savkin. *Rev. Sci. Instrum.* **75**, p.1704 (2004)

Stripping Target of 2.5 MeV 10 mA Tandem Accelerator

V.I.Davydenko, A.N.Dranichnikov, A.A.Ivanov, G.S.Krainov,
A.S.Krivenko, V.V.Shirokov

Budker Institute of Nuclear Physics, Novosibirsk, 630090, Russia

Abstract. An electrostatic tandem-accelerator with 2.5 MeV 10 mA proton beam is under development at BINP. One of the important accelerator parts is a target that converts the half energy accelerated negative hydrogen ions into the proton beam. In the tandem accelerator an argon stripping target with 1 cm tube diameter and 40 cm tube length will be used. To reduce argon flux from the target to accelerator gaps a gas recirculation by turbomolecular pump installed in high voltage electrode is provided. Processes of plasma production and ultraviolet emission due to target ionization by fast ions and stripped electrons are considered in the report.

INTRODUCTION

Neutron therapy facility based on 2.5 MeV tandem accelerator is being developed at BINP [1,2]. Figure 1 shows a view of the vacuum insulation tandem accelerator. Negative hydrogen ion beam produced in ion source and transported by low energy tract is injected into the tandem accelerator. Proton beam, produced by stripping of the negative ion beam in a gas target placed in high voltage terminal, is accelerated to double the voltage of the high voltage electrode. Set of intermediate electrodes of the tandem accelerator provides uniform potential distribution and prevents full voltage effect.

TARGET DESIGN FOR GAS STRIPPING

Different versions of the stripping target converting half-energy accelerated negative-hydrogen ions into a proton beam have been previously considered in detail [1,3]. As a result, argon stripping target was accepted for the tandem-accelerator.

Atomic gases do not dissociate by the energetic hydrogen beam. Argon has sufficiently large stripping cross-section and high pumping efficiency. Dependence of the hydrogen beam composition on the argon target thickness is shown in Fig.2. For the stripping target thickness of $2.1 \cdot 10^{16}$ cm^{-3} the proton fraction is 95% and for the target thickness $3.3 \cdot 10^{16}$ cm^{-3} the proton fraction is 99%.

The stripping cell is an oil-cooled copper tube with an inner diameter of 1 cm and a length 40 cm installed inside a high-voltage electrode. Argon is puffed into the middle

CP763, *Production and Neutralization of Negative Ions and Beams*, edited by J. D. Sherman and Y. I. Belchenko
© 2005 American Institute of Physics 0-7354-0248-5/05/$22.50

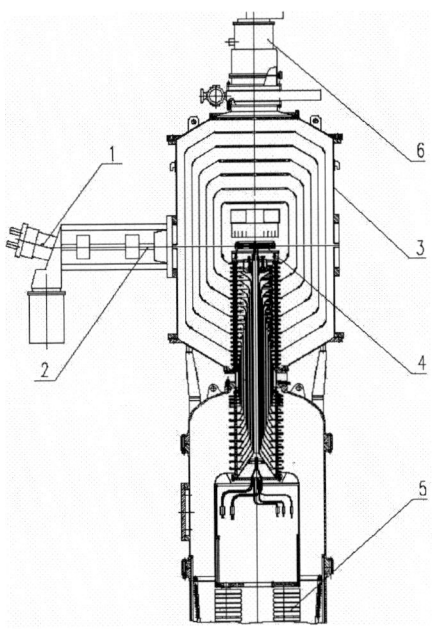

FIGURE 1. Cutaway view of the tandem accelerator. 1- negative hydrogen ions source, 2- low energy beam tract, 3- vacuum vessel, 4-stripping target, 5- sectionalized rectifier, 6- cryo pump.

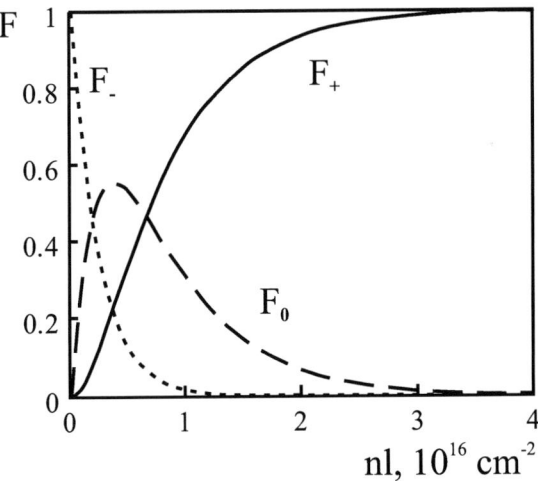

FIGURE 2. Dependence of beam fractions on target thickness.

of the stripping cell at a rate ~50 mTorr·l/s. At a pumping speed of 1500 l/s the argon pressure in the high-voltage electrode is ~$3.5 \cdot 10^{-5}$ Torr.

The stripping target actively influences tandem-accelerator operation. The target increases pressure in accelerator gaps and emits argon ions and ultraviolet radiation to the accelerator gaps. The influence of the stripping target can reduce electrical strength

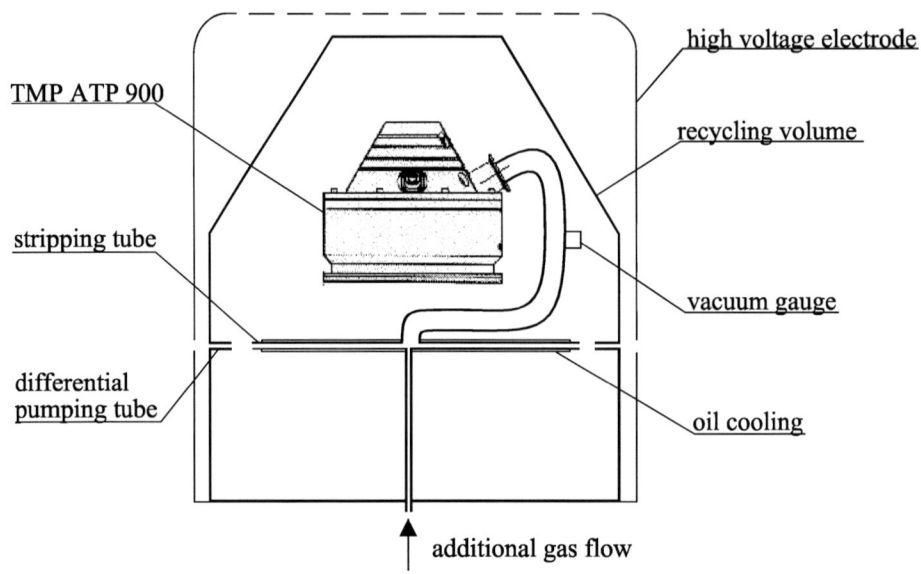

FIGURE 3. Stripping target with recirculation.

of the accelerator due to collisions of secondary particles with the electrodes and should be minimized.

To reduce the argon flux from the stripping target to the accelerator gaps, recirculation of argon in the target can be used. Figure 3 shows a scheme of the stripping target with recirculation. Inside the high-voltage electrode an additional recirculation chamber with small apertures for the beam is installed. Argon leaving from the stripping cell is intensively pumped by a turbomolecular pump mounted inside the chamber. Exit of the turbomolecular pump is joined with the stripping-cell tube and the pumped argon returns to the cell. Our simulations showed that for the presented geometry the recirculation reduces the argon flux to the accelerator gaps more than 10 times. The main problems of the recirculation scheme are operation of the turbomolecular pump inside the high-voltage electrode and high gas pressure at the entrance of the pump. The accepted scheme of recirculation differs from traditional [4,5] where turbomolecular pump is placed in atmosphere. Controller of the ATP 900 turbomolecular pump is installed in the rectifier volume filled with SF_6 gas.

PHYSICS OF GAS TARGET

The argon atoms in the target are intensively ionized by 1 MeV hydrogen particles which leads to plasma production in the stripping cell. There are two contributions to ionization of the target atoms. The first one is classical ionization losses. A hydrogen particle loses ~600 eV in the target and produces ~8 Ar^+ ions. The second factor is ionization of the argon atoms by two electrons with an energy ~500 eV stripped from an H^- ion. Estimation showed that these two electrons ionize ~10 Ar atoms. Thus, the number of ionizations in the stripping target produced by one H^- ion is $A \sim 18$. Rate of ion density production in the target is $\dfrac{dn_i}{dt} \approx \dfrac{AI_b}{\pi a^2 L}$, where I_b is the H^- current, a is the stripping tube radius and L is the stripping tube length. The ions leave the target at a rate $dn_i/dt \approx -n_i/\tau_i$, where $\tau \approx a/v_i$ is the Ar^+ ion life-time and $v_i \approx 5 \cdot 10^5$ cm/s is the velocity of the Ar^+ ions. In equilibrium the rates are equal and the plasma density in the target is estimated as $n_i \approx AI_b/\pi a L v_i \approx 3 \cdot 10^{10}$ cm^{-3}. The obtained value of the plasma density is much larger than the ion density in the hydrogen beam $n_b \approx I_b/\pi a^2 e v_b \approx 5 \cdot 10^7$ cm^{-3}. Assuming the electron temperature of the plasma target $T_e \approx 5$ eV, we obtain the Debye length $r_D = \sqrt{T_e/4\pi n_i e^2} \approx 10^{-2}$ cm and the plasma potential $\varphi \approx 3T_e/e \approx 15$ V. The arising plasma leads to emission of Ar^+ ions into the accelerator gaps. The total current of the Ar^+ ions produced in the target is $I_i \approx 2\pi a^2 n_i v_i \approx 40$ mA and the current of Ar^+ to the accelerator gaps determined by a solid angle is $I_{ac} \approx I_i b^2/s^2 \approx 40$ μA, where b is radius of high voltage electrode beam aperture, s is distance to the aperture.

After passing of the high-energy hydrogen particles through the stripping target, the Ar atoms are exited and a flux of ultraviolet photons with 10–15 eV energy from the target is generated. A number of photons produced in the target by one hydrogen particle is estimated as $B \approx 30$. The photon flux to the accelerator gaps is $S \approx BI_b \langle \Omega \rangle \approx 5 \cdot 10^{15}$ s^{-1}. For photoemission coefficient $\gamma \sim 0.1$, the current of secondary photoelectrons in the accelerator is $I_p \approx \gamma S_{ac} \approx 80$ μA. Coefficient of the ultraviolet reflection from metals is ~ 0.6 and after several reflections the ultraviolet photons can fall to the surface of a ceramic high-voltage insulator.

PLANNED EXPERIMENTAL PROGRAM

To reduce dangerous influence of the stripping target on the tandem-accelerator the following actions can be used. To minimize the ion and ultraviolet fluxes from the stripping target, the tube diameter should be taken as small as possible. To reduce the secondary-electron emission from the accelerator electrodes, the area near the beam apertures should be covered by metals with high work function – Mo, Ta, Nb. To avoid direct acceleration of the secondary electrons to the high-voltage electrode, small transverse magnetic fields near the beam apertures deflecting the electrons to the electrodes should be applied. Additional labyrinths shielding the high-voltage insulator can be useful.

To study the argon recirculation and the plasma production in the stripping cell an experimental test stand is under preparation. Figure 4 shows a general view of the test stand. At the initial stage of recirculation study the recirculation chamber will be pumped by a usual turbomolecular pump installed at the lower flange of the stand chamber. The turbomolecular pump can provide argon pumping from the stripping tube chamber without recirculation. At the next stage a turbomolecular pump ATR 900 will be mounted inside the recirculation chamber and the real recirculation

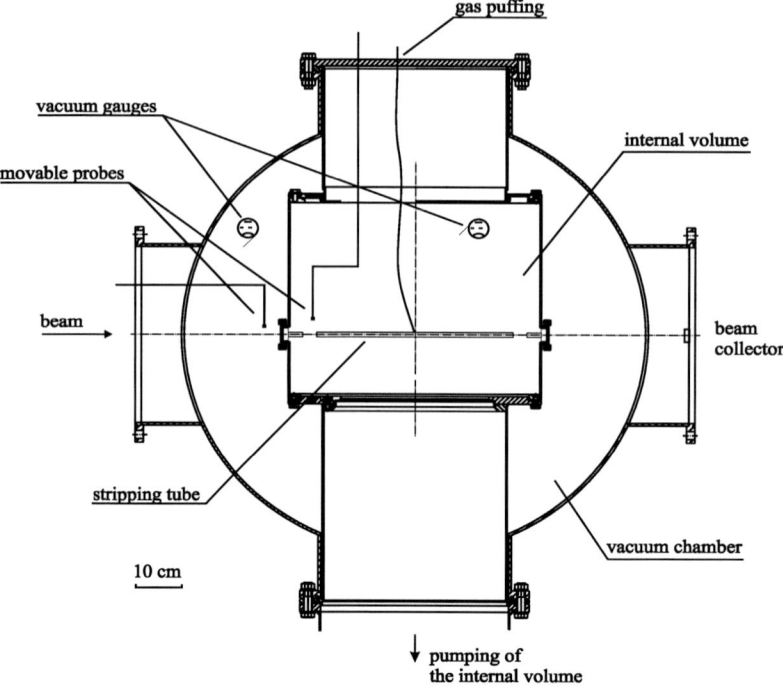

FIGURE 4. General view of experimental test stand.

scheme will be checked. To produce plasma in the target, a plasma diagnostic neutral-beam injector providing injection of a 10–100 mA, 10 keV hydrogen or helium beam into the stripping tube will be used. Also ionization of the target by 500 eV electrons will be studied by injecting a corresponding electron beam. Parameters of the target plasma will be measured by Langmuir probes, and ultraviolet emission will be registered by semiconductors detectors. At present a version of the stripping tube without the recirculation is prepared for mounting inside the high-voltage electrode. Assembling of the experimental test stand is being completed.

REFERENCES

1. B.F. Bayanov et al. Accelerator based neutron source for the neutron-capture and fast neutron therapy at hospital. *Nucl. Instr. and Meth. in Phys. Res.* **A 413** (1998) 397.
2. V. Dolgushin et al. Status of high-current tandem accelerator for the neutron therapy facility, *Proc. of Intern. Symp. on BNCT,* July 7-9, 2004, Novosibirsk, Russia
3. G.E. Derevyankin et al. Charge-exchange target for 40 mA, 2 MeV tandem accelerator. Preprint Budker INP 2001-23, Novosibirsk, 2001.
4. G. Bonani et al. Efficiency improvements with a new stripper design. *Nucl. Instr. and Meth. in Phys. Res.* **B 52** (1990) 338.
5. S.A.W. Jacob et al. Ion beam interaction with stripper gas – Key for AMS at sub *Mev. Nucl. Instr. and Meth. in Phys. Res.* **B 172** (2000) 235.

Symposium Program

Monday, September 13, 2004

Day of arrival
15:00-19:00   Registration
19:00   Welcome party

Tuesday, September 14, 2004

9:00-9:20		Opening of the Symposium
9:20-9:55	R.Hemsworth	Development of the Long Pulse Negative Ion Source for ITER
9:55-10:25	O.Kaneko	Status of Negative-Ion-Based Neutral Beam Injectors in LHD
		Coffee Break
10:55-11:25	L.Svensson	Experimental Results with the New ITER-like 1 MV SINGAP Accelerator
11:25-11:55	O.Fukumasa	Isotope Effect of H^-/D^- Volume Production in Low-Pressure H_2/D_2 Plasmas - Negative Ion Densities Versus Plasma Parameters
11:55-12:25	K.Tsumori	Correction of Beam Distortion in Negative Hydrogen Ion Source with Multi-Slot Grounded Grid
12:25-12:55	Yu.Belchenko	Plasma Injection from Several Cesiated-Hollow Cathodes into the 1/3rd Scale Large H^- Ion Source
		Lunch
14:30-15:00	M.Capitelli	Twenty Five Years of Vibrational Kinetics and Negative Ion Production in H_2 Plasmas: Modelling Aspects
15:00-15:30	A.Laricchiuta	Progress in Elementary Processes for Negative Ion Source Modeling
15:30-16:00	T.Sakurabayashi	Effects of a Weak Transverse Magnetic Field and a Spatial Potential on Negative Ion Transport in Negative Ion Sources
		Coffee Break
16:30-17:00	K.Benmeziane	2D PIC-MCC Code for Electron-Hydrogen Gas Interaction Study in H^- Ion Sources
17:00-17:30	V.Dudnikov	Relevance of Volume and Surface Plasma Generation of Negative Ions in Gas Discharges
17:30-18:00	M.Stockli	Emittance Analysis and Ghost Signal Busting in Allison Emittance Scanners
18:00-18:30	O.Fukumasa	Extraction Probability of Negative Ions from Hydrogen Ion Sources - Effect of Filter Magnetic Field and Gas Pressure

Wednesday, September 15, 2004

9:00-9:30	**M.Bacal**	ECR-driven Multicusp Volume H⁻ Ion Source
9:30-10:00	**V.Dudnikov**	Peculiarities of Compact Surface-Plasma Sources Operation (Practical Aspects)
10:00-10:30	**V.Ivanov**	Volume Production of High Negative Hydrogen Ion Density in Low-Voltage Cesium-Hydrogen Discharge
		Coffee Break
11:00-11:30	**J.Peters**	Finding the Optimum Frequency and the H⁻ Distribution in the HERA RF-Volume Ion Source
11:30-12:00	**V.Baturin**	H⁻ Source with Volumetric – Plasma Formation of Ions
12:00-12:30	**V.Baturin**	Flash Chamber of a Quasi-Continuous Volume Source of Negative Ions
12:30-13:00	**D.Faircloth**	Advances in the Ion Source Research and Development Program at ISIS
		Lunch

14:30   Excursion in Kiev city
20:00-21:30   General discussion

Thursday, September 16, 2004

9:00-9:30	J.Sherman	Physical Insights and Test Stand Results for the LANSCE H⁻ Surface Converter Source
9:30-10:00	D.Kuechler	H⁻ Source Developments at CERN
10:00-10:30	Yu.Belchenko	Study of Direct Current Negative Ion Source for the Medicine Accelerator
		Coffee Break
11:00-11:30	G.Kuznetsov	Source of Negative Hydrogen Ions with Hot Cathode
11:30-12:00	R.Gebel	Polarized Negative Light Ions at the Cooler Synchrotron COSY/Juelich
12:00-12:30	R.Gobin	Status of the Negative Hydrogen Ion Test Stand at CEA Saclay
12:30-13:00	M.Taniguchi	Acceleration of 100 A/m^2 Negative Ion Beam by Vacuum Insulated Beam Source
		Lunch
14:30-15:00	I.Soloshenko	Space Charge Lens for Focusing of Negative Ion Beams
15:00-15:30	V.Davydenko	Stripping Target of 2.5 MeV 10 mA Tandem Accelerator
15:30-16:00	D.Moehs	Sub Microsecond Electromagnetic Beam Notching at Low Energy Utilizing a Magnetron Ion Source with a Split Extractor
		Coffee Break
16:30-17:00	R.Becker	Mathematical Formulation and Numerical Modelling of the Extraction of H⁻ Ions
17:00-17:30	R.Welton	Ion Source Development at the SNS
17:30-18:00	V.Baturin	Neutralization of H⁻ Ions on the Plasma Target
18:00-18:30	M.Hanada	Energy Spectrum of Atoms Stripped from Negative Ions in a Multi-Stage Negative Ion Accelerator
18:30-19:00		Closing of the Symposium

20:00 Banquet

Friday, September 17, 2004

Day of departure

LIST OF PARTICIPANTS

ARTAMONOV Vladimir
MMPP "SALUT" MKB "GORIZONT"
16 Prosp. Budenogo
Moscow 105118
RUSSIA
Phone: 551-51-66
Fax: 551-51-66
E-mail: sasha-laron@mtu-net.ru

BACAL Marthe
Laboratoire LPTP, Ecole Polytechnique
Palaiseau 911218
FRANCE
Phone: +33-1-69-33-32-52
Fax: +33-1-69-33-30-23
E-mail: bacal@lptp.polytechnique.fr

BALAKIN Vladimir
LPI Physics-Technics Center
Protvino, Moscow region
RUSSIA

BATURIN Vladimir
Institute of Applied Physics
58 Petropavlovskaja St.
Sumy 40030
UKRAINE
Phone: +38 0542 223760
Fax: +38 0542 223760
E-mail: baturin@ipflab.sumy.ua

BECKER Reinard
IAP, Universitat Frankfurt/M
Fach 180
Frankfurt/M D-60054
GERMANY
Phone: +49 69 798 23488
Fax: +49 69 798 28510
E-mail: rbecker@physik.uni-frankfurt.de

BELCHENKO Yuri
Budker Institute of Nuclear Physics
11 Lavrentiev Ave.
Novosibirsk 630090
RUSSIA
Phone: 383-330-24-90
Fax: 383-330-71-63
E-mail: belchenko@inp.nsk.su

BENMEZIANE Karim
LPGP, 4 rue du pont de Pierre
Pantin 93500
FRANCE
Phone: 00 33 6 11 05 10 94
E-mail: karim.benmeziane@tiscali.fr

CAPITELLI Mario
Dipartimento di Chimica,
Universita di Bari,
via Orabona 4
Bari 70126
ITALY
Phone: +39-080-5442088
Fax: +39-080-5442024
E-mail: cscpmc05@area.ba.cnr.it

DAVYDENKO Vladimir
Budker Institute of Nuclear Physics
11 Lavrentiev Ave.
Novosibirsk 630090
RUSSIA
Phone: 7-3832-39-45-71
Fax: 7-3832-34-21-63
E-mail: davydenko@inp.nsk.su

DUDNIKOV Andrey
Budker Institute of Nuclear Physics
11 Lavrentiev Ave.
Novosibirsk 630090
RUSSIA
Phone: (3832)333268
E-mail: dudn@ict.nsc.ru

DUDNIKOV Vadim
BTG
202 Townehouse dr.
Coram NY 11727
USA
Phone: 631 696 3058
Fax: 631 941 9178
E-mail: dvg43@yahoo.com

FAIRCLOTH Dan
Rutherford Appleton Laboratory
Room R2 D-08 ISIS
Chilton Oxon OX11 0QX
UK
Phone: 44 (0)1235 446195
Fax: +44 (0)1235 445720
E-mail: Dan.Faircloth@rl.ac.uk

FUKUMASA Osamu
Yamaguchi University
2-16-1 Tokiwadai
Ube, Yamaguchi Prefecture 755-8611
JAPAN
Phone: +81-836-85-9445
Fax: +81-836-85-9401
E-mail: fukumasa@plasma.eee.yamaguchi-u.ac.jp

GEBEL Ralf
Institute for Nuclear Physics – COSY/ Juelich
Forschungszentrum Juelich IKP Leo Brandt Strasse
Juelich 52428
GERMANY
Phone: (49) 2461 61 3097
Fax: (49) 2461 61 2854
E-mail: r.gebel@fz-juelich.de

GOBIN Raphael
CEA/Saclay
DSM/DAPNIA/SACM, Bat 123
Gif sur Yvette 91 191
FRANCE
Phone: (33) 1 69 08 27 64
Fax: (33) 1 69 08 14 30
E-mail: rjgobin@cea.fr

GORETSKY Victor
Institute of Physics of NAS Ukraine
46 Prosp. Nauki
Kiev-39 03650
UKRAINE
Phone: +38(044)2650910
Fax: +38(044)2651589
E-mail: gorets@iop.kiev.ua

HATAYAMA Akiyoshi
Department of Applied Physics and Physico-Informatics
Faculty of Science & Technology
Keio University
3-14-1, Hiyoshi, Kohoku-ku
Yokohama 223-8522
JAPAN
Phone: +81-45-566-1607
Fax: +81-45-566-1587
E-mail: akh@ppl.appi.keio.ac.jp

HEMSWORTH Ronald
DRFC/CEA Cadarache
St Paul lez Durance 13108
FRANCE
Phone: +33 442256345
Fax: +33 442256233
E-mail: ronald.hemsworth@cea.fr

IVANOV Igor
Budker Institute of Nuclear Physics
11 Lavrentiev Ave.
Novosibirsk 630090
RUSSIA

IVANOV Vladimir
A.F.Ioffe Physical-Technical Institute
26 Polytechnicheskaya Str.
St.Petersburg 194021
RUSSIA
Phone: 7-812-515-19-69
Fax: 7-812-247-10-17
E-mail: baksht@mail.ioffe.ru

KANEKO Osamu
National Institute for Fusion Science
322-6 Oroshi Toki
Gifu 509-5292
JAPAN
Phone: +81-572-58-2211
Fax: +81-572-58-2634
E-mail: kaneko.osamu@lhd.nifs.ac.jp

KUECHLER Detlef
AB Department/CERN
Geneva CH-1211
SWITZERLAND
Phone: +41227676691
Fax: +41227679145
E-mail: detlef.kuchler@cern.ch

KUZNETSOV Gennady
Budker Institute of Nuclear Physics
11 Lavrentiev Ave.
Novosibirsk 630090
RUSSIA
Phone: 3832-394310
Fax: 3832-342163
E-mail: M.A.Batazova@inp.nsk.ru

LARICCHIUTA Annarita
c/o Dipartimento di Chimica
Universita di Bari
via Orabona 4
Bari 70126
ITALY
Phone: +39-080-5442104
Fax: +39-080-5442024
E-mail: cscpal38@area.ba.cnr.it

LITVINOV Peter
Institute of Applied Physics
58 Petropavlovskaja St.
Sumy 40030
UKRAINE

MOEHS Douglas
Fermi National Accelerator Laboratory
P.O.Box 500
Batavia IL 60561
USA
Phone: 630-840-4490
Fax: 630-840-2170
E-mail: moehs@fnal.gov

PETERS Jens
DESY
Notkestr. 85
Hamburg 22607
GERMANY
Phone: 49 40 8998 2495
Fax: 49 40 8998 4364
E-mail: Jens.Peters@desy.de

PIUNOV Ilya
Budker Institute of Nuclear Physics
11 Lavrentiev Ave.
Novosibirsk 630090
RUSSIA

RYABTSEV Andrew
Institute of Physics of NAS Ukraine
46 Prosp. Nauki
Kiev-39 03650
UKRAINE
Phone: +38(044)2650910
Fax: +38(044)2651589
E-mail: ryabtsev@iop.kiev.ua

SAKURABAYASHI Toru
Department of Applied Physics and Physico-Informatics
Faculty of Science & Technology
Keio University
3-14-1, Hiyoshi, Kohoku-ku
Yokohama 223-8522
JAPAN
Phone: +81-45-566-1607
Fax: +81-45-566-1587
E-mail: sakura@ppl.appi.keio.ac.jp

SCHEPANYUK Tadeush
LPI Physics-Technics Center
Protvino, Moscow region
RUSSIA
E-mail: tadeush56@mail.ru

SHCHEDRIN Anatoliy
Institute of Physics of NAS Ukraine
46 Prosp. Nauki
Kiev-39 03650
UKRAINE
Phone: +38(044)2652329
Fax: +38(044)2651589
E-mail: ashched@iop.kiev.ua

SHERMAN Joseph
Los Alamos National Laboratory
MS H838 LANSCE-2
Los Alamos NM 87545
USA
Phone: 505-667-3511
Fax: 505-665-2509
E-mail: jsherman@lanl.gov

SOLOSHENKO Igor
Institute of Physics of NAS Ukraine
46 Prosp. Nauki
Kiev-39 03650
UKRAINE
Phone: +38(044)2650925
Fax: +38(044)2651589
E-mail: solosh@iop.kiev.ua

STOCKLI Martin
UT-Battelle, LLC
Oak Ridge National Laboratory
P.O. Box 2008, MS-6471
Oak Ridge, TN 37831-6471
USA
Phone: 01-865-241-8817
Fax: 01-865-241-9831
E-mail: stockli@sns.gov

SVENSSON Lennart
DRFC/CEA Cadarache
St Paul lez Durance 13108
FRANCE
Phone: +33 442256169
Fax: +33 442256233
E-mail: lennart.svennson@cea.fr

TANIGUCHI Masaki
JAERI
801-1 Naka-machi
Naka-gun, Ibaraki-ken
JAPAN
Phone: +81-29-270-7552
Fax: +81-29-270-7559
E-mail: tanigucm@fusion.naka.jaeri.go.jp

TARASENKO Alexander
Institute of Physics of NAS Ukraine
46 Prosp. Nauki
Kiev-39 03650
UKRAINE
Phone: +38(044)2650910
Fax: +38(044)2651589

VLADIMIROV Alexander
"SALYUT"
Moscow,
RUSSIA

WELTON Robert
UT-Battelle, LLC
Oak Ridge National Laboratory
P.O. Box 2008, MS 6471
Oak Ridge TN 37831-6471
USA
Phone: 865-241-9479
Fax: 865-241-9831
E-mail: welton@ornl.gov

ZAVALOV Alexander
Institute of Physics of NAS Ukraine
46 Prosp. Nauki
Kiev-39 03650
UKRAINE
Phone: +38(044)2650910
Fax: +38(044)2651589
E-mail: zavalov_alexandr@ukr.net

Kondo, T., 17, 35, 47
Krainov, G. S., 332
Krivenko, A. S., 332
Krylov, A., 3
Küchler, D., 267
Kursanov, Y. V., 229
Kuznetsov, G. I., 275

L

Lara, P., 254
Laricchiuta, A., 81
Leitner, M., 145
Litvinov, P. A., 229, 235, 317
Longo, S., 66

M

Massman, P., 3, 28
Meyer, B. J., 254
Moehs, D. P., 145, 189
Mori, S., 57
Murray, S. N., 296

N

Naffziger, P., 254
Nagaoka, K., 17, 35, 47
Nishida, R., 159

O

Oka, Y., 17, 35, 47
Osakabe, M., 17, 35, 47

P

Pagano, D., 66
Pelletier, J., 203
Peters, J., 214
Piunov, I., 325

R

Rouillé, C., 203
Rouleau, G., 254
Rutigliano, M., 81

S

Sakamoto, K., 168
Sakurabayashi, T., 96
Sato, M., 17, 35, 47
Schaller, S. C., 254
Scrivens, R., 267
Seki, K., 168
Sherman, J. D., 107, 254
Shirokov, V. V., 332
Shkol'nik, S. M., 138
Sidlow, R., 243
Soloshenko, I., 176
Steiner, T., 267
Stevens, Jr., R. R., 254
Stockli, M. P., 145, 296
Svarnas, P., 203
Svensson, L., 3, 28

T

Takeiri, Y., 17, 35, 47
Taniguchi, M., 168
Thomason, J. W. G., 243
Tsumori, K., 17, 35, 47

V

von Rossen, P., 282

W

Watanabe, K., 168
Welton, R. F., 145, 296
Whitehead, M. O., 243

Z

Zaniol, B., 3
Zaugg, T., 254
Zavalov, A. M., 176